Astronomy is one of the oldest sciences, and one which has repeatedly led to fundamental changes in our view of the world. This book covers the history of our study of the cosmos from prehistory through to a survey of modern astronomy and astrophysics, itself sure to be of interest to future historians of twentieth-century astronomy!

It does not attempt to cover everything in depth, but deliberately concentrates on the important themes and topics. These include the Copernican revolution, which led to the challenge of ancient authorities in many areas, not just astronomy, and seventeenth- and eighteenth-century stellar astronomy, at the time subordinated to the study of the solar system, but the source of many important concepts in modern astronomy.

Based on the widely acclaimed *Cambridge Illustrated History of Astronomy*, this book is beautifully illustrated throughout, and follows a similar structure and style. However, it is focused to meet the needs of final year undergraduates or beginning postgraduates. This is an essential text for students of the history of science and for students of astronomy who require a historical background to their studies.

Michael Hoskin taught the history of astronomy at Cambridge University for thirty years, and is editor of *The Journal for the History of Astronomy*.

The Cambridge concise
history of astronomy

The Cambridge concise history of astronomy

Edited by
Michael Hoskin

CAMBRIDGE
UNIVERSITY PRESS

PUBLISHED BY THE PRESS SYNDICATE OF THE UNIVERSITY OF CAMBRIDGE
The Pitt Building, Trumpington Street, Cambridge, United Kingdom

CAMBRIDGE UNIVERSITY PRESS
The Edinburgh Building, Cambridge CB2 2RU, UK
40 West 20th Street, New York, NY 10011-4211, USA
10 Stamford Road, Oakleigh, VIC 3166 Australia
Ruiz de Alarcón 13, 28014 Madrid, Spain
Dock House, The Waterfront, Cape Town 8001, South Africa

http://www.cambridge.org

© Cambridge University Press 1999

First published 1999
Reprinted 2001

Printed in the United Kingdom at the University Press, Cambridge

Typeface Trump Medieval 9.5/12pt   *System* QuarkXPress® [SE]

A catalogue record for this book is available from the British Library

ISBN 0 521 57291 6 hardback
ISBN 0 521 57600 8 paperback

Contents

The contributors

Michael Hoskin is a Fellow of Churchill College, Cambridge. He recently took early retirement after three decades of teaching history of astronomy to Cambridge undergraduates, part of the time as Head of the Department of History and Philosophy of Science. In 1970 he founded *The Journal for the History of Astronomy*, which he has edited ever since. He is the General Editor of the multi-volume *General History of Astronomy*, being published by Cambridge University Press under the auspices of the International Astronomical Union and the International Union for History and Philosophy of Science.

J. A. Bennett is Keeper of the Museum of the History of Science, University of Oxford, and a Fellow of Linacre College. He was previously Curator of the Whipple Museum of the History of Science at Cambridge. His books include *The Mathematical Science of Christopher Wren* and *The Divided Circle*.

Christopher Cullen is Senior Lecturer in the History of Chinese Science and Medicine in the Department of History at the School of Oriental and African Studies, University of London, and Chairman of the Publications Board of the Needham Research Institute, Cambridge. He is the author of *Astronomy and Mathematics in Early China: The Zhou Bi Suan Jing*.

David Dewhirst is a Fellow of Corpus Christi College, Cambridge. He recently retired as Astronomer and Librarian of the Institute of Astronomy, Cambridge University.

Owen Gingerich is Professor of Astronomy and History of Science at the Harvard-Smithsonian Center for Astrophysics in Cambridge, Massachusetts.
He is Chairman of the editorial board for Cambridge University Press's *General History of Astronomy*. His interests cover a wide historical span, but he has concentrated

particularly on the Copernican revolution. He has published two anthologies of articles: *The Great Copernicus Chase* and *The Eye of Heaven: Ptolemy, Copernicus, Kepler.*

Clive Ruggles is Senior Lecturer in Archaeological Studies at Leicester University, and until recently edited the *Archaeoastronomy* supplement to *Journal for the History of Astronomy.* He has written and edited/co-edited several books, including *Astronomy and Society in Britain during the period 4000–1500 BC, Megalithic Astronomy, Records in Stone, Archaeoastronomy in the 1990s, Astronomies and Cultures,* and *Astronomy in Prehistoric Britain and Ireland.*

The Editor thanks four colleagues who acted as Advisors to *The Cambridge Illustrated History of Astronomy*: J. A. Bennett (Museum of the History of Science, University of Oxford), Owen Gingerich (Harvard-Smithsonian Center for Astrophysics), Simon Mitton (St Edmund's College, Cambridge), and Curtis Wilson (St John's College, Annapolis, Maryland). He is also grateful to David A. King (Johann Wolfgang Goethe University, Frankfurt) and Noel M. Swerdlow (The University of Chicago) for their help, and to Owen Gingerich for assistance in obtaining illustrations.

Preface

University teachers believe that in order to learn a subject, you have first to teach it. The thread running through this book is the history of astronomy as I learned it in three decades of lecturing to Cambridge undergraduates.

All teachers eventually convince themselves that they have seen the wood for the trees. I am no exception, and so I have elected to discuss at length, sometimes at considerable length, those few issues I believe to be of fundamental importance. To make room, questions that other historians might consider important, as well as innumerable lesser topics, are mentioned in passing, if at all.

We concentrate on the development, in the Near East and Europe, of the science of astronomy as the whole world knows it today. Other traditions, such as astronomy in China, and the sophisticated astronomies developed in the New World before the arrival of the *conquistadores*, occupy the attentions of respected historians of astronomy; but here they are described only briefly.

Readers sometimes come to the history of astronomy expecting the discussion to focus on 'who first got it right'. In the present work these expectations will be fulfilled very imperfectly, and this for two reasons.

First, 'getting it right' assumes that science is an onward and uninterrupted accumulation of truth, with theory approximating ever closer to reality. At the factual level, there is something in this. It is difficult to imagine that the claim, dating from Antiquity, that the Earth is roughly spherical will ever be abandoned, or that we shall one day discover that Venus is in fact closer to the Sun than is Mercury. But at a deeper, theoretical level, the development of science is immensely more complex. What has been termed 'normal science' often consists in the gradual clarification and elaboration of what is at first confused, with contributions from many hands. But there are sometimes dramatic and disturbing developments. A century after Isaac Newton's death it was generally believed that he

alone of the whole human race had been privileged to
announce the fundamental truths of the physical universe –
that this announcement had been made once, in 1687, and
that the feat could never be repeated. But this complacent
view was destroyed by Einstein's root-and-branch reform of
the most fundamental Newtonian concepts of space, time,
gravitation, and so forth. Yet it would be a poor historian
who declared Newton simply to have been 'wrong' and his
work therefore unworthy of attention.

Second, today's historians of astronomy see it their duty
not to award medals to past astronomers whose opinions
coincided with those of their modern counterparts, but to
take their readers on an exciting journey. This journey
introduces them to lands that are conceptually foreign – to
past cultures, that sought as we do to make sense of the
heavens, but did so by asking questions often very different
from those that we take for granted, and who looked for
answers strange to our way of thinking. Historians invite
their readers to venture with them into these alien ideas,
leaving behind modern assumptions as to the nature and
purpose of astronomy, and putting much of our modern
knowledge of the heavens onto 'hold'.

For example, Plato's contemporaries observed that the
heavens were rotating night after night with constant
speed. They saw that there were myriads of 'fixed' stars
which, while sharing in this rotation, preserved their positions relative to each other without change; but they also
saw, moving among the fixed stars in puzzling fashion,
seven 'wanderers' or 'planets': the Sun, the Moon, Mercury
and so forth. If, therefore, we are to understand astronomy
in the nineteen centuries between Plato and Copernicus,
we must put on one side the modern concept of 'planet',
and accept the Sun and the Moon as planets. More important still, we must put on one side what we nowadays think
of as the job of astronomers, for we are studying cultures in
which their job was to contrive, for each of the seven wanderers, a geometrical model from which accurate tables of
its future positions could be calculated.

This meant that for nearly two millennia, astronomy
was applied geometry. The culmination of this Greek
program came with the publication in 1543 of Copernicus's
De revolutionibus, in which the otherwise-conservative
author found himself compelled to make the Earth into a
planet in orbit about the Sun. In the early decades of the
seventeenth century, Kepler explored the physical implications of this claim – the *forces* at work in the solar system –
and he thereby transformed astronomy, moving it from

kinematics to dynamics. Not surprisingly, the new concepts developed by Kepler, Galileo, Descartes and their contemporaries were at first vague and confused, and
clarification came only in 1687, with the publication of
Newton's *Principia*, in which the author claimed that the
law of gravitational attraction was the key to understanding
the physical universe.

The test of this claim was whether or not the law, when
applied to the dauntingly complex solar system, could
account for the observed motions of the planets and their
satellites, and of the comets. During the eighteenth century
and beyond this question occupied the attentions of a tiny
band of mathematicians of outstanding genius; and how to
deal with their work is a problem for the historian of
astronomy. But while their conclusions were of the keenest
interest to astronomers, they were not themselves
astronomers but mathematicians working in the service of
astronomy, and so we can disregard the details of their calculations with a clear conscience.

These 'celestial mechanicians', like their ancient and
medieval precursors, were preoccupied with the solar
system. The stars were still little more than an unchanging
– and therefore uninteresting – backdrop to the movements
of the planets, and there was little to be done about them
beyond the cataloguing of their positions and brightnesses.
Even as late as 1833, the leading authority on the stars and
nebulae, John Herschel, published *A Treatise on Astronomy*
in which he dealt with these bodies in a single chapter.
With rare exceptions, his contemporaries, professionals and
amateurs alike, were preoccupied with just one star –
namely the Sun – and its satellites.

But since then the balance has tilted sharply in the opposite direction, and today's historian sees that the pioneering
eighteenth- and nineteenth-century investigations into
stars, nebulae, and 'the construction of the heavens' were to
have a profound influence on future astronomical thinking.
This book therefore gives more space to early explorations
beyond the confines of the solar system than would have
seemed proper to astronomers alive at the time.

One issue recurs throughout our account of astronomy
in recent centuries: distances. The observer sees the celestial bodies as spread out on the surface of the heavenly
sphere; the evidence, that is, is two-dimensional. To theo-
rize about the three-dimensional universe, observers must
investigate the third co-ordinate, that of distance.

The story of this investigation is an exciting one, for the
successful measurement of the distances of unimaginably

remote objects is one of the astonishing achievements of astronomy – even the nearest stars are so far away that their light takes years to reach us. But this remoteness of celestial bodies brings an unexpected bonus, for we see them, not as they are now, but as they were when their light set out on its journey through space. This enables the astronomer to do the seemingly impossible, and look back in time. The more distant the object, the further back in time its light takes us; and today the distances studied are sometimes so great that the objects involved are cited in evidence, for and against cosmological theories of how the universe appeared in its infancy.

When does history end and science begin? Historians are themselves too close to contemporary astronomy to be able to offer a considered perspective. But although it is too soon to see 'Astronomy Today' with historical eyes, astronomy has clearly been transformed in recent decades, and the changes are too dramatic and too exciting for the historian simply to ignore. We therefore end our historical journey by looking around us, at how things stand today in the quest we share with our ancestors both ancient and modern: to understand the universe in which we find ourselves.

The present text is based on that of *The Cambridge Illustrated History of Astronomy*. But whereas the *Illustrated History* was intended for 'the general reader' with a minimal grasp of mathematics and physics, the present text includes certain materials more suitable for those who have studied these subjects at school.

Astronomy before history

Clive Ruggles and Michael Hoskin

Most historians of astronomy spend their days reading
documents and books in libraries and archives. A few
devote themselves to the study of the hardware – astrolabes, telescopes, and so forth – to be found in museums
and the older observatories. But long before the invention of
writing or the construction of observing instruments, the
sky was a cultural resource among peoples throughout the
world. Seafarers navigated by the stars; agricultural commu-
nities used the stars to help determine when to plant their
crops; ideological systems linked the celestial bodies to
objects, events and cycles of activity in both the terrestrial
and the divine worlds; and we cannot exclude the possibility that some prehistoric and protohistoric peoples possessed a genuinely predictive science of astronomy that
might have allowed them, for example, to forecast eclipses.

This *History* will concentrate on the emergence of the
science of astronomy as we know it today. The historical
record shows this development to have taken place in the
Near East and, more particularly, in Europe. We therefore
begin by asking if anything is known of how prehistoric
Europeans viewed the sky, and whether there is any evidence of predictive astronomy. Because it is all too easy for
us to fall into the trap of imposing our Western thought-
patterns and preconceptions onto the archaeological
remains, we also look, by way of comparison, at members
of two other groups who viewed or view the sky with
minds untouched by Western ideas: the peoples who lived
in America before the Spanish conquest, and peoples living
today who pursue their traditional ways of life in relative
isolation from the rest of mankind.

The celestial phenomena in the two regions most inten-
sively investigated by students of prehistoric and protohistoric astronomy – northwest Europe and the American
tropics – are very different. In the tropics the Sun and the
other celestial bodies rise and set almost vertically, and for
people living there the two times in the year when the Sun

passes directly overhead often have special significance. At
the higher European latitudes the celestial bodies rise and
set along a slanting path and culminate in the south.
Around midsummer the days are long, but thereafter the
Sun's rising and setting points move steadily further south
and the days get shorter and colder: a pattern that threatened disaster, unless the Sun could be persuaded to turn
back. Although modern 'Druids' gather at Stonehenge at
the midsummer sunrise, the monument's orientation in the
opposite direction, towards the midwinter sunset, may well
have held powerful symbolism for its builders.

The sky as a cultural resource in prehistoric Europe

Europeans living today enjoy at best the flimsiest of links
with the prehistoric peoples who occupied the region. Some
links may nevertheless exist. It has been maintained that in
Bronze Age Britain a calendar was in use whereby the year
was divided into four by the solstices and equinoxes, and
each of these four into two and then into two again, giving
in all sixteen 'months' of from twenty-two to twenty-four
days each; and it may be that vestiges of an eight-fold division of the year survived into Celtic times and hence into
the Middle Ages, where they were represented by the feasts
of Martinmas, Candlemas, May Day and Lammas in addition to the four Christianized solstices and equinoxes.

Again, legends associated with the huge passage tomb at
Newgrange in County Meath, Ireland, built around 3000
BC, make the omniscient god Dagda (or his son) dwell in
the monument. Dagda's cauldron was the vault of the sky,
and a connection with much earlier practices may be indicated by the modern discovery that the winter sunrise penetrated the furthest recesses of the tomb. From an entrance
on the southeastern side, a 62-foot passage leads to a central
chamber 20 feet high, from which three side chambers open
out. Some time after construction, when the bones of many
bodies had been placed in the tomb, the entrance was
blocked by a large stone. Yet although the living were
excluded, the light of the midwinter Sun continued to enter
via an otherwise-inexplicable 'roof-box', a slit constructed
above the entrance. For some two weeks either side of the
winter solstice, the Sun, on rising, shone down the length
of the entrance passage and illuminated the central
chamber – as it still does. That this should happen by
chance, and that the 'roof-box' has some other explanation,
is so unlikely that there is little doubt that Newgrange was

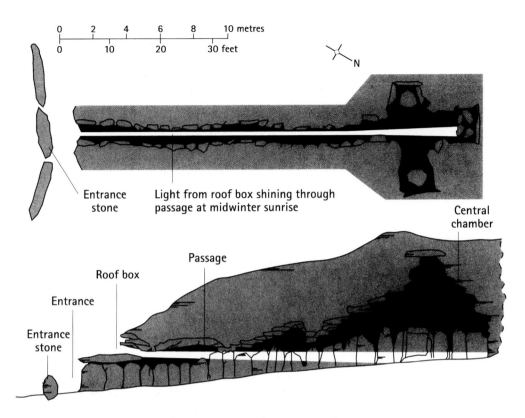

0 2 4 6 8 10 metres
0 10 20 30 feet
N

Entrance stone
Light from roof box shining through passage at midwinter sunrise
Central chamber

Passage
Roof box
Entrance
Entrance stone

deliberately constructed to face sunrise at the winter solstice. But we must note that the sunlight was intended to
fall upon the bones of the dead, not be seen by the living,
and that even a living occupant of the central chamber
would have learned only a very approximate date for the
solstice.

Even when no such direct links with the past exist, it
may be possible to identify with some confidence examples
of prehistoric monuments whose construction reflected a
concern for the heavens. In the Alentejo region of Portugal,
for example, to the east of Lisbon, there are numerous neolithic tombs. Each tomb has an axis of symmetry and an
entrance lying on this axis, and so there is a well-defined
direction in which the tomb may be said to 'face'. There are
scores of these tombs, scattered over a very wide area, yet
the directions in which they face all fall within the narrow
range of an octant or so – a uniformity that cannot have
occurred by chance.

How could the uniformity have been achieved? The
terrain is flat, and there is no mountain (for example) that
the builders could have used to determine the alignment
of the axis as they laid out a new tomb. Nor did these

Newgrange, diagrammed
from above (top) and in
cross-section (below),
showing the path of the
Sun's rays at midwinter
sunrise. The tumulus cov-
ering the tomb is some 250
feet across and over 30 feet
high.

neolithic peoples possess the compass. It seems, therefore, that the custom they were following must have involved the heavens, for only the heavens would have appeared the same from all places throughout this large region.

The orientation of such a tomb is something we can measure, and it is a matter of fact; the high degree of uniformity among the orientations of these tombs is likewise a matter of fact; and the involvement of the heavens in their layout is at least highly likely. On the other hand, we cannot interrogate the builders and they left us no written records, so we have to speculate on the meaning that the orientation of a tomb might have conveyed to its constructors and their contemporaries. Can the range of orientations shed any light on this?

It so happens that each tomb faced sunrise at some time of the year. The south-easterly limit of the tomb orientations coincided with the south-easterly limit of sunrise, at the winter solstice, but most tombs faced sunrise in the late autumn and early spring. The autumn is indeed a likely time of year for beginning the construction of a tomb, for then there would have been less work to be done in the fields and with the animals but the weather was still favourable. We know from historical records that many churches in England were laid out to face sunrise on the day construction began, and that one can calculate possible calendar dates for the beginning of construction from measurement of the orientation of the axis. It seems we can do the same for the Alentejo tombs, and so gain new insights into the annual rhythm of life in neolithic times.

We meet another example of the likely involvement of astronomy in the orientations of prehistoric monuments, in the *taula* sanctuaries on the Spanish Mediterranean island of Menorca, where a Bronze Age culture was at its height around 1000 BC. Such a sanctuary consisted of a walled precinct in the centre of which was the taula, a flat vertical slab of stone set into the ground, with a horizontal stone on top. The front face of the taula looked out through the entrance, nearly always in a southerly direction. Significantly, taulas were invariably located so that worshippers within had a perfect view of the horizon. Why was this important, when today there is nothing of interest to be seen away to the south?

We can find the probable answer by calculating backwards the effect of the wobble ('precession') of the Earth's axis caused by the pull of the Sun and Moon on the non-spherical Earth, which over the centuries alters the stars to be seen from any given location. We find that in Menorca in

1000 BC the Southern Cross was visible: it rose well to the
south, being followed shortly by the bright star Beta
Centauri, and then by Alpha Centauri, the second brightest
star to be seen from the island. This prominent star grouping has been (and is) of great importance in many cultures,
and not only in navigation. If, as seems probable, it was
associated with the rituals in the taula sanctuaries, we
learn something of the religion of the prehistoric people of
Menorca; and it may well be that they had links with
Egypt, where constellations were routinely identified with
deities.

The involvement of the heavens in prehistoric ritual in
Europe therefore seems well established. But was there also
a quasi-scientific astronomy of precise observation, perhaps
even leading to the prediction of astronomical events? In
Britain the suggestion that megalithic monuments, now
known to have been built in the third and early second millennia BC, incorporated alignments chosen for astronomical
reasons goes back to the eighteenth century, while at the
beginning of the twentieth century an astronomer of the
calibre of Sir Norman Lockyer could write: 'For my own
part I consider that the view that our ancient monuments
were built to observe and to mark the rising and setting
places of the heavenly bodies is now fully established.'

The subject came to popular attention in the 1960s with
the publication of a book on Stonehenge in which the
author – himself an astronomer – claimed that in addition
to the well-known phenomenon of the midsummer Sun
rising over the Heel Stone, a great many other astronomical
alignments were built into the configuration of the monument. He showed that, given regular observations extending
over many years, it was technically possible to use elements of Stonehenge to keep track of the solar calendar, to
study the more complex cycles of the Moon, and even to
predict eclipses. And this, the author insisted, had indeed
been one of its purposes.

If Stonehenge had been one among many similar monuments, these other monuments could have been examined
to see if they displayed the same features. Unfortunately
Stonehenge is without parallel anywhere in the world – it
was an object of wonder even in Antiquity. Its explanation
is further complicated by the fact that it was constructed,
modified, and reconstructed, over a period of some two millennia. Moreover, the stones we see today may not be
exactly in the position they occupied when first erected;
and when erected, they may not have been exactly in the
position the builders intended. As we cannot interrogate

the builders, and as they left no written records of their intentions, we are forced to fall back on probability: we must ask ourselves how likely it is that an arrangement of the stones, that to our eyes is of astronomical significance, has occurred by design rather than by chance. That is, the study of Stonehenge involves us in statistics – and for statistical investigation a unique monument is unsatisfactory.

The least contentious statement that can be made about Stonehenge is that the general orientation of the axis of the monument at various stages in its development was towards sunrise at the summer solstice in one direction, and towards sunset at the winter solstice in the other, and that this may well have been deliberate. A precision equivalent to, at best, two or three solar diameters is involved: the popular notion that the Heel Stone defined the direction of solstitial sunrise more precisely is quite unsupportable, because the supposed observing position (the centre of the monument) cannot be defined precisely enough, while the Heel Stone is too near to provide an accurate foresight and the horizon behind it is featureless.

Most students of Stonehenge have identified certain features at the site and tried to invent a theory to 'explain' them. Even when this is done impartially there are grave dangers in imposing astronomical (and geometrical) frameworks onto what is a very limited sample of the features at this much-altered site – those that today are superficially obvious, those that happen to have been excavated (while large areas of the site are still unexplored), and so on. For example, the Heel Stone is now known to have had a companion, long since destroyed, whose existence was discovered during rescue operations in 1979.

Some of the most famous astronomical theories regarding Stonehenge depend upon statistical arguments that the number of astronomical alignments between pairs of points selected are of possible significance. These arguments fall down on many different grounds: lack of prior justification for the points chosen, and archaeological doubts about some of those that were chosen; numerical flaws in the probability calculation; and, perhaps most importantly, the non-independence of data (for example, except in hilly regions, a line that roughly points towards midsummer sunrise in one direction will automatically point towards midwinter sunset in the other). When these errors are taken into account, no evidence whatsoever remains for preferred astronomical orientations of this sort.

One writer has pointed out that the 56 Aubrey Holes (named after their seventeenth-century discoverer, John

Aubrey) could have been used as an eclipse predictor, if markers were moved around from hole to hole. The problem here is that while this undoubtedly represents a way in which a modern astronomer could use a structure at Stonehenge to predict eclipses, there is ample archaeological evidence to suggest that the prehistoric users of Stonehenge did no such thing. There are in fact dozens of circular enclosures and so-called henge monuments (monuments that resemble the first phases of Stonehenge, before it acquired its distinctive structures of Bluestones and Sarsens) where rings of postholes or ritual pits inside a ditch have been found, and in these the holes vary in number from under twenty to over 100.

On the other hand, in the region around Stonehenge there appears to have been a shift from lunar to solar symbolism as development progressed from the Neolithic into the Bronze Age. This is reflected in the directions in which the burial cairns from each period are aligned, and also in the apparent shift in the axis of Stonehenge from lunar alignment in the earlier phases to solar alignment in the later. A group of post-holes situated in the northeastern 'entrance' – a gap in the ditch between the Aubrey circle and the Heel Stone – may represent evidence that the original construction of the axis was oriented on an extreme rising position of the Moon, though this interpretation remains controversial.

In short, there is good reason to think that the construction of Stonehenge and related monuments embodied astronomical symbolism, but we have as yet no convincing evidence that what we might think of as scientific astronomy was practised there.

While Stonehenge was attracting popular attention (and controversy) in the 1960s, Alexander Thom (1894–1985), a retired Oxford professor of engineering, was quietly continuing the mammoth task he had set himself, of surveying to professional standards the many hundreds of stone rings and other megalithic monuments that survive in Britain, Ireland and northern France. Thom was a collector of facts, and most collectors of facts shy away from speculation. Not so Thom. He maintained, not only that these megalithic monuments were constructed according to complex geometrical designs and laid out using carefully-determined units of measurement (one of which he termed 'the megalithic yard'), but that the prehistoric builders had anticipated an idea later proposed by Galileo and had precisely located their monuments in order to facilitate astronomical observations of great accuracy.

In 1632, in his *Dialogue on the Two Great World Systems*, Galileo has one of his characters relate how he found himself making an accurate determination of the summer solstice, with an instrument provided by Nature free of charge:

> From a country home of mine near Florence I plainly observed the Sun's arrival at, and departure from, the summer solstice, while one evening at the time of its setting it vanished behind the top of a rock on the mountains of Pietrapana, about 60 miles away, leaving uncovered a small streak of filament of itself towards the north, whose breadth was not the hundredth part of its diameter. And the following evening, at the similar setting, it showed another such part of it, but noticeably smaller, a necessary argument that it had begun to recede from the tropic.

Thom believed that the constructors of the megalithic monuments he was studying had anticipated Galileo by three millennia or more. Some standing stones, he maintained, were astronomical backsights; their locations had been carefully selected so that, for example, the Sun at a solstice, or the Moon at one of its extremes, might be glimpsed setting behind a distant mountain, very much as Galileo describes. Priests with knowledge of the dates of these significant solar and lunar events, Thom suggested, might even have been able to predict eclipses and thus reinforce their privileged status in the community.

Not surprisingly, Thom became the centre of controversy: such prehistoric sophistication, especially among the inhabitants of regions remote from the supposed cradle of civilization in the eastern Mediterranean, appeared incredible to many archaeologists. To assess the plausibility of Thom's claims it was necessary to decide whether Thom had focused attention on a particular feature of the skyline as seen from the given site because he already knew it lay in a direction of astronomical interest. Objectors argued that if the skyline contained numerous mountain peaks, one of which was in the direction of (say) the winter solstice, then the alignment of this particular peak with the solstice may well have been accidental.

Thom's sites have since been re-examined under procedures carefully designed to ensure objectivity. The controversy continues, but the re-examination has greatly reduced the plausibility of his claims to have demonstrated the existence in prehistoric Britain of a science of predictive astronomy.

How does the debate now stand? A particularly interesting example of Thom's sites is Ballochroy in the Kintyre peninsula in Scotland. Here there is a row of three standing stones, two of which are thin slabs oriented across the alignment of the row. A few yards away is a rectangular burial cist; this is aligned with the stones, and its longer sides are oriented in the same direction.

Around the solstices, the Sun's rising and setting positions are changing almost imperceptibly: thus in the week before or the week after a solstice, the Sun's rising and setting positions at this latitude alter by only one-third of its diameter. This makes determination of the actual solstices difficult, and the solstices are basic to a knowledge of the annual cycle of the Sun. Thom, however, believed that at Ballochroy the prehistoric erectors of the stones had overcome this problem by the location they had contrived for the stones – one from which the Sun was to be seen at the winter solstice setting behind Cara Island which is on the horizon 7 miles away, and at the summer behind Corra Bheinn, a mountain more than 19 miles distant. Even though the Sun is then altering its setting position from one night to the next by only a few arc minutes, this change becomes apparent to the observer within a very few days of the actual solstice, because of the sensitivity of the vast measuring instrument that Nature has provided. According to Thom, the direction of midwinter sunset was indicated by the alignment of the stones, and that of midsummer sunset by the flat faces of the central stone.

One problem with testing such a theory arises from our ignorance of when, to within several centuries, the stones were erected. Although the directions of solstitial sunrise and sunset at a given location alter only slightly from one millennium to the next, this is enough to make an important difference when we are observing with instruments tens of miles in length. At a site with distant mountains in roughly the right direction, it may well be possible to find a date for the site when it would have had the exceptional characteristics that Thom's theory requires. As to the 'indications' supposedly built into the stones themselves, these are of the kind that tend to be identified by the investigator after he has already convinced himself of the astronomical purpose of the site. It is then that he is likely (in this example) to focus attention on the middle slab (which points roughly in the 'right' direction) rather than on the northernmost (which does not), and to specify the 'intended' alignment of the stones themselves, to a precision quite unjustified for a despoiled (and originally longer) row of three closely-placed, large and irregular stones, two of which are slabs set across the axis. At Ballochroy there is the additional difficulty that the cist would have been covered by a cairn in prehistoric times, and this cairn would have obscured the view towards the midwinter sunset; indeed, the cairn is still to be seen in a seventeenth-century sketch of the site. All in all, then, while there is no doubt that what we may term Thom's Galilean method was feasible in prehistoric Europe (as elsewhere), the claims of this Scottish engineer to have discovered a prehistoric science of predictive astronomy at present merit the peculiarly Scottish verdict of *not proven*.

In conclusion we note that we must avoid a false dichotomy between ritual or folk practice on the one hand and high-level predictive astronomy on the other. Hesiod's description of an early Greek farmer's use of a constellation's heliacal rising (its reappearance at dawn after some weeks of absence lost in the glare of the Sun) to tell the season favourable to planting, is an example of prediction at a low level, and similar predictions are used by farmers in parts of Europe to this day. And since Galileian-type precision observations could have been recorded adequately by backsights consisting simply of poles inserted in the ground, then if stone monuments were indeed erected as backsights, they must also have served another and presumably ritualistic purpose.

Early astronomy in the Americas

The student of prehistoric Europe has virtually no written or oral evidence to guide him, and the monuments he studies are usually modest structures. The complex societies that developed in the American tropics have left a much richer legacy. Many of the buildings that have survived are of great sophistication; investigators have the opportunity to question living descendants; and we possess written records of various kinds – stone inscriptions and other meaningful carvings, documents such as the handful of Mayan bark books known as codices, and detailed accounts from the first Spaniards to come into contact with these cultures.

A strange aspect of Inca society that flourished in Peru at the time of the conquest (in the middle decades of the sixteenth century) has been revealed largely through the study of accounts written by Spanish settlers shortly thereafter. This is the system of *ceques*, conceptual straight lines radiating out from the Coricancha or Temple of the Sun, the central religious monument in the Inca capital of Cuzco. There were 41 ceque lines, along which sacred monuments were located and which served to divide society into different groups. Some ceque lines were oriented astronomically, for example on the rising position of the Sun on the day when it could be seen directly overhead (that is, in the zenith), while others were oriented upon sacred mountains on the horizon; still others were related to water flow and irrigation. Thus we see that such systems of radial lines related spatial divisions on the ground to the divisions in society, to geographical features, and to astronomical events. Astronomy was merely one component of a highly complex system covering many different aspects of society.

Such systems were also present, albeit in less complex forms, in other Inca cities. Indeed, the concept of radiality seems have existed in the Andes even earlier, in pre-Inca times, when systems of straight lines, radiating out from features such as hill tops and cairns, were given physical expression on desert pampas. The most famous such pampa is that at Nazca, in the coastal region of Peru, where there are several dozen 'line centres' with radiating lines constructed by brushing aside the thin layer of brown surface stones to reveal the bright yellow sandy soil beneath. Weather conditions in this region are so stable that these lines have survived to the present day.

Many of the radiating lines join one line centre to another; many are perfectly straight and run for several miles. It appears that they were sacred pathways, and that many factors may have influenced their orientation, just as was the case in the later ceques. Astronomy was one such factor; the direction of water flow was another. While the Nazca lines do not, as has been suggested, represent 'the largest astronomy book in the world', there is little doubt that astronomical symbolism, including alignments on the rising and setting Sun on significant days such as those of its passage through the zenith, features in the construction and use of the lines.

The significance of the Sun's zenith passage is easy to understand. At any given place in the tropics, the Sun spends part of the year to the north, and the rest of the year to the south; but at noon on the day when the Sun passes from north to south, or vice versa, the Sun stands directly overhead. Not surprisingly, zenith passages were also a focus of interest in Mesoamerica, many hundreds of miles to the north. The two days when this occurred could be pinpointed in a simple but very spectacular fashion by the use of so-called 'zenith tubes': when the Sun was directly overhead, its light shone down the tube onto the floor below.

One of the most extraordinary civilizations known to history was the classic Mayan, which flourished in parts of what are now southern Mexico, Guatemala and Belize. The Maya wrote in hieroglyphics, and although nearly all of their bark books were destroyed in the mid-sixteenth century by the invading Spaniards, a handful survived, including two that appear to be detailed astronomical (or rather, astrological) almanacs.

The Maya were obsessed with the passage of time, records of which were inscribed on every form of structure. They used three separate counts of days. The first was a year of 365 days, formed of eighteen months of twenty days each plus an (unlucky) additional period of five days. The second was a year of 360 days or *tun*, used in the calculation of very long periods of time. The third and

The 150-foot image of a spider seen in this aerial view is one of the famous stylistic depictions of animals and birds that occupy a small corner of the Nazca pampa. Much more extensive, however, are trapezoids, rectangles, spirals, and long straight lines, both narrow and wide, some of which can also be seen in the picture. The various features, some obliterating earlier ones, may well span several centuries. Some lines are astronomically aligned, but astronomy is only a small part of the symbolism that brought ritualistic order to an otherwise empty and infertile part of the landscape. Reproduced with kind permission, South American Pictures.

most significant was a sacred almanac of 260 days. Each
day of the almanac had a compound name, formed of one of
the numbers one to thirteen (which were taken in
sequence) and one of twenty names (also taken in
sequence).

The day 1 Ahau belonged to Venus, and it was on this
day that the cycle of revolutions of Venus had to begin and
end. As viewed from Earth, Venus appears to have a cycle
(its 'synodical revolution') of fractionally under 584 days,
and sixty-five times 584 days equals 146 sacred almanacs of
260 days each. Accordingly, the Venus table in the bark
book known as the Dresden Codex covers sixty-five synodical revolutions. However, since the period of 584 days is
roughly two hours too long, an adjustment was needed, and
this had to be one that somehow preserved 1 Ahau as the
beginning date of the Venus cycle. After the sixty-first
revolution, the error was between four and five days, the
revolution ending on the day 5 Kan which, as luck would
have it, was four days after 1 Ahau. The Maya therefore
took the opportunity to subtract four days at this point, so
that next cycle began again at 1 Ahau. Even this correction
left a residual error, and the Codex reveals how further
corrections of a similar form were made when the opportunity offered. The final sequence was accurate to some two
hours in 481 years.

The obsession with Venus was motivated by astrology:
its dawn reappearance after the period of invisibility while
it passed between the Earth and the Sun was a time of great
peril, and the tables would give forewarning of this and so
allow ceremonies to take place that might succeed in
warding off the threatened evil. There is no evidence that
the Maya took the least interest in other Venus events that,
to our geometrical outlook, are of equal if not greater
importance.

Immediately following the Venus table in the Dresden
Codex is a table that occupies eight pages. It covers some
11,960 days, which equals forty-six sacred almanacs of 260
days. Early in the present century it was noticed that the
numbers of days separating the pictures on the pages are
familiar to astronomers as intervals between solar eclipses,
and it emerged that one function of the table was to give
forewarning of these perilous events. The Maya did not
possess the knowledge needed to determine whether a solar
eclipse would actually be visible from their territory, but no
doubt those that were not seen had been averted by the ceremonies prompted by the table.

The table consists principally of totals of 177 (or 178)

days, which is the period occupied by six lunar-phase months, with occasional 148-day (five-month) periods. Eclipses take place only when Earth, Sun and Moon are (roughly) in line; seen from Earth, the paths of the Sun and Moon cross every 173.31 days, and a solar eclipse can occur only within a few days of this event. Mayan records must have shown that eclipses took place only during 'danger periods' that occurred every six months (177 days), but the four-day error required the occasional substitution of an interval of five months. The table makes 405 months equal to 11,958 days, which would imply a length of the month only eight minutes short of the modern value.

The table also provides the required calendar pattern involving the 260-day almanac, which in this case had to commence with the day 12 Lamat. The pattern depends upon a remarkable coincidence: three of these natural intervals of 173.31 days equal two of the Mayan 260-day almanacs to within a couple of hours. Occasional adjustments were made for this error, in a way similar to the treatment of Venus intervals.

It must be emphasised that we have been able to give no more than a taste of the tortuous complexities of Mayan calendrics. This unparalleled obsession with the interrelationship of time-intervals, some of them man-made and others supplied by Nature, would be well-nigh inconceivable if we did not possess the written record. But given this obsession, it would not be surprising to find that their buildings incorporated astronomical alignments. The investigator, however, once again faces the problem that each of the complex structures is unique in form, and it is difficult in any given instance to prove that the alignment is intentional rather than accidental. A good example is the Governor's Palace at the great Mayan city of Uxmal in Yucatan (Mexico). This vast building has a different orientation from other buildings on the site, and faces towards what seemed a bump on the horizon but proved to be a huge pyramid some 3 miles to the southeast. Measurements showed that the alignment pointed to the southernmost rising point of Venus, and the suggestion that this was the motivation for the orientation of the Palace found support in the Venus glyphs that are carved on the building. Many similar astronomical alignments have been proposed, and – given what we know of the Maya mentality – it would be surprising if all of them have come about by chance.

The sky as a cultural resource today

The Maya are an example of a highly sophisticated civilization whose interests in astronomy were alien to our own:
we shall go astray if we impose our own thought patterns
on those outside the European tradition.

The same lesson is taught us with equal force by living
societies that are close to nature. In Africa an example of a
people whose calendric preoccupations are very different
from ours is provided by the present-day Mursi of southwest Ethiopia. They depend for their subsistence on rain
cultivation, flood cultivation, and herding, and the timing
of their annual migrations from one region to another is
crucial. Yet they have no 'scientific' calendar such as we
might expect. Their year consists of thirteen months, and
so is eighteen days longer than the solar year. But their calendar keeps in step with the seasonal year, not by the occasional omission of a month, but by a process of
institutionalized disagreement with continuous adjustment. The balance between divergent opinions as to what
stage of the year has been reached is influenced by discussion of seasonal markers such as the appearance of birds,
the flowering of plants, and horizon observations of the
Sun, *all* of which are seen as inexact. However, one
crucial event – the annual flooding of the River Omo – is
monitored outside the calendar by the last sunset appearance for some weeks of four stars in Centaurus and the
Southern Cross. Thus to our eyes it would seem that in
this one instance where precision in timing is vital, the
Mursi do in fact use a precise rather than a haphazard
method of determining the time of year. But the Mursi do
not see it in this way. To them, there is a direct association
between these stars and the River Omo and certain
flowers and plants: the successive disappearance of the
stars in the morning sky is correlated with terrestrial
events such as the flooding of the river and the flowering
of the plants.

Such direct associations between the celestial and the
terrestrial are common among native societies. The
Barasana people of the Columbian Amazon, for example,
perceive a 'Caterpillar Jaguar' constellation which is the
father of caterpillars on Earth: as the constellation rises
higher and higher in the sky at dusk, so terrestrial caterpillars increase in numbers. To us this is the result of the coincidence whereby the constellation happens to be in the
eastern sky at dusk at the time of year when the caterpillars

pupate and come down from the trees on which they feed; but to the Barasana this is cause and effect.

To the people of the remote Andean village of Misminay, the association between earth and sky is stronger still. The Milky Way is regarded as a celestial river which is a reflection of the terrestrial Vilcanota river, and the two are perceived as parts of an integrated system that serves to circulate water through the terrestrial and celestial spheres. The Milky Way is directly overhead the village twice in every twenty-four hours, and it chances that the directions it then makes are at right angles to each other. This results in a conceptual quartering of the sky, which is reflected on the ground in the very layout of the village itself: radiating out from the central building (now a Catholic church) are four paths, together with irrigation channels, that divide the village into four quarters. For the villagers of Misminay, observations of the various celestial bodies are an integral part of their agricultural and pastoral activities and festivals. Some aspects of the system of practice and beliefs found here and in neighbouring villages can be traced back at least as far as Inca times.

Past traditions of the native Americans can sometimes be recovered, for their descendants are with us and can be questioned, and aspects of their practices still survive. A particular study has been made of the Hopi of Arizona. Their cardinal directions are not our north/south/east/west, but the directions to the points on the skyline where the Sun rises and sets at the solstices. The beginning of their winter solstice ceremony (Soyal) is decided by the Sun Chief and the Soyal Chief, who together observe the Sun from their village as it sets in a distant notch in the San Francisco peaks 80 miles away. Soyal lasts for nine days, and begins four days after the chiefs have made satisfactory observations. Calculations show that the Sun sets in the notch between one and two weeks before the solstice, so that the solstice occurs after the ceremonies are well under way. It is interesting that even when the notch in the skyline is so very distant, it has proved desirable to observe the Sun some time prior to the actual solstice, when it is still moving perceptibly from one day to the next.

The remaining chapters of this book will be devoted to the developments in the Near East and Europe that led to the science of astronomy as the whole world knows it today. They will rely mainly on written evidence, supplemented by what can be learned from the instruments that are preserved in museums and observatories.

Written evidence survives in quantity only from the last

centuries before Christ. In the present chapter we have seen
something of cultures that antedated those of Babylon,
Egypt and Greece, cultures that flourished in Europe in the
second, third and even fourth millennia BC. Our attempt to
infer what was in the minds of such prehistoric builders has
been based mainly on what is left of the stones they used in
their monuments. By looking at modern peoples whose cultures have been little affected by Western ideas – and by
seeing something of American cultures that preceded the
Spanish conquest – it has become evident that our pre-
occupations in studying the sky are by no means the only
ones, and that our attempts to interpret these silent stones
are fraught with danger.

As we now turn to the historical records, and read what
our predecessors actually wrote, we are on safer ground. But
the temptation to impose onto these writings our attitudes,
our interests, and our factual knowledge of astronomy, is all
the more insidious. It must be remembered that history of
astronomy is a journey back in time to cultures alien to our
modern thinking, and that, like good anthropologists, we
must try to see the world through the eyes and minds of
those cultures. What gives the history of astronomy its
special interest, is the fact that its object of investigation –
the sky that prehistoric, ancient and medieval cultures
sought to understand – is the same sky that modern
astronomers explore.

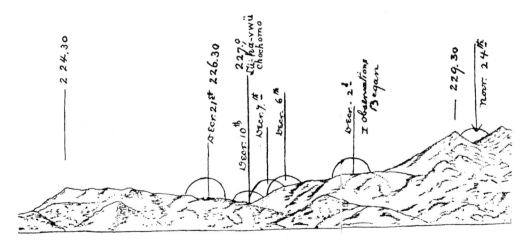

A Hopi horizon calendar, sketched by the anthro-
pologist Alexander M. Stephen who lived among
the Hopi in the 1890s. It shows a Sun priest's
observations for the purpose of timing midwinter
ceremonies. The Sun's setting in the notch that
it reached on 10 December (between peaks of the
San Francisco range near Flagstaff, Arizona) was
the signal to begin in four days the nine-day
celebration of the winter solstice. The solstice
occurs on or about 22 December. From A. M.
Stephen, *Hopi Journal*, Part II, Map 4, 1936, by
permission of the Syndics of Cambridge
University Library.

Astronomy in Antiquity

Michael Hoskin

From late Antiquity to the seventeenth century, astronomy had two related goals: to show that the movements of the planets were not haphazard but regular and therefore predictable, and to predict them with accuracy. All else that concerned astronomers was peripheral.

The first of the two goals was Greek, a product of the intense philosophical activity in Athens in the fourth century BC. The commitment to accuracy in prediction came from the quite different Babylonian tradition, which merged with Greek astronomy in the aftermath of the conquests of Alexander the Great in the third quarter of the same century.

The sources

The written materials, Babylonian and Greek, that have survived for the historian of astronomy to study are very limited. From Babylon we have clay tablets – mostly smaller than a man's hand – that have been recovered by excavation, legal or otherwise. The great majority are from the last seven centuries before Christ. Indeed, most were written in the 'Hellenistic era', after the death of Alexander in 323 BC, and in some cases even after the death in about 120 BC of Hipparchus, arguably the greatest of the Greek astronomers.

These Babylonian tablets, with their columns of numbers, have survived for over 2,000 years. The Greeks wrote on less durable materials, and most of their scientific writings have succumbed to the ravages of time and the pressure of economic forces. Some vanished because no copy was lucky enough to outlast the turmoil of later ages; but others were simply superseded, and thereafter the cost of copying them was no longer justified. Euclid's famous *Elements*, for example, incorporated – and therefore superseded – many of the mathematical works of his predecessors; these no longer survive, and the historian of

mathematics must reconstruct pre-Euclidean history, by
dismembering the *Elements* and relocating each section in
its original context.

The situation is especially difficult for the historian of
Greek astronomy, because the great work of synthesis –
Ptolemy's *Almagest* – was composed late in Antiquity, in
the second century *after* Christ, and half a millennium after
Aristotle was teaching in Athens. Ptolemy's monumental
achievement would have been awesome even if all the
earlier writings on which he drew – most notably those of
Hipparchus – had survived. But after the *Almagest* became
available, Hipparchus's works lost their value, and most of
them vanished. In their absence, the *Almagest* was without
peer, a daunting legacy to the Arabic and Latin civilizations
of the Middle Ages, whose astronomers were in awe of their
towering predecessor.

When we come to study the earliest Greek attempts to
make sense of the universe, we find that our sources are
fragmentary in the extreme. What little we have owes
much to the custom of Aristotle (384–322 BC), of citing his
predecessors before demolishing them. It seems that alongside the mythologies inherited from earlier times, there
gradually emerged a speculative interest in the natural
world, which led to efforts to make sense of nature – rather
than to attempts to answer quantitative questions, such as
led to the columns of numbers on the Babylonian tablets.
But what we would recognize as a mature, predictive
science of astronomy was to develop only in the Hellenistic
era, when these two approaches, the Greek and the
Babylonian, merged.

Astronomy in Babylon and Egypt

The city of Babylon lay on the left bank of the Euphrates
river, some 70 miles south of modern Baghdad. During
what is known as the Old Babylonian period (say 2000–1600
BC), the city was ruled by the dynasty of Hammurabi.
Babylon then fell to the Hittites, but was soon incorporated
into the Cassite empire, after which followed a long period
of Assyrian domination. This ended with the destruction of
Niniveh and its great library in 612 BC. After a period of
independence, Babylon came under Persian rule; but in 331
BC the city was conquered by the army of Alexander the
Great, and thereafter the cultures of Babylon and Greece
were in direct contact.

The tablets that survive from the Old Babylonian period
are more important for the history of mathematics than for

that of astronomy. They do, however, display a technique
that was to be crucial for the later development of astronomy: the employment of an efficient numerical notation.
To write the number 1, the Babylonian scribe pressed his
stylus into the clay edgeways on; to mark a 10, flatways.
Combinations of these two marks were used to write
numbers up to 59. For 60, however, the symbol for 1 was
again used, just as we use a '1' symbol in writing 10; and
similarly for 60×60, and 60×60×60, and so forth. Although
a zero appears late, and there was no symbol corresponding
to our decimal point, these minor limitations did not
detract seriously from the great merits of this 'place value'
system of writing numbers, which enabled the Babylonians
to perform elaborate arithmetical calculations with ease.
Our division of the hour into 60 minutes each composed of
60 seconds (each composed of 60 thirds, and so on), and our
similar division of the degree, reflect this Babylonian notation.

The early Babylonian skywatchers are often thought of
as astrologers, but if astrology is to be understood in the
Greek sense, as the study of the direct and unavoidable consequences for individuals that result from the configurations of heavenly bodies, this is a mistake. The Babylonians
were on the alert for unusual occurrences in every aspect of
nature – their examination of the entrails of sheep was one
example of this – and when they found something unusual
they interpreted this as an omen; not as the cause of a disaster about to happen, but as a warning sign of a disaster
that might yet be avoided by the appropriate ritual.

These interpretations were based on past experience,
and this experience was eventually codified in a series of
some seventy tablets with 7,000 omens, known from its
opening words as *Enuma Anu Enlil*. The collection, which
took definitive form before 900 BC, was in effect a list of
past signals that had been sent by the gods to the king,
indicating pleasure or displeasure. Learned scribes would
interpret new signals in the light of similar signals codified
in the *Enuma*, and it was hoped that their advice would
allow the king to take whatever avoiding action was
necessary.

The celestial body most frequently cited in the *Enuma*
was the Moon; and as their calendar was lunar, the cycle of
the Moon was doubly important. Calendars present every
culture with an awkward challenge because the natural
lengths of the day, the month, and the year are independent
of each other: no simple multiple of days makes a month
and no simple multiple of months makes a year. We see the

result of this today in the confusion of month-lengths that
we employ, and in our complex system of leap years.

Since the lengths of the seasonal year was more than
twelve lunar months and fewer then thirteen, from time to
time the Babylonians had to supplement the normal year of
twelve months with a thirteenth 'intercalary' month. For a
long time these intercalations followed no rule, but around
the fifth century BC the Babylonian astronomers realized
that nineteen solar years are very close to 235 lunar
months, and since 235=(19×12)+7, thereafter the intercalations followed a regular pattern of seven in every nineteen
years.

The recognition of this so-called 'Metonic cycle' was
made possible by the development in the eighth century BC
of a practice whereby the scribes of the *Enuma* began
systematically observing and recording astronomical (and
meteorological) phenomena. Their purpose was to improve
the efficiency of their prognostications, but the impact of
their efforts on the history of astronomy was to be incalculable. For seven centuries the recording went on; and just as
the *Enuma* was the expression of ominous experiences that
had been repeated so often that they had come to be recognized as regularities in the cosmic order, so cycles – regularities – in the behaviour of the Sun, Moon, and planets were
gradually identified and confirmed.

The need of astrologers to have tables ('ephemerides') of
the future positions of the Sun, Moon, and planets was,
until the seventeenth century, a driving force behind the
study of the movements of the heavenly bodies. Equipped
with ephemerides, the astrologer could ply his trade in fair
weather and foul. The Babylonian scribes learned to do this,
by using their sophisticated numerical system to take full
advantage of the cycles revealed by their observational
records.

For example, the speed of the Sun in its year-long
journey among the background stars constantly varies, and
the Babylonian techniques for coping with this illustrate
the ingenious methods their notation allowed them to
employ. For half the year the Sun's speed gradually
increases, until it reaches a maximum; thereafter, for the
next half-year, it slows down until it reaches a minimum.

The Babylonians devised two ways of approximating to
this. The first involved supposing that the Sun moved with
a *constant* speed for half the year, and for a different constant speed for the other half of the year. The second
involved supposing that the Sun increased its speed *uniformly* month by month for half the year, and then

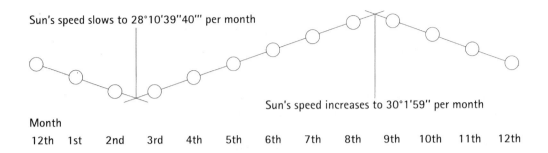

Sun's speed slows to 28°10'39"40''' per month

Sun's speed increases to 30°1'59" per month

Month

| 12th | 1st | 2nd | 3rd | 4th | 5th | 6th | 7th | 8th | 9th | 10th | 11th | 12th |

A representation in modern terms of the second Babylonian approximation to the speed of the Sun against the background stars, with the values found in a tablet for 133/132 BC. In each of six successive months the Sun was imagined to increase speed by the same amount, and thereafter its speed was imagined to decrease by the same amount for six months. In this tablet, the column of monthly velocities shows fluctuations between a calculated minimum of 28°10'39"40'" and a maximum of 30°1'59", even though these actual numbers are not listed for any given month.

decreased it uniformly month by month for the remaining half of the year. Clearly neither could have been intended as other than a highly artificial approximation to reality; but the resultant calculations were straightforward because of their efficient numerical notation, and the results were good enough.

Similar techniques were used to give control of the movements of the Moon and of the five planets, and this allowed the compilation of tables of extraordinary complexity. Some of these have come down to us, in the form of elaborate tablets with columns of numbers, whose underlying constructions historians patiently labour to unravel.

Whether the scribes had in mind any model of the universe, we do not know. What they handed on to astronomers who wrote in Greek were accurate arithmetical relationships involving time and angular distances (though the accuracy was due to the innumerable repetitions of the cycles involved, rather than to any precision of observation). Such relationships were just what was needed to turn the Greek speculative cosmologies into geometrical models from which reliable ephemerides could be calculated.

By contrast, the development of astronomy in neighbouring Egypt had been gravely handicapped by their failure to develop a place-value system of numerical notation. Instead, they had symbols for 1, 10, 100, . . ., and they simply repeated these as often as was necessary. For numbers less than 1 they used fractions that (with the exception of ⅔) always had unity in the numerator: ½, ⅓, and so on; other fractions had to be expressed in sums of these (for example, ⅖=⅓+¹⁄₁₅), while multiplication was achieved by successive doubling. Clearly, such a primitive arithmetic could permit only the most elementary grasp of celestial movements.

Society did however have certain requirements whose answers depended on astronomy. Nocturnal ceremonies had to be performed at the appropriate hour, and to tell the time at night astronomers selected from around the sky a sequence of 36 constellations or 'decans', which rose on any given night at regular intervals, like numbers on the face of a rotating celestial clock. More fundamental was the need for a convenient calendar. Egyptian life revolved around an annual event, when the river Nile mysteriously rose and covered much of the land in the valley; then, as the waters subsided, planting could be carried out; and finally there followed a period of growth and harvest. From the middle of the third millennium BC, we have records that show that the custom was already established of dividing the year into three seasons – the flooding, the subsidence of the river, and the harvest – each normally of four lunar months. However, since the lunar month lasts about twenty-nine and a half days, twelve such months fell well short of the average interval between one rising of the river and the next. From time to time, therefore, one of the three seasons would have to be given a fifth, additional ('intercalary') month. But how was this to be regulated?

The brightest star in the sky is Sirius. For part of the year the Sun is close to Sirius, and Sirius is then invisible, hidden from sight in the glare of the Sun. Then, one morning, in the eastern sky before dawn, Sirius briefly reappears. By chance, this 'heliacal rising' of Sirius took place in ancient Egypt around the time when the waters of the Nile began to flood, and some unknown genius saw that it was possible to control the calendar year by requiring that Sirius should always rise in the twelfth month, and by devising a rule to ensure this.

The seasonal year exceeds twelve lunar months by some eleven days. This meant that if Sirius rose in the current year in the last eleven days of the twelfth month, then – unless action was taken – a year later Sirius would rise in the first (and therefore wrong) month. To avoid this, whenever Sirius rose late in the twelfth month, an intercalary thirteenth month was announced for that same year.

This calendar was satisfactory for religious festivals. But Egypt had developed into a highly organized society, and to have months that were sometimes of twenty-nine and sometimes of thirty days, and years that were sometimes of twelve and sometimes of thirteen months, must have been very inconvenient. It was realized that the length of the seasonal year was close to 365 days, and another unknown genius proposed a system whereby every year was to contain exactly twelve months, and every month exactly 30 days divided into three 'weeks' of ten days each. At the end of the twelve months there were to be five extra days, to make 365 days in all, a total that never varied.

This administrative calendar was probably introduced
soon after 3000 BC, and it existed alongside the earlier, religious calendar. Because the administrative calendar allowed
of no exceptions whatever to the number of days in the
year, the interval between any two dates in this calendar
was easy to calculate; so convenient was this that
astronomers used the calendar until early modern times.
But the natural year is in fact a few hours more than 365
days (which is why we have our leap years). This meant
that before long the new administrative calendar began to
get seriously out of step with the seasonal year; but instead
of modifying the administrative calendar so that it fitted
nature, the bureaucratic Egyptians persisted with it (and
even invented a third calendar, a lunar calendar tied to the
administrative one). Only as late as 239 BC was an attempt
made to introduce a system of leap years so that the administrative calendar would keep in step with the seasonal
year.

Greek astronomy: the heavenly spheres

As society in Greece developed its characteristic structure
of the city state, the annual cycle traditionally used to regu-
late agricultural life, consisting of heliacal risings and set-
tings of star groups and events such as solstices, was no
longer adequate. Instead the various city states developed
formal calendars based on the lunar month, and this led to
the familiar problems as to which months were to be given
29 and which 30 days, and which years were to be given a
thirteenth, intercalary month. These decisions were left in
the hands of officials, whose competence varied considerably. The Athenian astronomer Meton, who lived in the
second half of the fifth century, proposed the adoption of
the cycle that we have already met in Babylon, whereby 19
solar years are equivalent to 235 lunar months. So far as we
know, he had no success in persuading the civil authorities
of the value of this reform; but astronomers were able to
use 'the Metonic cycle' to record the dates of eclipses and
similar events, in a form that – unlike the chaotic civil calendars – could later be used to determine the number of
days between two such events.

It is hard to believe that there were Greek records from
which the cycle could have been inferred, and so the cycle
must have reached Greece from Babylon, along with (for
example) the twelve signs of the zodiac. But such instances
of Babylonian influence were limited. For the time being,
the characteristic questions the Greek astronomers were

asking themselves, and the methods they were using to answer them, were altogether more abstract and generalized than the specific observational requirements that led to the tables we find on Babylonian tablets.

Such differences would persist even after the Greeks had learned to take observations seriously. In tackling the calendric problem of when new moons would occur, the Babylonians were using arithmetic to investigate special configurations of heavenly bodies – the times when Earth, Moon and Sun were aligned. This was in sharp contrast to the Greeks, who typically sought to develop geometric models of the movement of planets that would represent their observed positions at all times. Again, Greek philosophers might debate the status of a geometrical model that represented the motions of a planet – for example, whether it was to be considered true/false/possible, or merely accurate/inaccurate. But (as far as we can tell) the Babylonian astronomers would have been judged simply by results – either their predictions were good enough for their purpose, or they were not.

The first cosmologists known to us came from the prosperous Greek colonies in Ionia. Thales of Miletus (c. 625–c. 547 BC) is said to have taught that there was a material unity (which he identified as water) underlying the transient phenomena that our senses register – and, this being so, nature was more intelligible than the superficial appearance of endless variety might have suggested.

Anaximander (c. 610–c. 545 BC), also of Miletus, attempted to explain the form of the heavenly bodies in the context of his vision of worlds constantly coming into being from the Infinite, only to perish and be reabsorbed into the Infinite. If we can rely on accounts written many centuries later, he thought of the stars as wheel-like condensations of air filled with fire, with openings through which flames were discharged. The Sun was the highest (that is, most remote) of the heavenly bodies, with the Moon next below it, then the 'fixed' stars (those unchanging in their positional relations), and finally the planets. The Earth he believed to be a cylinder, on one of whose end surfaces lived mankind; it rested in the middle of the universe, remaining where it was because it had the same distance from everything. The limitations of this cosmology are evident, but a fundamental shift had occurred: earlier mythologies had been replaced by a nature in which an impersonal law was at work.

Not long after Thales had suggested that there was a material unity in nature, the members of a religious sect in the Greek colonies of southern Italy were finding the underlying unity of nature in structure. The sectarians looked on Pythagoras as their founder. Later myths were to obscure his personal achievements, but a significant legend has it that when listening to the sounds made by a blacksmith striking his anvil, Pythagoras was led to a perception of the relationship between arithmetic ratios and harmonic intervals in music – between abstract numbers and a phenomenon of nature. What is certain is that the Pythagoreans, generalizing on this, perceived number as the basis of all things.

These same Pythagoreans attained a very remarkable insight into the natural world: they recognized that the Earth is a sphere. The arguments they used are not recorded, but the proof given later by Aristotle – that the Earth's shadow cast on the Moon during eclipses is always circular – was cogent. In confirmation he pointed out that one sees different stars as one travels north or south, which showed that the Earth was in fact a sphere of no great size. For the rest of Antiquity and throughout the civilizations of the Middle Ages and the Renaissance, anyone with any pretensions to education knew that the Earth was round.

Indeed, the Greeks later arrived at an excellent appreciation of the actual size of our globe, an achievement due to Eratosthenes (c. 276–c. 195 BC), who was for many years in charge of the famous library that formed part of the Museum of Alexandria. He knew that, in the town of Syene (the modern Aswan) in Upper Egypt, the Sun was directly overhead at noon on midsummer's day: a vertical pointer cast no shadow, and the Sun's rays reached to the bottom of a well that had been dug for the purpose. One year, on midsummer's day in Alexandria, he measured the shadow cast by a vertical pointer at noon, and found it to be $\frac{1}{50}$th of a circle. Believing Alexandria to be some 5,000 stades due north of Syene, Eratosthenes concluded that the circumference of the Earth was fifty times 5,000 stades, or 250,000 stades. Although the data used by Eratosthenes are not quite accurate, and although the modern equivalent of 250,000 stades is a matter of debate – some put it at 29,000 miles – there is no doubt that the fit between Eratosthenes's 250,000 stades and our modern value for the circumference of the Earth of nearly 25,000 miles is good, and possibly excellent.

More generally, the Pythagoreans believed the natural world to be a *cosmos*. This term implied a rational order, but had in addition overtones of symmetry and beauty, and of the harmony that existed in a healthy organism. This

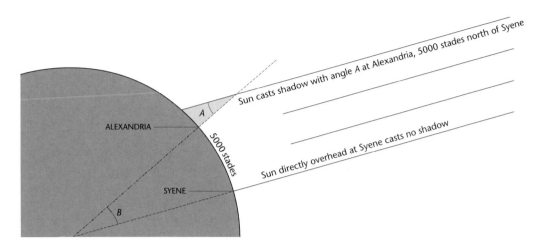

Sun casts shadow with angle A at Alexandria, 5000 stades north of Syene

A

ALEXANDRIA

5000 stades

Sun directly overhead at Syene casts no shadow

SYENE

B

intuition that the universe must be harmonious was to be a powerful driving force in astronomy in the Renaissance, when a Pythagorean stance became popular once more following the recovery of the writings of Plato, who had visited southern Italy and been much influenced by the doctrines he found there.

Plato (427–348/7 BC) was the second of the three great philosophers who graced Athens in the late fifth and the fourth centuries BC. His teacher, Socrates, wrote nothing, but lives on as a character in the dialogues of Plato. His pupil, Aristotle, who was a naturalist as well as a philosopher, wrote much, and a vast quantity of his writing survives; by chance, nearly all of this was available to the universities of the later Middle Ages (unlike most of Plato's dialogues), and this encouraged there an attitude to nature that emphasized the naturalist's sense of function and purpose.

Plato, in contrast to Aristotle, focused on the certainty to be found in mathematical reasoning, and to later civilizations he was to represent the mathematical view of nature: the Pythagoreans' insight, though shorn of their extreme mysticism. But Plato and Aristotle agreed that the world was a cosmos, a macrocosm whose parts corresponded to the organs of the individual living body or microcosm. It was this correspondence that would provide a theoretical basis for astrology and especially for astrological medicine.

Earlier philosophers had seen two fundamental and contrasting pairs of qualities in the objects people encountered in the natural world: hot versus cold, and moist versus dry. According to Aristotle, bodies that were cold and dry were mostly made of earth; those that were cold and wet, of water; those hot and wet, of air; and those hot and dry, of

Eratosthenes found by measurement that the angle A was one-fiftieth of a circle and, since A and B were equal, he concluded that the circumference of the Earth was fifty times 5,000 stades, or 250,000 stades.

fire. The elementary earth was to be found mainly in the
(roughly spherical) Earth that was at the centre of the
cosmos; around this was a shell of water (the seas), around
this a shell of air (the atmosphere), and around this a shell
of fire that terminated just this side of the Moon.

Within all this region – which constituted the terrestrial,
or sublunary, world – there were life and death, coming to
be and passing away. Any given body had a natural place – a
natural height, or distance from the centre of the Earth –
depending on the proportion of the elements in its makeup;
and unless prevented, the body would move towards this
natural place. This was why stones fell down and flames
moved up. Such 'natural' motions, we note, took place in a
straight line, either towards the centre of the Earth or away
from it; and they were transient, ending when the body
reached its natural place or got as near to it as circumstances permitted.

Granted that the Earth was a sphere, the Greeks found
there to be strong evidence that this spherical Earth was at
the centre of a universe bounded by a much greater sphere,
the sphere of the 'fixed' stars. For how else would people
invariably see half the heavens?

There was, according to Aristotle, a fundamental contrast between the terrestrial and the celestial regions,
between the imprecision and impermanence of what people
saw about them here on Earth and the geometrical perfection of the eternal heavens with their points and circles of
light. In the heavens, there was no life or death, no coming
to be or passing away. Instead, the heavenly bodies cycled
around eternally; they were all formed of a single fifth
element, or 'quintessence', and they expressed the permanence of their nature by the permanence of their uniform
circular motions. (Comets, which did indeed come to be
and pass away, presented no difficulty: their behaviour
showed them to be part of the terrestrial world – indeed
they were supposedly caused by the impact of the rotating
heavens on the shells of air and fire, and Aristotle discusses
them in his *Meteorology* (see page 96).)

It is important to recognize that this Aristotelian cosmology drew strength from being an intellectual formula-
tion that reinforced common sense, in contrast to modern
science, which contradicts what seems self-evident.
Aristotle tells us that the Earth is at rest beneath our feet,
that we are on *terra firma*, whereas today's astronomers
would have us believe that we are hurtling through space at
an almost unimaginable speed. Aristotle confirms that
there is indeed a fundamental contrast between Earth and

sky, whereas post-Copernican astronomy insists that we must disregard the plain evidence of our senses, and accept the Earth as a heavenly body like Venus and Mars and the rest. The history of cosmology is not the easy story of the rejection of absurd ideas in favour of what (perhaps after a little thought) is seen to be patently true, but the heroic saga of the hardwon rejection of the patently true in favour of the absurd. It is this that gives the history of cosmology its fascination.

We have an example of this in the physical proofs that the Earth was at rest rather than spinning on its axis. These proofs were to prove convincing to nearly everyone who considered the question, from the philosophers of Athens in the fourth century BC to Tycho Brahe in his observatory on the Baltic island of Hven two millennia later, and even today they retain a certain plausiblity. Let an archer fire an arrow vertically into the air; and suppose that while the arrow is in flight, the Earth moves, and therefore the archer with it. If so, then by the time the arrow has climbed verti-cally into the air, paused, and fallen vertically down again, the archer will no longer be underneath it. Yet experience shows that in fact it is hazardous for an archer to fire arrows vertically. *Ergo*, the Earth does not move.

But if the stability of the Earth was not in doubt, the status of the heavens as a cosmos where law prevailed was open to question, and would remain so until the move-ments of the 'wandering' stars, or 'planets', had been shown to be lawlike and regular. With seven exceptions, the innumerable heavenly bodies behaved in rational fashion; maintaining their positions relative to one another, the 'fixed' stars wheeled around the central Earth with perfect mathematical regularity, night after night, year after year. But the seven 'planets' – Sun, Moon, Mercury, Venus, Mars, Jupiter and Saturn – moved as individuals among the fixed stars, with varying speeds, seemingly as the fancy took them. Indeed, from time to time each of the five lesser ones actually stopped and went backwards (or 'retrogressed') for a while, before resuming its normal forward movement.

Plato (according to a late commentator) put the challenge to his contemporaries, to show that the planetary motions were in fact as regular as the rest (though not of course as simple). And the terms of an acceptable answer were clear: the stellar motions are uniform and circular, and so the planetary motions must be shown to be similar in nature; that is, to result from combinations of uniform circular motions.

It was Eudoxus of Cnidus (c. 400–c. 347 BC), Plato's

The hippopede produced by a pair of Eudoxan spheres. The planet is located on the equator of the sphere that has the inclined axis, whose poles are embedded in the sphere with the vertical axis and are carried round by it. The two spheres rotate with equal and opposite speeds. As a result of the combined motions, the planet moves on a figure-of-eight (which is in fact the intersection of the spheres with a cylinder that touches them internally).

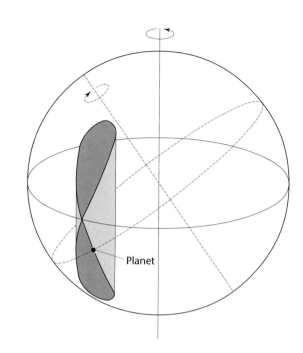

Planet

younger contemporary and one of the finest mathematicians of Antiquity, who first offered a solution to Plato's challenge. The occasional periods of backward ('retrograde') motion indulged in by the five lesser planets were the most extreme examples of non-uniformity, and Eudoxus's masterstroke was to achieve the representation of such an effect simply by means of a pair of concentric spheres. The astronomer was to imagine the planet as located at the equator of a sphere that was spinning uniformly. Projections protruded from its two poles, and these were embedded into a second sphere, outside the first and concentric with it. This outer sphere was also spinning but about a different axis, and as it spun it carried round with it the inner sphere. In consequence, the movement of the planet reflected the spinning of both spheres. Eudoxus realized that if the two spins were equal in speed but opposite in direction, and if the two axes were not very different, then the planet would move back and forth in a figure-of-eight – a 'hippopede', or 'horse-fetter', named from the hobble placed round the front feet of a horse to prevent it straying.

In each of the Eudoxan models of these five lesser planets, the outer of the two spheres was itself carried around by a third sphere located outside it, whose spin was chosen so as to reproduce the average (mean) west-to-east motion of the planet along the ecliptic (the apparent path of the Sun); and that sphere in turn was carried round by a

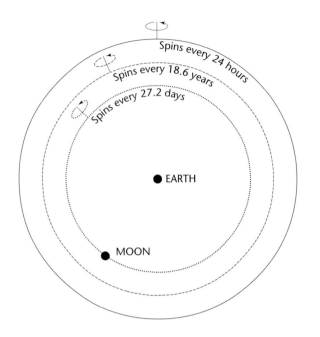

Spins every 24 hours

Spins every 18.6 years

Spins every 27.2 days

● EARTH

● MOON

The motion of the planet
Moon, according to
Eudoxus. The Moon is on
the equator of the inner-
most sphere, which spins
once a month. The poles of
this sphere are embedded
in the middle sphere,
which spins once every
18.6 years, a cycle familiar
from eclipse records; and
this in turn has its poles
embedded in the outermost
sphere, which spins daily.

fourth and outermost sphere that generated the east-to-west
daily journey of the planet around the Earth. These two
outer spheres thereby reproduced the principal movements
of the planet, while the two inner spheres served to replicate – qualitatively if not quantitatively – the occasional
retrogressions.

For the planet Moon Eudoxus proposed a nest of three
spheres. The outermost again generated the daily east-to-
west journey of the planet around the Earth. The speeds of
spin of the other two were responsible respectively for the
lunar month and for the 18.6-year cycle familiar from
eclipse records. The various quantities or 'parameters' could
be chosen so as to represent the Moon's motion with some
success. The nest of three spheres that Eudoxus proposed
for the planet Sun, however, was less satisfactory.

With four spheres for each of the five lesser planets,
three each for the planet Sun and the planet Moon, and one
for the fixed stars, Eudoxus required a total of twenty-seven
spheres in all. Eight of the spheres – one from each set –
simply supplied the identical daily movement, so that
overall the complexity was far from excessive.

We are not told how Eudoxus viewed the status of his
spheres, but it is probable that they were a mathematician's
solution to the problem of the wandering stars. They were
the equivalent of equations that set out to describe how
these bodies moved. There was need for further refinement
– Eudoxus's younger contemporary, Callippus, gave the

system greater flexibility by increasing the numbers of spheres – but Eudoxus's nests provided a reasonable basis for believing that the world was indeed a cosmos, that law prevailed even among the seven planets. This was after all what Plato, himself a mathematician, had called for.

But Aristotle was a naturalist who lived in the real world, and who insisted on knowing the physical causes of the celestial motions. Fortunately he saw an easy way of achieving this, by adapting Eudoxus's nests of spheres appropriately. He made the spheres physically real, and combined the various distinct nests of three or four spheres into an integrated whole, with the nest for the Moon (the planet believed to be nearest the Earth) at the inside and the nest (or more exactly, the single sphere) for the fixed stars at the outside.

But this was not satisfactory as it stood, for in any such physical system of spheres, the motions of each sphere would be transmitted down through the system and so would affect every one of the spheres within. There was indeed one motion – the daily spin – common to every planet, and so Aristotle was able to eliminate all of the twenty-four-hour spheres except the outermost; this one sphere would be sufficient to impose the appropriate motion on all the seven planets within. But most of the spherical motions applied only to the planets for which they had been designed. Aristotle therefore interpolated appropriate new spheres between any given nest and the next inner nest. Each such additional sphere moved about the same axis and with the same speed as one of the spheres of the given nest, but in the opposite direction; its function was to cancel the motion of the sphere in the given nest, and so prevent a motion that was peculiar to that planet alone from being transmitted down to all the planets within.

These interpolated spheres served a physical purpose only. Although they brought about a nominal increase in the total number of spheres in the system, they did not add to its mathematical complexity. Aristotle's coherent system of concentric spheres, each spinning with a uniform angular speed, was to be a cornerstone of natural philosophy as taught in the later Middle Ages; and the attractive simplicity of his conception of the cosmos would contrast with the daunting complexity of the geometrical models developed by later mathematical astronomers.

These same astronomers, on the other hand, saw grave defects in concentric spheres, whether they were viewed as mathematical conceptions or as physically real. They

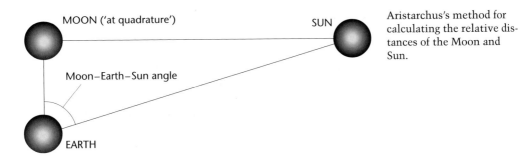

Aristarchus's method for calculating the relative distances of the Moon and Sun.

denied astronomers the flexibility they needed. In particular, nothing that could be done by way of adding spheres to a nest would permit the planet to vary its distance from the centre of the Earth. Yet some of the lesser planets altered in brightness in a manner that strongly suggested that they were varying their distances from Earth, while the Sun and Moon actually altered in apparent size and so demonstrably varied their distances.

Hellenistic astronomy: the merging of two traditions

Such quantitative considerations became more pressing after the conquests of Alexander the Great in the third quarter of the fourth century resulted in the spread of Greek culture into the Near East. This brought speculative Greek geometric astronomy into contact with the pragmatic and observation-based arithmetical astronomy of the Babylonians. Few astronomical writings survive from the early centuries of the Hellenistic era; but we need no books to tell us why the Eudoxan models, for all their geometrical elegance, would have been unacceptable to astronomers who had learned from the Babylonians the fundamental importance of quantitative observation.

Yet meanwhile, whatever its limitations, the more adventurous side to Greek astronomy retained its intellectual excitement. This is well exemplified by the work of Aristarchus (c. 310–230 BC) of Samos, an island just off the coast of Asia Minor. In the only treatise of his that is extant, Aristarchus correctly shows how (in principle, at least) we may calculate the relative distances of the Moon and Sun. When the Moon is half full ('at quadrature'), the angle Earth–Moon–Sun is a right angle. If we then measure the angle Moon–Earth–Sun, we learn the shape of the triangle joining the three bodies, and hence the ratio of any two sides.

In practice, the time of quadrature of the Moon is very

hard to determine accurately, and it is also difficult to measure the small difference that then exists between the angle Moon–Earth–Sun and a complete right angle. Aristarchus takes the angle to be 3° short of a right angle, whereas the true difference is only ¹⁄₁₈th of this. As a result, his conclusion that the Sun is some 19 times further than the Moon is less than ¹⁄₂₀th of the true value. But his treatise is modelled on works in pure mathematics, and the 3° may have been no more than a convenient value with which to illustrate his method: some Greek astronomers were more fascinated with finding out *how* to answer a question, than with answering it.

Aristarchus also dared to discuss the possibility that the Earth was in motion. He was not the first Greek to do so. According to one tradition, Philolaus, a Pythagorean living in south Italy in the second half of the fifth century BC, held that the Earth, together with an anti-Earth, the Sun and Moon, and the five lesser planets, orbited around a central fire, 'the hearth of the universe'. Another tradition tells us that Hicetas of Syracuse, a contemporary of Philolaus, believed the Earth was spinning on its axis, and the same teaching is ascribed to Heraclides of Pontus (c. 390–post 339 BC), a pupil of Plato. Aristarchus, however, anticipated Copernicus by teaching that the Earth is in orbit about the Sun. According to Archimedes (c. 287–212 BC), 'His hypotheses are that the fixed stars and the Sun are stationary, that the Earth is borne in a circular orbit about the Sun, which lies in the middle of its orbit . . .'.

But if Earth-based astronomers observed the stars from a moving platform, why did not the stars appear to move, in reflection of the Earth's motion? Because, Aristarchus suggested (and as Copernicans were later to maintain), the radius of the Earth's orbit was negligible compared to the distances of the stars, whose apparent movements were in consequence too small for astronomers to detect.

We know of only one person who was convinced by Aristarchus's speculation, the Babylonian Seleucus of Seleucia, who lived in the mid-second century BC and who tried to prove the hypothesis true. This lack of support is not surprising: Aristarchus's proposal, however perceptive it can seem to modern eyes, belonged to a tradition of speculative cosmology that was giving place to an astronomy that was more concerned with quantitative observations, and preoccupied with devising geometric models that would replicate these observations.

Nevertheless, belief in uniform circular motions as the key to the secrets of the cosmos remained strong in Greek

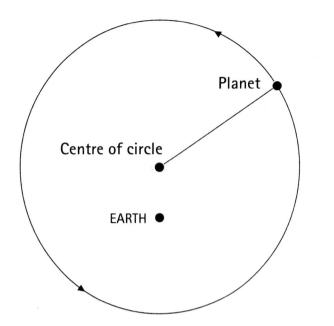

In an eccentric circle, the planet moved as usual with uniform speed around the circle; but since the Earth was not at the centre of the circle, the planet appeared from Earth to move with varying speed (slower when at the top of the circle and faster when at the bottom).

minds; and so it seemed that the way forward was to keep faith with the tradition of circular motions, but to employ them in a more flexible manner, and with parameters taken from Babylonian sources or calculated from Babylonian observations. By about 200 BC significant progress had been made in developing two alternative and versatile forms of circular motions, and these devices were investigated around that time by Apollonius of Perga, who is famous for his mastery of the geometry of conics and who is known to have lived for a time in Alexandria.

In the first of these devices, the planet moved around the Earth uniformly on a circle. However, the Earth was not at the centre of the circle, but to one side. As a result, in moving on this *eccentric* circle the planet varied its distance from Earth, and therefore its apparent speed across the sky.

In the second device, the planet moved uniformly on a small circle or *epicycle*, whose centre was carried uniformly on a large 'carrying' circle, or *deferent*, that had the Earth at its centre. If the motion of the planet on the epicycle was sufficiently rapid in relation to the motion of the epicycle on the deferent, then the planet would appear from time to time to move backwards (or 'retrogress'). It was not difficult to see that simple motion on an eccentric circle is mathematically (and observationally) equivalent to a special case of the motion produced by epicycle and deferent; and such equivalences would one day lead to the

How an epicycle might explain the apparent backward motion of a wandering star. The Earth is assumed to be at the centre of the large circle or *deferent*, on which travels the centre of the small circle or *epicycle*, which carries the planet. Each circle rotates with uniform speed. The path of the planet as seen from Earth depends upon the values chosen for the various parameters, but the figure indicates how the backward (retrograde) motion of a planet could be produced.

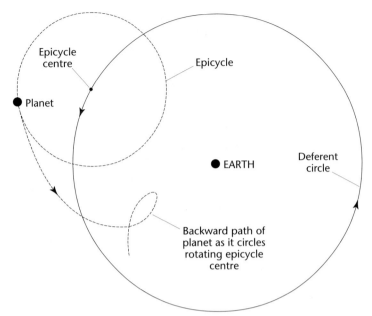

philosophical question of how one could decide which of two such models was 'true', if the observed effects were necessarily identical.

Apollonius investigated just when a planet moving on an epicycle would appear to halt its forward movement and begin to retrogress. His result is preserved for us by Ptolemy in Book 12 of the *Almagest*.

With hindsight we can see that the introduction of epicycles and deferents set Greek astronomy onto a most promising path. For since in fact Mercury, Venus, Mars, Jupiter and Saturn all circle round the Sun (in near-circular ellipses), and the Sun appears to circle round the Earth (also in a near-circular ellipse), a combination of two circles (epicycle and deferent, each doing duty for a near-circular ellipse) could provide an encouraging first approximation to the observed motion of one of these five planets.

Yet these planetary orbits are not circles but ellipses, and so no combinations of eccentres, epicycles and deferents could ever reproduce the observed movements perfectly. Astronomers – whether in Antiquity, the Middle Ages, or the Renaissance – who limited themselves to circles would never finally close the gap between prediction and reality, try as they might. And so the struggle continued down the ages, until at length in the seventeenth century the unprecedented observational accuracy that had been achieved by Tycho Brahe would force his assistant and successor, Johannes Kepler, to abandon circles and try alternatives.

We know little of the use, if any, Apollonius attempted
to make of these devices, for his astronomical works are
lost. So too are all but one of the writings of Hipparchus,
who was born in Nicaea in northwestern Asia Minor, and
who made observations at Rhodes between 141 and 127 BC.
Most of what we know of his astronomy we owe to the
many references Ptolemy makes to him in the *Almagest*.

Until now, Greek astronomy had been largely descriptive, while Babylonian astronomers had concentrated on
making predictions of specific phenomena. In the work of
Hipparchus we find the first of the geometric models that
would permit the prediction of the position of a planet *for
all times*. Using Babylonian astronomical records and parameters to develop quantitative geometrical models for the
motions of the Sun and Moon (the lesser planets he left to
posterity), Hipparchus showed how immensely fruitful was
to be the union of the contrasting Babylonian and Greek
traditions in astronomy.

In particular, Hipparchus compiled a list of lunar eclipses
observed at Babylon from the eighth century, and this
involved not only linguistic translation, but also conversion
from the chaotic Babylonian dating into a calendar that
could almost have been designed for the convenience of
astronomers, namely the Egyptian year with its unvarying
365 days. These records of past eclipses were crucial for the
study of the motions of the Sun and Moon (and so for the
prediction of future eclipses), for whenever an eclipse
occurs the Sun and Moon are directly in line with the
Earth. Hipparchus adapted for Greek use the sexagesimal
system of numbers that had been fundamental to the
Babylonian calculations, along with arithmetical techniques based on this system, and he adopted 360° as the
division of the ecliptic and other circles. But the trigonometry needed when geometry was to be used for prediction
did not exist, and had to be developed by Hipparchus
himself.

From the *Almagest*, we can reconstruct both
Hipparchus's model for the Sun (which Ptolemy was happy
to take over largely unchanged), and his much more
complex model for the Moon (which Ptolemy improved and
elaborated). In the elegant and simple model of the motion
of the Sun, the Sun orbited the Earth on a circle, moving
about the centre of the circle with a uniform angular speed
that took it round in 365¼ days. But Hipparchus knew that
the seasons are of different lengths: in particular, as seen
from Earth, the Sun moved the 90° from spring equinox to
summer solstice in 94½ days, and the 90° from summer

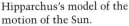

Hipparchus's model of the
motion of the Sun.

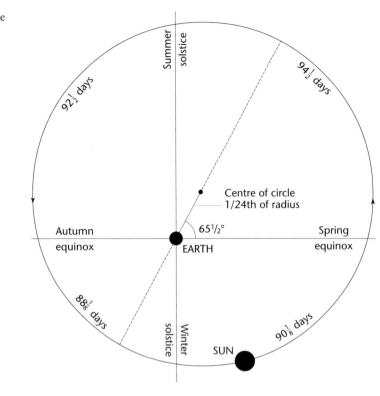

solstice to autumn equinox in 92½ days. Since both these
intervals were longer than one-quarter of 365¼ days, the
Sun appeared to the observer to be moving across the sky
with a varying speed.

The Earth therefore could not be located at the centre of
the circle, but must be 'eccentric'. By how far, and in which
direction? To generate these longer intervals, the Earth had
to be displaced from the centre of the circle in the opposite
direction, so that the corresponding 90° arcs seen from
Earth were each more than one-quarter of the circle, and it
therefore took the Sun intervals of more than one-quarter of
365¼ days to traverse them. Hipparchus's calculations
showed that the eccentricity needed to be ¼₄th of the radius
of the circle, and that the line from Earth to centre had to
make an angle of 65½° with the spring equinox.

Granted these parameters, the model was completely
defined. Luckily it reproduced the remaining two seasons
well enough, and so this single eccentric circle was sufficient to reproduce the motion of the Sun.

The Moon presented Hipparchus with a more challeng-
ing problem, and the simple model he derived gave good
results around new and full moons but was less successful

elsewhere. In the model the planet was thought of as moving uniformly on an epicycle, whose centre was moving uniformly about the Earth on a deferent circle. The values Hipparchus uses for the speeds of rotation of the two circles are credited to him by Ptolemy, but modern scholars have encountered these values (explicitly or implicitly) on tablets found at the site of Babylon.

To determine the relative sizes of the two circles and an initial configuration, Hipparchus contrived an ingenious geometrical method using the positions of the Moon at three eclipses. For each of these eclipses, however, the Babylonian records listed, not the position of the Moon, but the time of the mid-eclipse. Since an eclipsed Moon is aligned with the Sun and Earth, Hipparchus calculated the position of the Sun at the time (possibly from his model of the Sun but more probably by Babylonian methods) and added 180°. This gave him the required positions of the Moon on these three occasions. When we study this sophisticated Hipparchan model, it is hard to believe that little more than two centuries had passed since the speculative cosmologies of Eudoxus and his contemporaries.

Hipparchus was a dedicated observer, and he compiled a catalogue of the stars, apparently because he suspected that one of the stars may have moved, and he wished to bequeath to his successors data against which any future suspected movements might be tested. His single most important discovery was that of the precession of the equinoxes, the slow movement from east to west among the stars of the equinoctial points (the places where the Sun crosses the celestial equator). The spring equinoctial point is used by astronomers in defining the celestial frame of reference, and the discovery of precession showed that this point in fact is steadily moving, so that the measured position of a star varies with the date of the measurement. According to Ptolemy, Hipparchus thought precession amounted to 1° per century (though in fact it is nearer to 1° in seventy years).

The three centuries that separate Hipparchus and Ptolemy form a dark ages of astronomy, a period of which we know little, partly because Ptolemy despised the work that was done and ignored it. We would know even less were it not for the limited aims and ultra-conservative nature of such astronomy as was practised in India, for writings have been handed down in Sanskrit that embody elements from the Greek works of this period from which they derive. But with Ptolemy the historian is on secure ground at last, for his works were cherished by later generations and much of his voluminous output has survived.

Of his career we know little. The observations he reports in the *Almagest* date from AD 127 to 141, so that he cannot have been born much later than the beginning of the century; and he wrote several major works after the *Almagest*, so that his death may have been around 165 or 170. The only location he gives for his observations is Alexandria, and he very probably spent his adult life in that great centre of Hellenistic civilization.

The original title of the *Almagest* is *Megale syntaxis*, or *Mathematical* (that is, astronomical) *Compilation*; it became known in late Antiquity as *The Greatest Compilation*, and Arabic translators rendered 'the greatest' as *al-majisti*, which became *almagestum* in medieval Latin. It is indeed a magisterial work, one that provided geometrical models and related tables by which the movements of the Sun, the Moon and the five lesser planets could be calculated for the indefinite future. It also contained a catalogue of over a thousand stars arranged in forty-eight constellations, with the longitude, latitude and apparent brightness (magnitude) of each. Some historians have argued that Ptolemy simply took the positions from Hipparchus's catalogue and reduced each longitude by $2\frac{2}{3}°$ to allow for precession in the interval between the two men. Whole forests have been sacrificed to the resulting dispute, and it is now evident that straightforward plagiarism by Ptolemy is altogether too simplistic an explanation.

Ptolemy later published a modestly revised version of the tables with an introduction explaining their use, under the title *Handy Tables*; its format became the norm in the Middle Ages for both Islamic and Christian compilers of astronomical tables. He also published a digest of the *Almagest* which he called *Planetary Hypotheses* and in which (as we shall see) he added a physical dimension to the geometrical models of the *Almagest*.

Ptolemy wrote treatises on a wide range of other mathematical sciences. They included a major work on astrology, the *Tetrabiblos*, and his studies in astronomy were motivated in part by the need, which he shared with fellow-astrologers, for tables of planetary positions that were accurate and yet could be computed without excessive labour. The techniques of eccentre, epicycle and deferent he had inherited from Apollonius and Hipparchus. But he found that to compute planetary positions both accurately and conveniently, he needed to resort to another device, the *punctum aequans* or equant point.

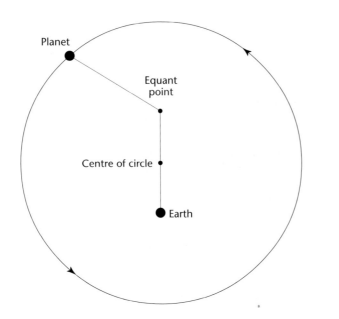

In a circle with an equant point, the Earth is located to one side of the centre, and the equant point is symmetrically positioned on the opposite site. The planet moves with a variable speed in such a way that it appears from the equant point to be moving uniformly. The planet therefore moves slower when near the equant point (at the top of the figure) and faster when distant from it (at the bottom).

Suppose the Earth is located at a distance from the centre of a given circle. Ptolemy defined the equant point as the mirror image of the Earth, on the opposite side of the centre and at an equal distance from it. This point he then used to define motion on the circumference of the circle. A point on the circle was required to move, not with uniform speed, but with a speed that varied in such a way as to appear uniform to an observer at the equant point.

Why was the equant so convenient that, to avail himself of it, Ptolemy was prepared to sacrifice the most fundamental principle of astronomy – that circular motions must be uniform? To understand this we must take advantage of our modern knowledge of the planetary motions. The first of the two laws published by Johannes Kepler in 1609 tells us that a planet moves around the Sun in an elliptical orbit with the Sun at one focus. Kepler's second law prescribes the speed of the planet in its orbit: a line (the 'radius vector') from the Sun to the planet traces out equal areas in equal times.

In obedience to the second law, the planet moves through space with a slower velocity than usual when furthest from the Sun (the radius vector then being longer than usual), and faster when nearest to the Sun (the radius vector then being shorter). How would the planet's movements appear to a hypothetical observer located at the focus of the ellipse *not* occupied by the Sun – the 'empty focus'? When the planet in its orbit is furthest from the Sun (and moving through space more slowly than usual), it is nearest to this

Kepler's first two laws tell us that a planet orbits the Sun in an elliptical path with the Sun at one focus, and that the line from the Sun to the planet traces out equal areas in equal times. As a result, the planet moves fastest when nearest the Sun and slowest when furthest from the Sun. The equant was a useful tool for Ptolemy because of the analogy between the equant point and the empty focus of the ellipse. (In the figure the eccentricity of the ellipse has been greatly exaggerated: with the exception of Mercury which is difficult to observe, the planetary orbits are in fact nearly circular.)

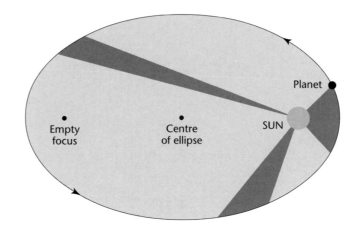

observer at the empty focus; the slower velocity is therefore masked by the proximity of the planet to the observer. Similarly, when the planet is nearest to the Sun (and moving through space more quickly), this too is concealed from the observer because the planet is now furthest from him. As a result, viewed from the empty focus, the planet appears to move around the sky with almost uniform angular velocity.

The consequence is that the motion of the planet in its orbit about the Sun – it matters little whether this orbit is a nearly-circular ellipse or a true circle – is well represented by a model in which the planet moves with uniform angular velocity as seen from the empty focus (in the case of an elliptical orbit); or, if the orbit is taken as circular, from an equant point similarly located, that is, on the opposite side of the centre to the Sun and equidistant from the centre. Since this is true in particular of the Earth, it follows that the same is true of the motion of the Sun relative to the Earth. It was of course the motion of the planet relative to the Earth (rather than to the Sun) that Ptolemy wished to represent; but this is compounded from the motion of the planet about the Sun, and the motion of the Sun relative to the Earth.

With the benefit of hindsight, then, we can see that Ptolemy's equant point was so useful because it was closely related to the empty focus of a Keplerian ellipse.

No doubt Ptolemy was prepared to use a device that violated the centuries-old principle that the heavenly motions were uniform, because he was more concerned with accuracy and mathematical convenience than with questions of truth. Yet, as time would show, Ptolemy had created a serious point of philosophical tension, one that disturbed many Islamic astronomers, and became a regular matter of

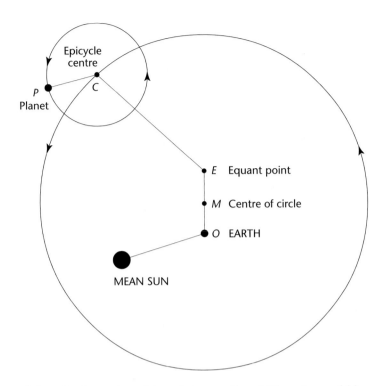

Epicycle
centre

P
Planet

C

E   Equant point

M   Centre of circle

O   EARTH

MEAN SUN

The technical ingenuity that Ptolemy's displays in the *Almagest* is illustrated by his model of the motion in longitude (that is, around the Sun's path or 'ecliptic') of one of the planets Mars, Jupiter or Saturn. The Earth is at O, and M is the centre of the large (deferent) circle. C is a point on the deferent circle, and is itself the centre of the epicycle on which travels the planet P. E is a point (the equant point) on the line OM (extended), such that $OM=ME$. C moves around on the deferent circle with a variable angular velocity, such that, viewed from the equant point E, it *appears* to be travelling uniformly. P moves on the epicycle in such a way that CP is always parallel to the line from O to the 'mean Sun' (reflecting the fact that the Earth-based observer is in reality in motion around the Sun). The ratio of the radii of the circles, the 'eccentricity' OM, and the two angular velocities, are then chosen to fit the motion of the particular planet.

debate in the universities of the medieval West. It would be difficult enough for university students of Arts to reconcile eccentres and epicycles with what they were taught in their natural philosophy about the uniform motions of the Aristotelian concentric spheres; to reconcile the equant was quite impossible. For this very reason, Copernicus in the sixteenth century would see it as a matter for self-congratulation, that his models needed no equants.

In the *Almagest*, Ptolemy provided geometric models that would, with fair accuracy, predict the motions of every one of the seven wanderers. He showed how the all-important parameters in his models could be derived from observations, though his general strategy was to work in the opposite direction: to present the models and then to 'test' them by comparing deductions from them against observations. With the help of the *Almagest*, mathematical astronomers (and astrologers) would be able to calculate tables of the positions of a planet in longitude and latitude for the indefinite future.

There were of course problems. He had found that the model for the Moon as adapted from Hipparchus represented well enough positions of the planet when Earth, Sun and Moon were in a straight line – unsurprisingly, in view of the use Hipparchus had made of Babylonian eclipse

records in deriving his model – but was unsatisfactory elsewhere. To remedy the defect, and so account for the Moon's
irregularity that later became known as 'evection', he introduced a 'crank' mechanism that varied the distance from
Earth of the Moon's epicycle. When Earth, Sun and Moon
were aligned, the crank was fully extended and its presence
made no difference; but elsewhere the crank pulled in the
epicycle towards the Earth, most of all when the angle
Sun–Earth–Moon was a right angle. The resulting model
was satisfactory as far as computing lunar positions was
concerned, but it implied that the height of the Moon above
the Earth varied between 33 and 64 Earth radii. This ought
to have resulted in its apparent diameter varying by a factor
of nearly two; yet such variations were not in fact observed
in the real Moon. How much this bothered Ptolemy's
readers depended on whether they were looking for something more than accurate tables of position.

Another problem stemmed from a curious fact of
observation. Two of the lesser planets – Mercury and Venus
– are never to be seen far from the Sun: they rise and set
with the Sun, unlike Mars, Jupiter and Saturn which may
be seen in the sky at any time of night. Ptolemy replicated
this in his models for Venus and Mercury by aligning the
centres of their epicycles with the 'mean Sun', so that all
three had the same period of one year. Some of his readers,
however, were unhappy with such an *ad hoc* device.

But these were details. Ptolemy's *Almagest* achieved
what it set out to do: to provide a set of geometrical models
for calculating accurate tables of the future positions in the
sky of each of the seven planets. It marked the culmination
of centuries of effort.

Ptolemy made no attempt in the *Almagest* to present an
integrated 'Ptolemaic System' of the heavens, but he did
permit himself a few sentences on the order of the planets,
in terms of their respective heights above the central Earth.
It seemed plausible to make the fixed stars the outermost
(as indeed Eudoxus and Aristotle had assumed half a millennium before), and to place nearest to these stars those
planets whose motions most closely imitated the motion of
the stars. On this basis Saturn – whose motion differed
from that of the stars by only one circuit every thirty years
– was the highest of the planets, with Jupiter (one in twelve
years) and Mars (one in two years) next. At the other
extreme, the Moon (one each month) was to be placed
closest to the Earth.

That left him with three planets, Sun, Mercury and
Venus, whose order was as yet unassigned. Unfortunately,

they kept company with each other as they wheeled about
the central Earth, and so all their motions differed from that
of the stars by the same amount: one circuit in one year.
Which then was nearest the Earth and which furthest?

As the Sun was of unique status, Ptolemy decided to
follow earlier astronomers in placing the remaining planets
above it and below in equal numbers. Indeed, in the Latin
Middle Ages the Sun would be seen as a king on a royal
progress through his dominions attended by three planetary
courtiers on either side. Since the previous argument had
placed Saturn, Jupiter and Mars above the Sun and only the
Moon below, Ptolemy evened things up by locating Venus
and Mercury below the Sun. The order he selected for these
two, putting Mercury below Venus, was based on little
more than the toss of a coin. And so, by reasoning that
varied from the plausible to sheer guesswork, Ptolemy
finally arrived at the order: Moon, Mercury, Venus, Sun,
Mars, Jupiter, Saturn, fixed stars.

This enabled Ptolemy, in his later *Planetary Hypotheses*,
to combine his planetary models into a unified and physically real system, one that later craftsmen attempted –
imperfectly, of course – to imitate in metal. Knowing the
order of the planets, he made the assumption that all the
possible heights in the heavens were shared out among
them; that each planet had its own range of heights that it
occupied from time to time, and these ranges neither over-
lapped nor left any gaps.

This being so, he could exploit the fact that (according to
the *Almagest*) the maximum height of the Moon was 64
Earth radii. Because the range of heights occupied from
time to time by the Moon adjoined the range occupied by
the next planet, Mercury, he inferred that the minimum
height of Mercury was also equal to 64 Earth radii.
Knowing the ratio of Mercury's epicycle to its deferent, he
could now calculate the maximum height of Mercury. This
he put equal to the minimum height of Venus; and so on,
until finally he put the fixed stars at the maximum height
of the outermost planet, Saturn.

By this means Ptolemy arrived at a universe whose
radius was 19,865 times the radius of the Earth, or about 75
million miles. Some modern writers dismiss this figure as
hopelessly wrong, pointing out that it is less than the true
distance from the Earth to the Sun. But to think in this way
is unhistorical. Rather, it was in the work of Ptolemy that
the universe first became too large for the human mind
truly to comprehend.

Ptolemy's astronomical writings brought to a triumphant

conclusion the opening, Greek phase of the campaign to
devise geometrical models that would replicate the motions
of the seven planets. Future planetary positions could now
be predicted encouragingly well, and there seemed no
reason to doubt that adjustments made by later
astronomers would bring prediction and observation ever
closer. This programme was to preoccupy astronomers for
the next fourteen centuries – indeed, the radical revision of
Ptolemaic astronomy carried out in the Renaissance by

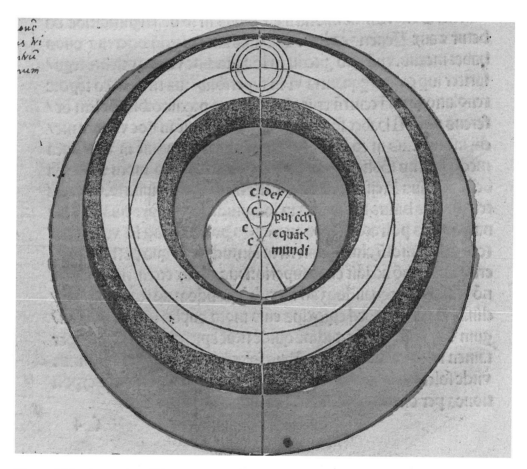

The model for the motion of Mercury proposed
by Georg Peurbach in the fifteenth century.
Peurbach popularized an idea transmitted by
Islamic astronomers, that the details of the planetary motions could be realized within a system
of solid spheres. Here the epicycle carrying
Mercury is indicated by the small, triple circle at
the top of the figure, and this epicycle moves in
the unshaded groove, about the central Earth.
Being the innermost planet of the solar system,
Mercury was to ancient observers invisible most
of the time, lost in the glare of the Sun. Ptolemy
made assumptions to supplement the observations available to him, and these implied that the
planet came close to Earth, not once but twice in
each orbit. As a result, the Ptolemaic model for
Mercury was more complex than those of the
other planets. From G. Peurbach, *Novae theoricae planetarum* (1460) 1482, by permission of the
Syndics of Cambridge University Library.

Copernicus was inspired by a commitment to the uniform circular motions of the Greek natural philosophers even greater than Ptolemy's.

Circular motions would finally succumb to the precision of Tycho Brahe's observations. But meanwhile, after the collapse of the *pax romana*, astronomers of civilizations unborn in Ptolemy's day would struggle to master the intricacies of the *Almagest*, against a background of philosophical disquiet at the devices Ptolemy had found it necessary to employ.

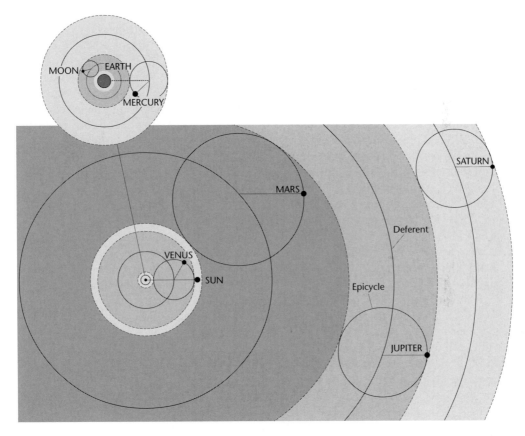

The Ptolemaic system in outline. Note that every distance from the Earth is occupied from time to time by one, and only one, planet. Note also that the centres of the epicycles carrying Mercury and Venus are aligned with the Sun, while the radii of the epicycles of Mars, Jupiter and Saturn are parallel to the line from the Earth to the Sun. This unexplained involvement of the Sun in the geometry of the other planets was to puzzle later astronomers. Following Copernicus, we realize that it reflects the fact that we Earth-dwellers observe the planets from a platform that is in orbit about the Sun.

Astronomy in China
Christopher Cullen

The traditional Chinese term for the inhabited world was *tianxia*, 'That which is below Heaven'. In astronomical terms the Chinese were under the same sky as everybody else, but the ways in which they interpreted what they saw in the sky were often unique.

There was in China no single subject of 'astronomy'. On the one hand there was *lifa*, 'calendrical methods', a discipline that aimed to master the regularities of celestial phenomena through careful measurement, record-keeping, and mathematical calculation. On the other there was *tianwen*, 'celestial patterns', whose practitioners watched for unpredictable and transient celestial phenomena and tried to interpret their significance for the world of human beings.

There was however one important factor linking *lifa* and *tianwen*: both were regarded as of vital importance to the state, and the practitioners of both disciplines were nearly all members of the imperial civil service. From around 1000 BC both the moral and the natural order of the cosmos were seen as embodied in a somewhat impersonal entity known as *Tian*, 'Heaven', whose will it was that the world below should be well governed. If the ruler misconducted his government or his personal life, this would be expected to cause disturbance in the natural world as an expression of Heaven's displeasure. Such disturbances might include flood, famine and plague, but disorder in the sky was the most ominous of all and might be a warning of

other troubles to come. It was therefore prudent for the imperial government to maintain a staff of sky-watchers, whose job it was to record, report and try to interpret all unusual celestial events. This was the province of specialists in *tianwen*. Meanwhile, specialists in *lifa* worked to detect and codify all discernible regularities in celestial phenomena, with the aim of providing the emperor with all predictable phenomena, so that he could demonstrate his own effectiveness in maintaining the orderliness of the cosmos by promulgating an accurate astro-calendrical almanac.

Detailed records of the activities of *lifa* practitioners go back to 104 BC, when a new system of mathematical astronomy was inaugurated by Emperor Wu of the Han dynasty. This was to be followed by forty-seven other new systems up to the seventeenth century. All such systems had as their core requirement the need to run a luni-solar calendar, in which a civil year of twelve lunar months had to be kept in step with the cycle of the seasons by the addition of extra 'intercalary' months, *run yue*, at appropriate intervals. Lunar months began with conjunction (the 'astronomical' new moon), and the basic mark point for the seasons was the winter solstice. But in addition to predicting conjunctions and solstices, it became customary to include in officially promulgated astronomical tables, methods for predicting the apparent motions of the planets, and for predicting lunar eclipses.

Solar eclipses could never be accurately predicted by the available methods, and hence remained to a large extent within the domain of *tianwen*. The increasingly sophisticated mathematical methods used by *lifa* practitioners were numerical rather than geometric, and exploited the recurrence of long-term cycles without laying stress on the motions of celestial bodies in three-dimensional space. Unlike the ancient Greeks and medieval Europeans, Chinese *lifa* specialists were not much concerned with cosmographical disputes. For most of the imperial period it was found satisfactory to assume the inhabited world was a small central region near the centre of a flat Earth, whose surface was a diametric plane of a celestial sphere rotating about an inclined axis. The celestial bodies moved on the inner surface of this sphere; the means by which they did so were not discussed.

The activities of *tianwen* specialists can be traced back to the late second millennium BC, when inscribed 'oracle bones' are thought to mention an observation of a nova (a new star). From around 200 BC a continuous stream of data is available from official sources. Apart from novae, officials recorded meteor showers, comets, solar eclipses, sunspots, and phenomena such as unusual clouds or the aurora borealis. In many cases such records are detailed and accurate enough to be useful to modern astronomers interested in such matters as identifying supernova remnants or tracing early appearances of Halley's Comet. Like the specialists in *lifa*, the observers of such phenomena had access to large and well-made armillary spheres and other graduated instruments that enabled them to record the apparent positions of celestial bodies to a degree of naked-eye accuracy not surpassed in Europe until the work of Tycho Brahe.

3

Islamic astronomy

Michael Hoskin and Owen Gingerich

It had taken Greek geometrical astronomers nearly half a millennium to achieve their goal of predicting with tolerable accuracy the positions of the seven 'planets' – the Sun, the Moon, Mercury, Venus, Mars, Jupiter and Saturn – which moved as individuals against the background of the 'fixed' stars. By the time Ptolemy's *Almagest* was written, in the second century after Christ, the golden age of Athenian culture was a distant memory, and the power of Rome itself was showing signs of decline. Some elements of Babylonian and Greek astronomy had already percolated into India, whence they were eventually to be restored to the mainstream tradition by Arabic authors. But although many Greek works (including the *Almagest*) would survive into the later Middle Ages in the original language, as manuscripts gathering dust in libraries in Constantinople and elsewhere, active study of the natural world was to continue within the cultural context of a religion not yet born, that of Islam.

In the mid-fourth century AD, a school was founded in Edessa in Mesopotamia by St Ephrem. The language spoken there was Syriac, but Greek was studied and some writings of Aristotle and of Ptolemy's distinguished medical contemporary, Galen, translated into Syriac. In 489, the school was closed by the Emperor Zeno, and some of the teachers moved further east, to Jundishapur. There they translated further medical and other writings from Greek into Syriac: modest in quantity, but providing a sample of the intellectual riches on offer to those who could gain access to works originally written in Greek.

The year 570 saw the birth of the prophet Muhammad at the trading centre of Mecca. Soon after his death in 632, the religion he had founded began to spread with astonishing speed, throughout the Middle East, and across north Africa and into Spain. In 762 his successors in the Middle East founded a new capital, Baghdad, by the river Tigris at the point of nearest approach of the Euphrates, and within

reach of the Christian physicians of Jundishapur. Members
of the Baghdad court called on them for advice, and these
encounters opened the eyes of prominent Muslims to the
existence of a legacy of intellectual treasures from
Antiquity – most of which were preserved in manuscripts
lying in distant libraries and written in a foreign tongue.
Harun al-Rashid (caliph from 786) and his successors sent
agents to the Byzantine empire to buy Greek manuscripts,
and early in the ninth century a translation centre, the
House of Wisdom, was established in Baghdad by the
Caliph al-Ma'mun. There Syriac- and Arabic-speaking
scholars, under the leadership of the Christian Hunayn ibn
Ishaq al-Ibadi (808–73), collaborated to translate the Greek
works, either from the original or from a Syriac version,
into Arabic.

That Arabic should emerge as an international scientific
language is itself surprising. Hitherto it had been the language of poetry, and of the Koran and of studies related to
the Islamic religion. Wherever Islam ruled, Arabic would be
understood, and the translation of these Greek works into
Arabic ensured their widespread diffusion, not only
throughout the Middle East but across north Africa and into
Islamic Spain. Eventually, mainly as a result of the
Christian advance in Spain against the Moors in the twelfth
century, these works would come into Christian hands, and
many would be translated into Latin. Although Greek writings that made their way into Latin through these cir-
cuitous routes (possibly via Syriac, Arabic and Castilian, for
example) would arrive somewhat the worse for having been
subjected to successive translations, they nevertheless provided an important continuity in traditions of astronomical
thought.

Astronomy and Islamic practices

Most of the 10,000 or so astronomical manuscripts written
in Arabic, Persian or Turkish that have come down to us
continue to rest undisturbed on the shelves of libraries.
However, it is clear that a large number of these writings on
astronomy are Islamic in character, in the sense that they
were the work either of religious scholars dealing with folk
astronomy, or of astronomers dealing with mathematical
techniques, each group addressing the same problems of
Islamic practice. Other extant astronomical writings are relatively few in number, though of considerable historical
significance; some attempted to perfect the planetary
models of the *Almagest* (which was among the works

translated, and later retranslated, in Baghdad), while others continued the debate between philosophers and astronomers over the nature of the cosmos.

Long before translations began, a rich tradition of folk astronomy already existed in the Arabian peninsula. This merged with the view of the heavens presented in Islamic commentaries and treatises, to create a simple cosmology based on the actual appearances of the sky and unsupported by any underlying theory. Meanwhile religious practices generated three specific challenges to which mathematical astronomers attempted solutions, the complexity of which often went beyond the actual needs of the community.

The first challenge arose out of the lunar calendar inherited in a modified form from pre-Islamic times. (The usual twelve lunar months in the year had previously been supplemented at intervals with an extra or 'intercalary' month to keep the year roughly in step with the seasons; but Muhammad apparently taught against such intercalations, so the Muslim 'year' was, and still is, consistently some eleven days short of the seasonal year. One result of this is that Ramadan, the sacred month of fasting, may – unlike Lent – occur in any season.) Each month began with the new moon – understood not astronomically (when the Sun, Moon and Earth were shown by calculation to be in a straight line), but practically, when the lunar crescent was first sighted in the evening sky. On the appropriate evenings witnesses would be sent to suitable locations with instructions to watch the western sky; if the lunar crescent was not seen one night, they would try again the following night.

This simple procedure led to difficulties. The skies were not always clear; and even if they were, the watchers of one town might see the new moon on an evening when those of the next had been unsuccessful, so that the two towns would begin their months on different days. Early Muslim astronomers followed a criterion they found in Indian sources, whereby the new moon would be seen if the difference in the setting times of the Sun and Moon on the local horizon was at least forty-eight minutes. Ptolemy's theory of the lunar motion was tolerably accurate around the time of new moon, but it described the Moon's position with respect to the Sun's path (the ecliptic), and this is inclined to the horizon. To relate this to motion with respect to the horizon was therefore a problem in spherical geometry. Later Muslim astronomers devised more elaborate conditions and compiled sophisticated tables to assist in the resulting calculations, leading to the production of almanacs with information on the possibility of sightings around the beginning of each month. But even today the problem remains a challenge to the Muslim world.

The second religious requirement that involved astronomy concerned the times of prayer, the number of which was established as five: sunset, nightfall, daybreak, midday and afternoon. The timing of the latter two, plus that of a voluntary midmorning prayer, corresponded to the ends of the third, sixth and ninth of the (variable) hours of daylight, and an approximate formula of Indian origin was used to relate these 'hours' to increases in shadow lengths. The problem of the translation of these rules into uniform hours and minutes aroused the interest of astronomers, though the muezzins who actually made the calls to prayer were more likely to follow a simple folk astronomy. Early in the ninth century a member of the House of Wisdom, al-Khwarizmi (a corruption of whose name gives us the word *algorithm*), compiled prayer-tables for the latitude of Baghdad, while the first tables for finding the time of day from the altitude of the Sun or the time of night from the altitudes of bright stars appeared in Baghdad soon after. Solving any of these problems involved finding the unknown sides or angles of a triangle on the celestial sphere from the known sides and angles. Thus the time of day could be derived from a triangle whose vertices were the zenith and the celestial North Pole (so that the line joining them was in the meridian), and the Sun's position. The observer had to know his geographical latitude (and hence the angle between his zenith and the Pole), and the Sun's position on the ecliptic for the time of year. He measured the altitude of the Sun, and the time was then given by the angle between the meridian and the Sun's hour circle (the arc from the Pole to the Sun). The necessary formulae were derived from two-dimensional projections of the celestial sphere, using techniques adapted from Indian sources. Later, spherical trigonometry was also used.

The method Ptolemy used to solve spherical triangles had been devised around the end of the first century AD by Menelaus of Alexandria. It involved a clumsy procedure whereby five segments had to be known in order to determine the unknown sixth quantity, but it sufficed to solve all the basic problems of spherical trigonometry. Unfortunately, however, this meant that telling the time from the Sun's altitude involved many applications of the theorem.

By the ninth century, the medieval equivalents of the six modern trigonometric functions had been recognized,

Calculating the time of day from the altitude of the Sun. The procedure involved finding the angle at the Pole in the triangle (shaded) whose vertices are the Sun, the observer's zenith, and the celestial North Pole. The zenith–Pole distance is the observer's latitude, the zenith–Sun distance is the complement of the Sun's altitude, and the Sun–Pole distance is (by definition) the complement of the Sun's 'declination' on that particular date. The angle at the Pole is proportional to the time.

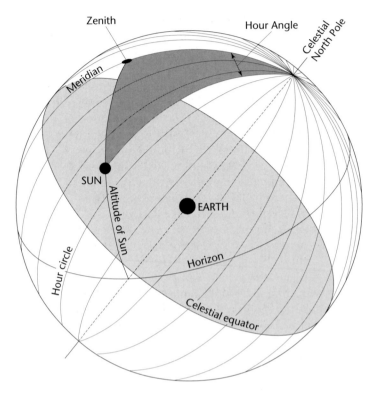

whereas Ptolemy had operated with only a single chord function. The concept of the sine of an angle was introduced into Islam from India, as were the tangent and cotangent functions important in calculations involving shadow lengths. Islamic astronomers discovered basic trigonometric identities that greatly simplified calculations involving triangles on the celestial sphere.

Eventually, the office of *muwaqqit* or timekeeper was instituted for mosques. This at last gave competent astronomers an institutional haven within one of the central structures of society, and the development resulted in a rapid increase in the quantity and quality of astronomical writings. Because of the hostility of Islam to astrology, astronomers who became muwaqqits were denied the financial rewards that could be earned by astrologers, but they were compensated by their secure and respected position in the community.

The third challenge posed by Islam to astronomy stemmed from the requirement that many acts of a religious nature, and more especially the orientation of mosques, should be directed towards the sacred shrine at Mecca known as the Kaaba. In the first two centuries of Islam many mosques, from Spain in the west to central

Asia in the east, paid token respect to this requirement by
facing south, on the grounds that when at Medina (which is
north of Mecca) the Prophet had faced south when praying.
In other places the *qibla*, or sacred orientation, was
assumed to be the direction taken by pilgrims leaving for
Mecca. Others again adopted an orientation of the rectangular shrine itself, whose major axis faced the rising of the
star Canopus and whose minor axis was aligned with the
summer sunrise and winter sunset. But in later centuries,
as muwaqqits and other professional astronomers applied
their minds to the problem of how to determine the qibla
mathematically using the available geographical data, formulae in spherical trigonometry were developed and tables
calculated from them. One outstanding achievement,
which seems to date from the eleventh century, was the
development of cartographic grids for Mecca-centred world-
maps, from which one could read off the qibla and distance
to Mecca directly. The culmination of this activity was a
table prepared at Damascus in the fourteenth century by
the muwaqqit al-Khalili, which gives the qibla for each
degree of latitude from 10° to 56° and for each degree of longitude from 1° to 60° east and west of Mecca, with the vast
majority of the entries correctly computed from complicated (and accurate) formula.

The emergence of observatories in Islam

The Koran declares that 'Nobody but God can know the
future', and Islamic religious leaders were as resolute as
their Christian counterparts in condemning astrology – and
as ineffective. Ruler and people alike saw astrology as of
great practical use, and they were prepared to pay for the
information they wanted. Astrologers who practised in the
market-place were little more than fortune-tellers, but in
princely courts and elsewhere were to be found astrologers
who were also astronomers and who based their astrological
predictions on tables of planetary positions.

Early Arabic works on planetary astronomy had been
eclectic, drawing on pre-Ptolemaic writings in Greek as
well as importations from Persia and India. The translation
of the *Almagest* transformed the situation by demonstrat-
ing the incomparable superiority of the Alexandrian
astronomer. However, the effectiveness of Ptolemy's planetary models depended not only on the geometric configurations themselves but also on the accuracy of the parameters
employed; and by now several centuries had passed and
improvement in many of these parameters was clearly

necessary. Ptolemy had shown his readers how to derive the parameters from observations and this was a lesson his Islamic successors learned well: we find that the observations they record are mostly designed for the improvement of the values used for such quantities as the eccentricity of the solar orbit and the obliquity of the ecliptic plane (the angle at which it is inclined to the celestial equator). These observations called for instruments of increased precision.

At first the instruments in question were small and portable, but in time the quest for greater accuracy led to a call for large and fixed instruments. Here and there, a ruler or other powerful patron would pay for the construction of such instruments; as these were no longer portable they required permanent homes, and these institutions marked the beginnings of astronomical observatories. But hostility to astrology on the part of the religious authorities made it all the more likely that the death of a patron, or even his loss of nerve in the face of criticism, would bring observing to an end.

Two of the larger observatories were actually demolished. In Cairo in 1120 construction of an observatory was begun on the order of the vizier of the Fatamid caliph, and when the vizier was murdered the following year work continued under his successor. But in 1125, when the instruments were constructed but the building was not yet complete, the new vizier was killed by order of the caliph, his alleged crimes including communication with Saturn. The observatory was demolished, and the personnel forced to flee for their lives.

A similar fate befell the observatory built in Istanbul for the astronomer Taqi al-Din by the Sultan Murad III. Its construction began in 1575 and was completed in 1577, which made it an exact contemporary of the first major observatory of northern Europe, that of Tycho Brahe (see pages 98–100). In addition to the main building, there was a smaller observatory, where

> fifteen distinguished men of science were in readiness in the service of Taqi al-Din. In the observations made with each instrument five wise and learned men cooperated: there were two or three observers, and the fourth was a clerk, and there was also a fifth person who performed miscellaneous work.

Building work finished just in time for observations of the bright comet of 1577. Taqi al-Din interpreted the apparition as boding well for the Sultan in his fight against the Persians, but the success of the Turks proved to be far from

complete, and other misfortunes befell the Sultan, including an outbreak of plague and the deaths of several notable
figures. In 1580 religious leaders persuaded him that
attempts to pry into the secrets of nature would only bring
misfortune, and he ordered the observatory to be destroyed
'from its apogee to its perigee'. Tycho's observatory did not
long outlive it: in 1588 Tycho's royal patron died, and soon
his observatory was in decline.

Only two Islamic observatories enjoyed more than a brief
existence. The first was at Maragha, the present-day
Maragheh in northern Iran. It was built for the great Persian
astronomer Nasir al-Din al-Tusi (1201–74) by the Mongol
ruler of Persia, Hulagu, who was addicted to astrology.

Construction began in 1259. The observatory (whose
foundations still survive) was built on the flattened top of a
hill, and included an extensive library; no doubt it was
there that the students were given part of their systematic
instruction. The instruments were set up in the open air.
They comprised a mural quadrant (an instrument fixed in
the north–south direction, for measuring altitudes) no less
than 14 feet in radius, an armillary sphere (a representation
of the principal astronomical circles, used for other
measurements of position) with circles of about 5 feet
radius, and numerous lesser instruments. With their help,
the team of astronomers (known to history as 'the Maragha
School') completed in 1271 a *zij*, or collection of astronomical tables with instructions for their use, composed in the
tradition of Ptolemy's *Handy Tables*.

Three years later al-Tusi left Maragha for Baghdad, where
he died. His departure from the observatory marks the end
of its creative period, though observations continued into
the next century.

The other major Islamic observatory was built at
Samarkand in central Asia, by Ulugh Beg (1394–1449), who
was a provincial governor long before he succeeded to the
throne in 1447. For once there had been no need to petition
for royal patronage: the most enthusiastic, and perhaps
most knowledgeable member of staff was Ulugh Beg
himself. Construction of a three-story building took place
in 1420. The major instrument was mounted out of doors,
in a trench, between two marble walls aligned north–south
and separated by some 20 inches. The instrument itself was
a form of sextant, its range being chosen so that the Sun,
Moon, and other planets could all be observed. The radius
of the sextant was over 130 feet, an illustration of Islamic
astronomers' mistaken conviction that increased size
would inevitably lead to increased accuracy.

The great achievement of Ulugh Beg's observatory was a set of astronomical tables that included a catalogue of over 1,000 stars. Many of its star positions had been established by the Samarkand astronomers themselves, making this the one important star catalogue of the Middle Ages. Ulugh Beg was murdered in 1449, by which time the observatory had enjoyed three decades of existence. As so often with the death of the patron, the end of observing soon followed.

Arabic planetary astronomy

An early example of a zij had been composed in the House of Wisdom at Baghdad by al-Khwarizmi. It derived from a Sanskrit astronomical work that had been brought to the Bagdhad court around 770 by a member of an Indian political mission, and most of the parameters used in the zij derived from Hindu astronomy, as did many of the computational procedures. In a version by a later astronomer, it was to be translated into Latin by Adelard of Bath in the twelfth century, and so became one of the ways in which Indian methods reached the medieval West.

One of the many astronomers who respected the Ptolemaic models but amended certain of his parameters was Muhammad al-Battani (c. 850–929), who spent most of his working life at al-Raqqa on the Euphrates. His zij, with an improved treatment of the orbit of the Sun in relation to the Earth, reached Christendom via Muslim Spain. The invention of printing eventually gave it wider circulation, and it was much used by Copernicus, who mentions its author in *De revolutionibus* no fewer than twenty-three times.

An astronomer whose work, by contrast, was unknown in the West in the Middle Ages was Abd al-Rahman ibn Yunus, who lived in Cairo in the late tenth century. He composed a major astronomical handbook called the *Hakimi Zij*, which he prefaced with a series of more than one hundred observations, mostly of eclipses and planetary conjunctions. His timekeeping tables were still in use in Cairo in the nineteenth century. The *Toledan Tables* prepared by the eleventh-century Moorish astronomer al-Zarqali (Azarquiel to the Latins), on the other hand, were translated early and had wide circulation; they became the model for the *Alfonsine Tables* that, from early in the fourteenth century, were to dominate this aspect of Latin astronomy into the Renaissance.

The first revision of the star catalogue in the *Almagest* was due to a tenth-century astronomer who worked in

Persia and in Baghdad, Abd al-Rahman al-Sufi (903–86). In his *Book on the Constellations of Fixed Stars* he gave improved magnitudes, and Arabic versions of the identifications, but left both the stars themselves and their (often inaccurate) relative positions unchanged. This is a symptom of the lack of commitment to observation that was common among Islamic astronomers: despite their industry in applying mathematics to astronomy, their workplace was often the study rather than the open air. Thus they were almost entirely oblivious of the supernova explosion of 1054 that gave rise to the Crab Nebula; only a single reference to this is known from Arabic sources. They had a versatile and convenient observing and calculating instrument in the astrolabe (see pages 62–7), but a single measurement with it was sufficient for most tasks. After all, astrology depended on the positions of planets, and for this the *Almagest* provided an excellent basis.

Imperfections in Ptolemaic astronomy of a more conceptual nature became clear from a comparison of the abstract, geometrical models of the *Almagest* with the same author's physical view of the cosmos as embodied in his *Planetary Hypotheses*. Already in the ninth century we find Thabit ibn Qurra (836–901), a Syriac-speaking scholar who worked in Baghdad, drawing attention to such inconsistencies, and by the tenth century texts appeared regularly whose subject matter was *shukuk*, or 'doubts', concerning Ptolemy. While we look in vain in Islam for a rebel who challenged the basis of the Earth-centred Aristotelian/Ptolemaic cosmology, there was philosophical unease over some of the aspects of the received planetary models, as there had been in Antiquity and as there would be in the Christian Middle Ages and Renaissance. The most obvious target for criticism was Ptolemy's device of the equant (see page 41), which led the planet to slow down and speed up and so violated a fundamental rule of Greek astronomy, that motions must be uniform. Eccentres (see page 35) were also objectionable in that they involved circular motions about centres other than the centre of the Earth, and even epicycles (see page 36) fell foul of Aristotelian fundamentalists.

The debate over eccentres and epicycles (which took place mainly in Islamic Spain) was essentially between philosophers and mathematical astronomers. An influential spokesman for the philosophers was the greatest Islamic exponent of Aristotle, the Andalusian Muhammad ibn Rushd (1126–98), who became known to the Latins as Averroës, or simply The Commentator. He accepted that the predictions from the Ptolemaic models might indeed save the appearances, but in his view only concentric spheres could compose the true universe. Most mathematical astronomers realized that to attempt planetary models on this principle was hopeless, but Averroës's contemporary and fellow Andalusian, Abu Ishaq al-Bitruji (Alpetragius), was bold enough to try. Not surprisingly, the results were grossly unsatisfactory; Saturn, for example, found itself deviating from the ecliptic by a monstrous 26° instead of the correct maximum of only 3°. Yet his attempt struck a sympathetic chord among Christian natural philosophers after his work was translated into Latin early in the thirteenth century.

Most attempts at reform of the layout of the planetary models were the work of Eastern mathematical astronomers, and were driven by theoretical considerations rather than the need to improve predictions. One of those who rejected the Ptolemaic equant was Ibn al-Haytham (965–c. 1040), later known to the Latins as Alhazen, who worked in Cairo in the early eleventh century. In his *On the Configuration of the World* he tried to adapt the planetary models of the *Almagest* so that they could take on physical reality. He saw the heavens as formed of concentric spherical shells, within whose thicknesses other shells and spheres were located. In this way he tried to assign a single spherical body to each of the simple motions of the Ptolemaic models. In the thirteenth century his work was translated into Castilian in the court of Alfonso the Wise, and not long after from Castilian into Latin. The concept of separate celestial spheres for each component of the planetary motions was to gain currency in the later fifteenth century through the work of Georg Peurbach.

The equant was enough to arouse misgivings even among observational astronomers. In the thirteenth-century the Maragha astronomer al-Tusi, in his *Tadhkira* or *Memorandum*, succeeded in devising a satisfactory substitute that used only uniform circular motion: two small epicycles were to be added to each of the planetary models, a complication that seemed a price worth paying.

Eccentres, though less shocking than equants, still represented a violation of Aristotelian doctrine. The most competent attempt to purge the Ptolemaic models of all their objectionable features and to replace them with models that were acceptable both philosophically and observationally was made around 1350 by Ibn al-Shatir, muwaqqit of the Umayyad mosque at Damascus. Ibn al-Shatir's lunar model avoided the gross variation in the diameter of the Moon

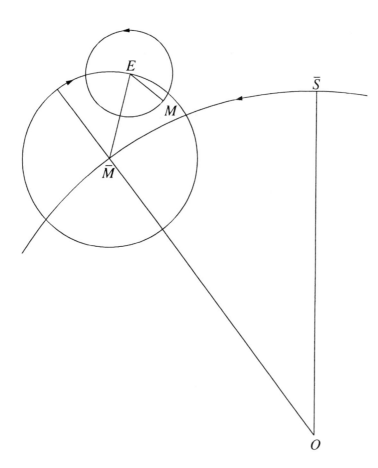

The model of the motion
of the Moon proposed in
the fourteenth century by
Ibn al-Shatir, an example of
his use of a pair of epi-
cycles in order to avoid the
equant of Ptolemy, which
he considered unacceptable
in principle. The Earth is at
O, the centre of the universe, and $\bar{S}$ is the 'mean
Sun'. $\bar{M}$ (the 'mean Moon')
moves uniformly from
west to east on the circle
centre O. The primary epicycle, centre $\bar{M}$, rotates in
the opposite direction, carrying the point E. The true
Moon M moves on the secondary epicycle whose
centre is E and which
rotates from west to east.
With the values chosen by
Ibn al-Shatir for the radii
and speeds of rotation of
the various circles, not
only is the motion of the
Moon across the sky well
reproduced, but the
changes in the distance of
the Moon from the Earth
are moderate. This avoided
a striking defect in
Ptolemy's lunar theory, in
which the distance of the
Moon (and hence its diam-
eter as it should appear
from Earth) varied by a
factor of 2.

that Ptolemy's model had implied (see page 44); his solar
model was based upon new observations of the solar diameter; and all his planetary models were free not only from the
equant but also from eccentres. True, he could not do
without epicycles; but the existence of individual stars
demonstrated that Aristotle had gone too far in his insistence that the heavens were composed throughout of
uniform matter, and so even an Aristotelian ought to keep
an open mind over the possible existence of epicycles.

Though not the last mathematical astronomer to write
in Arabic, Ibn al-Shatir represented the culmination of a
movement that had been maturing for half a millennium.
But the moment had passed when Arabic astronomers
could influence the future course of planetary theory. The
enthusiasm in the Latin world for translations from the
Arabic had long since abated; the Latins were developing
their own astronomical tradition, and the writings of al-
Tusi and Ibn al-Shatir seem to have been virtually unknown
in the West.

But if so, historians have to accept a strange coincidence; for when Ibn al-Shatir's work was rediscovered in modern times, it was realized that it employed geometrical devices similar to those later used by Copernicus, who was likewise scandalized by the Ptolemaic equant. Copernicus took the radical step of turning the Earth into a planet, but when it came to developing detailed models, many of the problems he encountered were not so very different from those confronted by his predecessors. In the *Commentariolus*, a preliminary sketch of his Sun-centred theory that circulated in manuscript in the early years of the sixteenth century, Copernicus used an arrangement equivalent to Ibn al-Shatir's in order to eliminate the equant and generate the intricate changes in the Earth's orbit. In the fully developed *De revolutionibus* (1543), Copernicus reverted to the use of eccentric orbits, but he used a model that was the Sun-centred equivalent of one developed at the observatory founded by al-Tusi at Maragha in Persia. No Latin translation of these Arabic works has been found, nor is any Latin account of them known. A Greek translation of some of al-Tusi's writings found its way to Italy in the aftermath of the fall of Constantinople in 1453, and Copernicus studied in Italy from 1496 to 1503 and acquired a knowledge of Greek. But whether he owed to others his technique for replacing the equant, or whether he developed the method independently, is for the present moot.

The astrolabe
Michael Hoskin

By far the most sophisticated (and historically important) astronomical instrument of the Middle Ages was the planispheric astrolabe. It was one of four forms of astrolabe, but two of the other forms, the linear and the spherical, were very rare, and the mariner's astrolabe was a crude tool for use at sea and seemingly developed only towards the end of the Middle Ages.

The origins of the astrolabe go back to Greek Antiquity; it reached maturity in Islam (where it was to retain its popularity into modern times), and then received further refinements in the West. The basis of the instrument is a disc of brass which can be suspended from a ring. The back of the disc is essentially an observing instrument, being fitted with an observing bar or *alidade* that rotates about a pin in the centre of the disc and is used for measuring the altitude of celestial bodies. The observer simply suspends the instrument from the ring so that it hangs vertically, looks along the alidade towards the body in question, and reads off its altitude from a scale engraved around the circumference of the disc.

Also on the back of the instrument in

The back of an astrolabe. The observer holds the instrument suspended from the ring at the top, and measures the altitude of a celestial body by sighting along the observing bar and reading the angle on the outermost scale. The bar can also be used to determine the current position of the Sun on the ecliptic: the user aligns the bar with the current date on the circle engraved with the days of the year, and the position of the Sun is shown by the intersection of the bar with the circle engraved with the signs of the zodiac. Reproduced with kind permission of Paul Freestone Photography, and the Museum of the History of Science, Oxford.

A fourteenth-century astrolabe belonging to Merton College, Oxford. Reproduced with kind permission, of the warden and fellows of Merton College, Thomas Photos, Oxford.

its Western form are two circular scales which together give the position of the Sun on the ecliptic (its path against the background of the stars) for any date in the year. One scale is engraved with the days of the year, and the other shows the corresponding position of the Sun; the alidade can be used to line up corresponding points.

The front of the astrolabe is a calculating device that embodies representations both of the heavens (the individual stars together with the ecliptic, equator, tropics and so on), and of the local system of coordinates (angular altitude above the horizon, angle of 'azimuth' around the horizon, usually from south) in which the user will be making

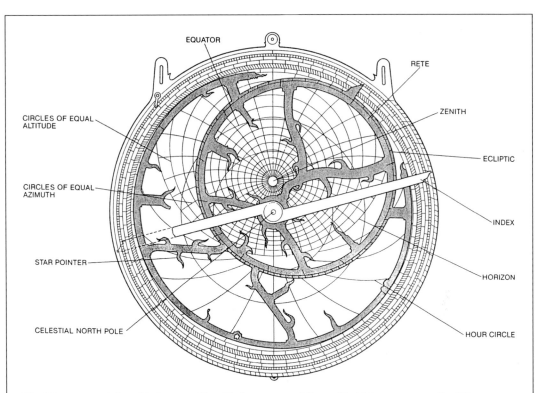

The Merton astrolabe with features identified. The hour circles were used to tell the time with the period of daylight divided into twelve equal 'hours' and the period of dark divided into twelve equal 'hours' (which were normally different from the daylight 'hours'). Copyright Scientific American, Inc. All rights reserved.

measurements of the positions of heavenly bodies.

To make an astrolabe, we need to transfer the sky onto our sheet of brass. Fortunately, we know from geometry how to set up a projection between the points of the celestial sphere and the points of an (infinite) plane, so that each point on the sphere projects into exactly one point on the plane. We simply imagine the line joining the celestial South Pole to any other celestial point, and take the projection of the point to be the place where this line intersects the plane that contains the celestial equator. Such a projection has the valuable and surprising property that angles between curves on the heavenly sphere are unchanged after projection, so that problems in spherical triangles can be converted into more tractable problems in plane trigonometry.

The brass disc of our astrolabe is a physical replica of the plane through the celestial equator, onto which we have projected (that is, represented) the celestial sphere. Unfortunately the disc is limited in size, and this means that in practice not all of the celestial sphere can be represented on it. But as the Muslim and Christian users of astrolabes never saw the skies near the celestial South Pole, no harm was done if the southern skies were left unrepresented. This was the reason behind the choice of the South Pole as the centre of projection.

At the very centre of our disc of brass is the representation of the celestial North Pole. Around it, three concentric circles represent respectively the Tropic of Cancer, the equator, and the Tropic of Capricorn. We choose to make this latter circle the outer limit of our disc, so that it is the skies that lie further south of the Tropic of Capricorn that are not represented.

The observations that the user makes with the alidade of the astrolabe are of the angular altitudes of heavenly bodies, and these are angles within a co-ordinate system in which the horizon is 0° and the zenith 90°. To enable us to make convenient use of these observations, circles of equal altitudes (0°, 10°, up to 90°) must be represented in the projection, along with circles of equal azimuth. The problem is that these circles depend upon the latitude of the place from which the observations are being made, and so therefore do their projections. The solution adopted was to supply the astrolabe with discs for a range of latitudes, each engraved with the projection of the appropriate coordinate circles. These *climates* were stored in the instrument one above the another, and the user would simply select the most appropriate one available and place it at the top of the pile.

So far, every feature of the representation we have mentioned is fixed and static. But we now need to project the individual stars of the heavenly sphere (other than those too far south, of course); and since the heavens rotate, the projection must rotate. Had transparent plastic been available in the Middle Ages, no doubt the projected stars would have been engraved on a plastic disc that rotated about the central point that represents the celestial North Pole;

through the plastic the user would have been able to see the co-ordinate circles engraved on the climate below. Instead, the medieval maker took another disc of brass (the *rete*), and marked on it the projections of the important moving features of the sky (the ecliptic path of the Sun and 'pointers' to indicate the positions of the principal stars). Retaining these, he cut away as much of the rest of the brass as possible, thereby exposing the climate below with its coordinate grid.

The heavens spin: in modern parlance, the heavenly sphere has just one degree of freedom. Therefore, only a single observational fact is sufficient to locate all the stars in the position they occupy at the time of the observation. If the astronomer, for example, has measured the altitude of Sirius with the back of his astrolabe (and has verified whether the star is rising or setting), he may then retire to his study and rotate the rete until the representation of Sirius is appropriately located over the corresponding altitude circle in the coordinate grid. This done, he now has not only Sirius correctly located, but all the other stars as well; he can determine their current positions, he can tell which of them are about to rise and which have just set, and so on.

A basic use of the astrolabe was to tell the time. The pair of scales on the back of the astrolabe tells the user the position of the Sun for the day in question, and so he knows where to mark its position on the projection of the ecliptic. In a Western astrolabe an outer circle on the front was engraved with twenty-four equal divisions, to represent the hours taken by the heavens to complete one rotation (Islamic astrolabists followed a related procedure). An *index* bar allowed the user to align the projected position of the Sun with this outer circle and so read off the time. In other words, the astrolabe is a twenty-four-hour clock that permits the time to be determined from a single observation – of the Sun in the daytime, or of a star at night.

Alternatively, the astrolabe can be used to predict the time when an astronomical event will happen. So, for example, if the user wishes to know the hour when the Sun will rise, he need only rotate his rete until the Sun is on the eastern horizon, and he can then use the index to read off the time when this will happen.

These are only a few of the applications of this very remarkable instrument, which in the Islamic and the Christian Middle Ages was fundamental to astronomy, astrology, and astrological medicine alike.

4

Medieval Latin astronomy

Michael Hoskin and Owen Gingerich

In the second century BC Roman soldiers had conquered the Greek city states, but it was Greek culture that had triumphed over the Roman mind. Works written in Latin scarcely featured in our discussion of astronomy in Antiquity, because within the *pax romana* the language of cultivated men was Greek: among writers on astronomy Latin was mainly reserved for low-level didactic works.

The final disintegration of the *pax* came about in the second half of the fifth century AD, and thereafter everyday life in the regions of western Europe became increasingly precarious. For the few to whom scholarly writings still mattered, access to such works became ever more difficult, and an ability to read the Greek in which they were written became a rarity.

One scholar made a desperate attempt to render the outstanding works of Greek philosophy into Latin before it was too late: Anicius Manilius Severinus Boethius (c. 480–524/5), a high official in the Roman Gothic kingdom. His most ambitious plan was to 'translate and comment upon as many works by Aristotle and Plato as I can get hold of'. Sadly, he angered his master Theodoric by his refusal to stand silently by while injustices were perpetrated, and he was tortured and executed, his grand task scarcely begun. But what he did achieve was to prove significant. In particular, he translated or paraphrased a number of Greek writings on logic and assembled them along with logical writings by Latin authors such as Cicero, so providing students living centuries later and in happier times with a corpus of secular writings that they might compare and contrast, discovering in the process how to display an independence of mind. Logic would become an obsession in the medieval universities, most of whose Arts teachers would be more concerned with the status of Aristotelian spheres or Ptolemaic equants than with the effectiveness of the planetary models constructed out of them.

By his own writings and by his translations and commentaries, Boethius codified the existing tradition of mathematical studies into what would later become the standard university format of the *quadrivium*: arithmetic (the abstract theory of discrete quantities); harmony (the same discipline applied to nature); geometry (the abstract theory of continuous quantities); and astronomy (the same discipline applied to nature). However, it so happened that his own quadrivial writings that survived were mainly on arithmetic and harmony, rather than astronomy.

The 'Dark' Ages

One possible definition of the Middle Ages is the period between the fall of Rome and the fall of Constantinople (in 1453). Another is the period when few Western scholars knew Greek. For the time being this ignorance of Greek scarcely mattered, since Greek works were rarely accessible in the original. Sadly, neither were they available in translation. Aside from the legacy of Boethius, the most significant translation from the Greek was due to Calcidius (it is a sign of his disturbed times that we are not sure if he lived in the fourth or the fifth century). He rendered into Latin two-thirds of Plato's cosmological myth, the *Timaeus*, and wrote a lengthy and influential commentary. Remarkably, even in the later Middle Ages, when the whole of the vast Aristotelian corpus had been translated into Latin and was dominating the learned world, Calcidius's version of the *Timaeus* was still the most influential of the tiny handful of Plato's dialogues available.

Writings composed in Latin in the middle centuries of the first millennium AD bear testimony to the dire state of secular studies. The best Latin treatise on astronomy from this period appeared in *The Nuptials of Philology and Mercury* by Martianus Capella of Carthage (c. 365–440). This work was an allegory of a heavenly marriage in which seven bridesmaids presented a compendium of astronomy and the other six liberal arts. In particular, it contains a statement of the theory, nowadays often ascribed to Heracleides of Pontus (c. 390–c. 310 BC), that Venus and Mercury always appeared close to the Sun because they circled the Sun while all three circled the Earth. In the Middle Ages, the Capellan tradition became entangled with another from Pliny, sometimes with the result that diagrams showed the orbits of Mercury and Venus curiously interlinked as they circled the Sun. In any event, diagrams and passages such as these would ensure that in the medieval universities, dissenting voices would always be heard, and Martianus Capella was to be cited with approval (albeit some puzzlement) by Copernicus.

Ambrosius Theodosius Macrobius, who was apparently also from North Africa and who lived in the early fifth century, was the author of a commentary on the *Dream of Scipio* of the Latin author, Cicero. Macrobius took the opportunity to expound a cosmology deriving ultimately from Plato and the Pythagoreans, in which numbers underlay all things. The commentary also incorporated a popular handbook on astronomy. A spherical Earth lay at the centre of the spherical universe, encircled by seven planetary spheres and, at the outside, a starry sphere; this rotated

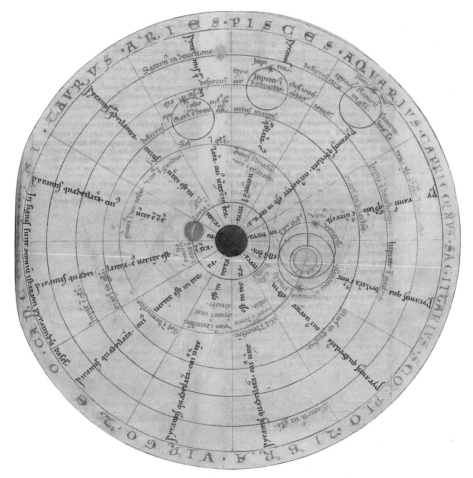

A thirteenth-century manuscript showing a curious variant of the system of Martianus Capella (c. 365–440). Venus and Mercury are carried around the Earth by the Sun, but the paths along which they circle the Sun are interlinked. Notice that the three superior planets (Mars, Jupiter, and Saturn, in the upper part of the diagram) have the customary Ptolemaic epicycles. Reproduced with kind permission of the Manuscript Department, The Royal Library, Copenhagen Denmark. (MS GKS 277, 2°, fol. 143r.)

daily from east to west, dragging along the planetary
spheres, though each of these also had its own motion in
the opposite direction. Significantly, Macrobius left the
order of the planets vague, because his sources – Plato and
Cicero – differed on the place of the Sun in the sequence.
Like Martianus Capella, Macrobius related the theory of
Crates of Mallus (second century BC) that the Earth was
divided by oceans into four quarters each of which was
inhabited, and he reported Eratosthenes's value (see page 26)
for the circumference of the Earth.

As in Islam, so in Christendom religious requirements
posed questions whose answers demanded some study of
the heavens. Monasteries needed to know the time at night
so that the community could rise and sing the office
psalms; this was often done by observation of the stars, the
monks having to resort to more mundane methods – water
clocks or rhythmic chanting – when the stars could not be
seen. Of wider significance was the determination of the
appropriate date for the great festival of Easter; in conformity with its origins in Jewish observances, Easter was
linked with the first full moon after the spring equinox, so
that its dating depended on both the lunar month and the
solar year. Church authorities needed to publish Easter
dates for some years ahead, so that the feast could be celebrated on the same day throughout Christendom. One
possibility would have been for the bishop of Alexandria to
invite local astronomers, as the inheritors of accurate
values for the month and year that had been handed down
from the Babylonians, to carry out calculations of the date
of the spring equinox, year by year. Instead, the authorities
took the more practical course of looking for a cycle in
which a certain number of solar years would equal (with
sufficient accuracy) a number of lunar months: the dates of
Easter would then be calculated within the cycle, and
simply repeated in succeeding cycles. From the third
century onwards various cycles were tried ranging from 8
years to 84, until the Greek Metonic cycle (see page 24) of
19 years was recognized as the most suitable. 19 years differ
from 235 lunar months by only one-twelfth of a day, and so
such a cycle departs from nature by only one day in 12×19
years, or more than two centuries.

The definitive treatise on the Easter problem was written
in 725 by the Venerable Bede (672/673–735), monk of
Jarrow in the north of England. In his *On the Divisions of
Time*, he set out for the date of Easter a cycle of
$19 \times 28 = 532$ years. Under the Julian calendar, a leap year
occurred every fourth year (without exception). Then as

now, the weekday on which a given day of a given month
fell normally advanced by one day each year, but by two
when the extra day of a leap year intervened. Accordingly,
in 4 years, the weekday invariably advanced a total of 5
days, and in $4 \times 7 = 28$ years it advanced by 5×7 days or
exactly 5 weeks, thereby repeating itself. Since the Metonic
cycle of 19 years dealt adequately with the solar/lunar component of the Easter dating, and any cycle of 28 years
repeated the day of the week, a compound cycle of 19×28
years took into account the requirement that Easter Day be
a Sunday.

The revival of astronomy in the West

Writers such as Macrobius had helped preserve fundamental concepts of Greek astronomy such as the sphericity
of Earth and its location at the centre of a spherical universe, and the distinction between the seven planets and
the 'fixed' stars. But compared to the sophistication of the
Almagest, knowledge of astronomy among the Latins in the
second half of the first millennium was primitive in the
extreme. The tenth century, however, witnessed the first
contacts with Islamic centres of learning, and monasteries
on the southern slopes of the Pyrenees became centres for
the transmission of Arabic culture. One influential scholar
to visit Spain and study there was Gerbert of Aurillac (later
Pope Sylvester II; c. 945–1003). After his return, Gerbert
was given charge of the cathedral school of Reims, which
soon began to attract students from far and wide. Gerbert
may well have brought from Spain knowledge of the astrolabe, that sophisticated and versatile instrument for both
measurement and calculation. Certainly the instrument
itself had appeared in the West by about 1025, and soon
thereafter two Latin treatises on it were composed (or
adapted from the Arabic) by an Austrian monk, Hermann
the Cripple (Hermannus Contractus, 1013–54).

The astrolabe allowed astronomy to become once more a
mathematical science, for with it an astronomer could
measure the angle between the horizon and the position of
a heavenly body. As in Islam, astrology had always retained
a certain popularity despite forthright expressions of dis-
approval from religious leaders, and this popularity would
increase with the passage of time, most notably in the
demoralization that followed the terrible Black Death of
1348–50. Nor was astrology wholly without a basis in
reason. So, for example, the Aristotelian analogy between
the macrocosm (the world in the large) and the microcosm

The Aristotelian analogy between the world in the large and the individual living body enabled physicians to use astrology as a guide to the treatment of patients. This fifteenth-century figure instructs them in the correspondence between the various organs of the human body and the planets and 'houses' of the zodiac. Reproduced with kind permission, Wellcome Institute Library, London.

(the individual living body) enabled medical treatment of a patient's organs to be planned in the light of the favourable or unfavourable disposition of the corresponding heavenly bodies. The teachers in university medical faculties would include instructors in astronomy/astrology, and some knowledge of the influence exerted by the planets, and therefore of the celestial configurations that generated this influence, would be expected of every practising physician. The English poet Geoffrey Chaucer says of the physician among the pilgrims on their way to Canterbury, that 'he was grounded in astronomy', and so 'well could he fortunen the ascendent'.

Astrology therefore generated an incentive to master the movements of the planets, one that went far beyond the disinterested pursuit of astronomy for its own sake. Accordingly, the tables that had been compiled by Islamic astronomers and the rules for their use – and the theories that had underlain them – aroused keen interest among the translators attracted to Spain in the twelfth century, as the tide of Islam receded and left in Christian hands manuscripts that embodied the philosophical and scientific achievements of both Greece and Islam. The fall of the great intellectual centre of Toledo in 1085 opened the floodgates. Gerard of Cremona (c. 1114–87) was one of the translators who flocked there, and his companions were later to list no fewer than seventy-one astronomical and other works translated by him. Surprisingly, modern scholars agree that even this list is not exhaustive, and credit Gerard with translating additional works, including the rules (*Canons*) for the use of the *Toledan Tables* of al-Zarqali. These tables were adapted for various longitudes in western Europe and enjoyed an immense success. Their availability stimulated the study of astronomy, for it was now possible to compute planetary positions for any moment in time, and to compare these predictions with observations. However, the underlying theories remained a mystery until the *Almagest* itself became available – though to minds accustomed to astronomical treatises of the level of Boethius and Martianus Capella, it was at first wholly indigestible. In Sicily, which had been ruled in turn by Greeks, Muslims and Latins, translations likewise got under way. Communications with Constantinople were restored, and from the libraries there came another copy of the *Almagest*, this time in the original Greek.

It is hard for us today to comprehend the impact of the translations of works from Antiquity and the Islamic Middle Ages, as they descended upon the newly-fledged universities of the Latin West in the twelfth century. When a teacher in a Faculty of Arts first laid hands on the *Almagest*, it must have represented a level of technical virtuosity exceeding by several orders of magnitude any astronomical writings he had previously known.

Paris University soon established itself as the great intellectual centre of Christendom. In a period when all educated men could write and speak Latin, there were few obstacles (other than financial) to prevent the best teachers, and the best students, gravitating towards a recognized centre of excellence, and Paris in the thirteenth century attracted an astonishing galaxy of talent. At Paris, as elsewhere, the role of the Faculty of Arts was simply to provide students with the basic education that was all that most of them would need. The Faculty's teaching was structured initially around the seven Liberal Arts: the literary *trivium* of grammar, rhetoric and logic, and the mathematical *quadrivium* of arithmetic, harmony, geometry and astronomy. The students of Arts were mostly young – much younger than is normal today – and the Arts teachers in Paris were acutely conscious of the prestige of the higher faculties (medicine, law, and especially theology), to which the more ambitious students might progress after becoming

Masters of Arts. They were therefore exhilarated to find
that most of the non-medical works that were now arriving
in translation clearly belonged to the Faculty of Arts. In particular, Aristotle's writings owed nothing to the Hebraeo-
Christian revelation that was the subject-matter of the
Faculty of Theology. The new works therefore provided the
Masters of Arts with a weapon to use in their struggle to
improve their status *vis-à-vis* the theologians.

   This political dimension added to the inevitable difficulties of assimilating a pagan Aristotelian world-picture into
a culture that had been dominated for a millennium by
Christian theology, and as a result the University of Paris
was in turmoil for much of the thirteenth century. The synthesis eventually achieved there by the Italian Dominican
friar, Thomas Aquinas (1225–74), propagated an immensely-
powerful cosmology, in which the teachings of Aristotle
concerning the natural world – so many of them plain
common sense anyway – seemed to acquire an aura of
revelation.

   Most of the Aristotelian works were eventually incorporated into the last years of the Arts course, where they were
divided between the three 'philosophies' of metaphysics,
ethics, and natural philosophy. This meant there was now
less time for the quadrivium, where a very rudimentary
astronomy had long been taught; but this loss was more
than offset by the teaching, in the courses on natural philosophy, of Aristotelian material relating to astronomy. As
to the *Almagest* and the major Arabic works in mathematical astronomy, no young student in Arts could think for a
moment of attempting to master them; manuscripts, even
if available, were rare, prohibitively expensive, and
subject to copyist's errors, and in any case the conditions
under which students worked precluded the study of anything other than the most elementary treatises of astronomy.

   Furthermore, research – the enlargement of the frontiers
of knowledge – was not the function of a university, whose
duty was rather to educate. This was done through the
systematic study of approved texts of recognized quality,
and through intensive training in the skills of logical reasoning. The typical medieval Arts student was no
mathematical astronomer, and certainly there was no place
for observation in his astronomy course; but on the other
hand, once introduced to epicycles or equants, he was likely
to form the most decided opinions about their reality or
validity.

   He might also have opinions about the heavens beyond

the planets, a region in which both astronomers and theologians had a stake. How did the multiple motions of the sphere of the fixed stars come about? What was the relationship between this sphere and 'the heaven' created on the first day according to the story in Genesis – and between that and 'the firmament' brought forth on the second day and made visible on the fourth? And what was to be made of 'the waters above the firmament'?

Everyone agreed that in addition to their daily motion, the fixed stars performed the slow movement that in fact results from the wobble of the Earth's axis ('precession'). Many astronomers wrongly believed that the rate of precession itself varied, and that this involved a third movement known as 'trepidation'. To generate these three movements, writers often thought it necessary to postulate three spheres. Albert of Saxony (c. 1316–90), for example, a leading figure first at Paris and then at the University of Vienna, is typical in assigning precession to the eighth sphere, trepidation to the ninth, and the daily motion to the tenth.

But how did these astronomical spheres relate to the heavenly realms described in Genesis? The 'firmament' was frequently identified with the eighth sphere, in which lay the fixed stars themselves. 'The waters above the firmament' were often thought of as being as hard as crystal, and as forming the ninth sphere (or the ninth and tenth). 'The heaven' created on the first day was the outermost (usually the eleventh) of all the spheres, the motionless *Empyreum*, which served a purely theological purpose. The Empyreum was the ultimate container of the universe, and the dwelling place of God and the elect.

The translated works in astronomy were of a quality that soon made the introductory texts previously in use no longer acceptable. The first to fill the gap was John of Holywood, better known by the Latinized form of his name, Sacrobosco, who taught in Paris in the mid-thirteenth century. He provided students with three short works that together supplied their needs in astronomy: a *Compotus* or introduction to time-reckoning, an *Algorismus* that taught the arithmetic needed in astronomical computation, and a *Tractatus de sphaera* (or *Sphere*).

Sacrobosco's *Sphere* contained four 'books'. Book I dealt with the heavenly sphere, its revolution, the sphericity of the Earth, and its location at the centre; Book II defined the celestial equator, the ecliptic, the zodiac and zodiacal constellations, the meridian, the altitude of the pole, and the division of the Earth by means of the tropics and polar

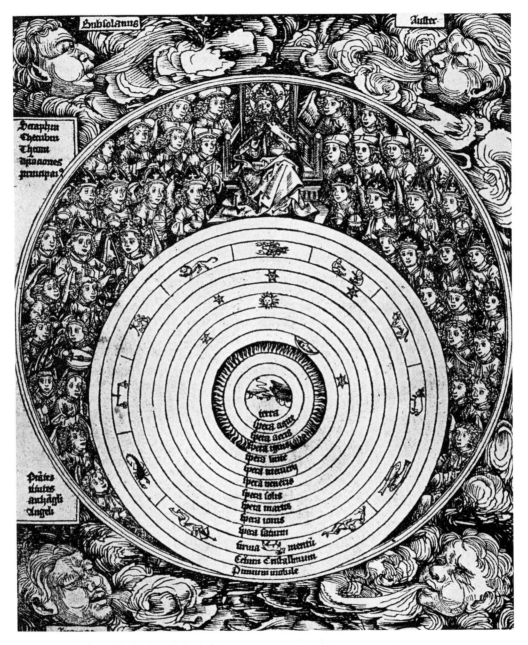

The Christianized cosmos as represented in the *Nuremberg Chronicle*, a world history published in 1493. At the centre is the region of the four elements, surrounded by the seven planetary spheres, after which come the firmament, the crystalline heaven, and the First Mover postulated long ago by Aristotle. At the outside we see God enthroned, with the nine orders of angels. The corners are decorated with four winds. From Hartmann Schedel, *Liber chronicarum*, 1493.

circles; Book III dealt with the rising and setting of celestial bodies, and the length of day and night at different latitudes and 'climates'; and Book IV gave a brief description of the motion of the Sun, Moon and planets, and an outline theory of eclipses.

While the first three books were of a tolerable standard, if only just, the fourth was hopelessly inadequate, and teachers needed something better. The most popular of the texts written to fill the gap was the anonymous *Theory of the Planets*, written in the second half of the thirteenth century. It began with Hipparchus's theory of the Sun, and went on to the Ptolemaic theories of the Moon and the outer planets. The author then had chapters on Venus and Mercury, and on the direct and retrograde motion of planets. He described the geometrical models briefly but well, with definitions clearly stated. He was, however, less satisfactory on the other topics he tackled, such as eclipses.

In time other authors were to write treatises to make good the shortcomings in the *Theory*, just as the *Theory* made good shortcomings in the *Sphere*. The historian of medieval astronomy is able to monitor the steady improvement in the technical quality of teaching in astronomy, by studying the works that were grouped together in surviving manuscripts to form the set of treatises that a student needed: the quality of the sets steadily improved as the less satisfactory works dropped out and were replaced by better ones.

Other works that we find in such collections are treatises on the calendar, and the *Alfonsine Tables* of the planetary motions, named after the thirteenth-century patron of astronomy, King Alfonso X of Castile (though some modern historians believe the tables are in fact of French origin). These tables soon replaced the old *Toledan Tables* inherited from the Arabs. Modern computer analysis has shown that they were calculated from Ptolemaic models of the planetary motions, with only occasionally modified parameters.

Another striking feature of the manuscript collections of the time is the new interest in astronomical instruments. These instruments were, like the astrolabe, small and portable. There was therefore no need for purpose-built observatories, which were unknown in the Latin West and would remain so until the time of Tycho Brahe in the late sixteenth century, any observations being taken from private homes. So, for example, an astronomer of Roskilde in Denmark in 1274 used an astrolabe to measure the noon altitude of the Sun every day of the year; from this could be calculated the length of daylight.

One of the instruments, the 'old quadrant', was described in an encyclopedia compiled by Alfonso's astronomers. This brass quarter-circle could be used to measure altitudes, the observer sighting the target object along one edge of the instrument and then reading the altitude from the position of a plumb-bob on the scale on the circumference. But its sophistication lay in the engraved, curved hour lines and adjustable scale, which enabled it to be used as a sundial for any latitude. It was supplemented at the end of the thirteenth century by the much more complex 'new quadrant' (a combination of quadrant and astrolabe) of the

A navigator learns how to use a cross-staff to measure the angle between the Moon and a particular star. Elsewhere in this woodcut of 1533, surveyors are using the instrument to determine the height of a tower. To use a cross-staff to measure the angle separating two objects, the observer slides the cross bar towards or away from himself until each object is in line with one or other end of the bar; the angle can then be read off from the scale on the central member. A version of this instrument was later in widespread use among seamen, although it suffered from the defect that in moving the cross bar, the observer affected the alignment of both objects: it was not feasible to align one object and then adjust for the other. From J. Werner and P. Apianus, 'Introductio geographia', in Doctissimas in Verneri annotationes, 1533.

Montpellier astronomer Profatius the Jew (Jacob ben Mahir, c. 1236–c. 1304). Another Jewish astronomer, Levi ben Gerson (1288–1344) of Provence, invented the cross-staff, which allowed the angle between two objects (such as stars) to be measured.

The cross-staff was to be one of the instruments used by European navigators who, as the Middle Ages drew to a close, began to venture ever more frequently out of sight of land. Mistakes in position could easily lead to shipwreck, and so navigators usually kept to basics: they followed a course that brought them to the required latitude well to the west (or east) of their destination, confirmed their latitude by observation, and then travelled east (or west) along the parallel of latitude.

Measurement of latitude was therefore of great importance. In the daytime, latitude could be determined from the altitude of the Sun at noon (that is, when it was highest in the sky). To measure this, the navigator would use a cross-staff or some other instrument such as a mariner's astrolabe. He would then consult a table that gave the position of the Sun above (or below) the celestial equator for the time of year; combining his noon observation with the angle given in the table, he obtained the altitude of the celestial equator as seen from his ship, and hence the ship's latitude.

Had the Pole Star been located *exactly* at the celestial pole, the navigator at night could simply have used the altitude of the star as the ship's latitude. Unfortunately this was not the case – in 1500 the star was some 3° from the pole – and some adjustment was necessary.

The current position of the Pole Star in relation to the pole could be inferred from the configuration of the other stars rotating around the pole. In practice navigators normally used the 'Guards', two stars in the constellation of Ursa Major (the Plough or Big Dipper) that happened to be in line with the Pole Star. This line turned in the sky like the hand of a 24-hour clock. There was however a complication: the clock ran on star ('sidereal') time rather than Sun time. We see the Sun and stars go round every day, but because the Sun also circles once a year among the background stars, the 'day' of a star is some 4 minutes shorter than that of the Sun, and this difference accumulates steadily throughout the year.

An ingenious instrument called a 'nocturnal' allowed the navigator either to calculate the required adjustment in altitude, or simply to tell the (conventional) time at night. It was fitted with two concentric circles. The outer circle, which was attached to the handle, was graduated with an annual calendar, the handle itself being located at the day when the two stars crossed the meridian at noon. The inner circle, which rotated freely, was graduated into 24 hours.

The navigator rotated the inner circle until the noon mark was aligned with the date on the outer circle; this took care of the difference between sidereal and Sun time. He then held the instrument above him so that it was roughly in the plane of the celestial equator. He looked at the Pole Star through a hole in the centre, and he then took hold of an index bar, which rotated about the centre and projected beyond the edge of the circles, and turned it until the bar touched the two stars. The time was then shown by where the index bar crossed the hour circle.

To find out what adjustment to make to the altitude of the Pole Star in order to get the altitude of the true pole, and hence the latitude of the ship, the navigator simply consulted tables known as 'The Regiment of the Pole Star'. If he did not have such tables to hand, he could resort to using a scale on the reverse of the instrument.

A technological innovation of the Middle Ages that, centuries later, would solve the navigator's problem of longitude (see pages 150–4), as well as facilitating the compilation of star catalogues of precision, was the introduction of clockwork. The mechanical clock is recorded with certainty only from the fourteenth century, and historians have been intrigued and puzzled by the fact that it appeared, not in a simple and primitive form, but as a highly complex and sophisticated mechanism. Not surprisingly – since timetelling was derived from astronomy – the more sophisticated of these clocks were astronomical. Some were mechanical astrolabes, while others were mechanizations of astronomical demonstration instruments. Historians have been hampered in their investigations of earlier and simpler mechanisms by the fact that the Latin word *horologium* can refer either to a clock or to a sundial.

Richard of Wallingford (c. 1292–1336), Abbot of St Albans in England, built one of the most celebrated of these astronomical clocks in the early fourteenth century, and left a detailed technical account of its construction. The clock was a replica of the medieval universe. Its features included the phases and eclipses of the Moon (using differential gears), and an oval wheel gave an accurately varying velocity to the Moon as it moved around an astrolabe dial.

Even more famous in its time was the 'astrarium' of the professor of astronomy at Padua, Giovanni de' Dondi (1318–89). The machine, which he completed in Pavia in

1364, is also fully documented, and modern reconstructions have been made. It had a seven-sided frame with one dial for each of the planets (Sun, Moon, Mercury, etc.), each mechanism being in effect a geared Ptolemaic diagram. Below there were dials or displays showing the twenty-four hours, the times of sunrise and sunset, and much else. The astrarium was one of the marvels of the age; Regiomontanus (see page 85) visited it in 1462 and he records that prelates and princes flocked to see it as if they were to witness a miracle.

While these technical developments were in progress outside the universities, students of natural philosophy in the faculties of arts were applying their logical training to critiques of the Ptolemaic models. For the mathematical astronomers, these models (only rarely with updated parameters) were indispensable for their calculations, but the natural philosophers were more concerned to learn the truth about the cosmos. Most agreed with their Arabic counterparts, that the non-uniform motions implied by equants could not occur in the real world. In fact, in the eyes of many, epicycles and eccentres were likewise unacceptable: the motions involved were indeed uniform and circular, but they did not take place around the Earth, that spherical aggregate of earthy bodies gathered at the centre of the cosmos.

But what of the possibility that the Earth itself was spinning on its axis? Such a motion is of course inherently implausible to us humans who believe ourselves to be on *terra firma*, but the arguments against came primarily from consideration of the motion of projectiles such as arrows – not from astronomy, nor from the relevant Scripture passages (to which the medievals were able to take a more measured approach than would later be possible in the post-Reformation period). As Aristotle had taught, the fact that an arrow fired vertically upwards on a calm day would fall to ground where the archer was standing, proved that the ground below had not moved during the period of flight.

However, the Aristotelian theory of projectiles was itself in disarray, for his explanation of why it was that the arrow continued to rise in the air after it had lost contact with the bowstring was clearly absurd. In continuing its upwards movement, the arrow, although composed mainly of the element earth, was getting still further from its natural place at the centre of the Earth; and this unnatural (or 'violent') movement must be the result of the action of an external force. Anxious to identify the source of this external force, Aristotle saw only one possible candidate, the surrounding air, for this was the only material body in contact with the arrow. Somehow, the air contrived for a time to push the arrow upwards, so preventing its beginning its fall to the ground the instant it left the bowstring.

Parisian masters of the mid-fourteenth century were not the first to point out the absurdity of this – for if Aristotle was right, how was it possible to fire an arrow into the teeth of a gale? – but they were the ones who elaborated an alternative theory, and then applied it to the possible spin of the Earth. Jean Buridan (c. 1295 – c. 1358) and Nicole Oresme (c. 1320 – 1382) both agreed with Aristotle that *some* force must be at work, but both saw the absurdity of making the air the origin of this force, and both proposed instead that the projector – in this case, the archer with his bow – might impart to the projectile an 'incorporeal motive force', which they termed *impetus*.

Buridan then used this concept to explain other puzzling phenomena, such as the acceleration of a freely falling body: impetus was responsible for the continuance of the downward motion of the body, while the natural downwards tendency produced the acceleration. Oresme meanwhile pointed out that if the Earth were to rotate, then the archer would be moving sideways with the Earth. Consequently, merely by holding the arrow prior to firing it, the archer would share with the arrow his sideways motion and hence impart to it a sideways impetus. This impetus would cause the arrow while in flight to keep pace with the sideways movement of the archer below, and hence to fall beside the archer. In other words, an arrow fired vertically would fall onto the archer, whether or not the Earth was in motion: one therefore could not decide whether or not the Earth was at rest, simply by firing arrows into the air. Nor, declared Oresme, could the question be decided by other arguments from natural philosophy, or by astronomical observation, or by the study of quotations from Scripture. He himself believed that the Earth is at rest; but there was no proof either way.

Buridan also used *impetus* to illuminate a long-standing debate concerning the causes of the rotations of the various heavenly spheres. Aristotelian natural philosophy required all motions to have a cause, and the motions of the planetary and starry spheres were no exception. The heavens were all made of a single substance, the fifth element or 'quintessence', and so their motions were, so to speak, frictionless and therefore effortless; but nevertheless these motions had to have causes. The ultimate cause was God, the Prime Mover, who acted directly on the outermost of the moving spheres. But God, in the opinion of many, had also assigned to each of the other moving spheres an immaterial, spiritual intelligence or angel, which was able, voluntarily and inexhaustibly, to move its sphere with the appropriate uniform circular motion.

This Buridan found unsatisfactory, and instead used the concept of *impetus* to give a more acceptable explanation of the continued motion of the heavenly spheres. There was, he pointed out, no mention of the angelic intelligences in Scripture; and if God had impressed an impetus on each sphere at the Creation, then in the absence of friction this impetus would itself have been sufficient to keep the sphere in motion indefinitely.

The development of the concept of *impetus* was but one example of the growing self-confidence of Masters of Arts of the fourteenth century, and of their readiness to contradict Aristotle. The twelfth century had been that of the translations, the thirteenth, that of assimilation; now the learned world was on the move forward. Soon a new technology was to become available that would revolutionize the study of the mathematical sciences, and especially geometry and astronomy: the invention of printing.

Printing and the rise of Humanism

Books are the most effective means whereby the knowledge of one generation is passed to the next. Before printing, books existed in the form of manuscripts, copied – one from another, at great expense of labour – by scribes. If the subject matter was expressed in words – if the text was by Aristotle, for example – then the scribe would have some grasp of what he was copying; he might even be able to correct slips of his predecessors that he found in the manuscript before him. But if the subject matter was expressed in symbols, as in a treatise on geometry or astronomy, then the scribe would seldom have any hope of noticing when one of his predecessors had introduced an error. These earlier errors would therefore be repeated, and no doubt in making this fresh copy our scribe would himself introduce new ones of his own. As a result, a would-be astronomer wishing to master the *Almagest* – even when he had managed to lay hands on a copy – would find himself struggling to make sense of an unreliable text.

The mid-fifteenth-century invention of printing with movable type transformed this situation. It now made sense for the scholarly editor of an ancient treatise to devote months, or even years, to establishing a text faithful to the intentions of the original author; for he could then correct printer's proofs and supervise the making of woodblock illustrations, and, when everything was to his satisfaction, hundreds of identical copies would be run off and distributed throughout Europe for sale at reasonable cost.

Nowhere was this development more significant than in astronomy. Furthermore, the invention coincided with an increasing appreciation of the achievements of the Greeks and Romans. Antiquity was now coming to be seen as a Golden Age of artistic, literary and philosophical achievement that had been followed by the decline of the 'Middle Ages'. This 'Humanist' movement was primarily concerned with the works of man rather than with the understanding of nature, but all ancient writings, even the scientific, were legacies to be treasured.

A leading figure of the Humanist movement was the Greek scholar Johannes Bessarion (c. 1395–1472), who had been educated in Constantinople. He had come to Italy to take part in moves to reunite the Greek and Latin churches; and when these proved unsuccessful, he had remained in the West, becoming a cardinal in 1439. In 1460 Bessarion arrived in Vienna as papal legate, and there he met the Austrian court astrologer, Georg Peurbach (1423–61), and his young collaborator in astronomy, Johannes Müller of Königsberg (1436–76), known as Regiomontanus from the Latinized name of his birthplace. Peurbach had already written a *Theoricae novae planetarum* (*New Theories of the Planets*) designed to supersede the thirteenth-century textbook of similar name. Regiomontanus was to see this work into print in about 1474, some years after Peurbach's early death, and it became a best-selling textbook, appearing in dozens of editions well into the seventeenth century. It described the Ptolemaic planetary theory that underlay the *Alfonsine Tables*. One feature of the work was its detailed descriptions of physically-real representations of Ptolemaic planetary models in terms of solid spheres, representations whose shortcomings may have been the motivation that led Copernicus to take up planetary astronomy. Eventually, Tycho Brahe's demonstration that the comet of 1577 had passed without hindrance through the planetary regions (see page 97) was to show that no such solid spheres existed.

Bessarion was himself an outstanding scholar and collector of manuscripts, and he was anxious to see the *Almagest* made more accessible to students, through the preparation of a systematic abridgement. This project appealed to the two astronomers, who set to work; within a year, however,

Peurbach died, with the abridgement completed of only the first six 'books'. On his deathbed he made his friend promise to complete the project, and this Regiomontanus did while with Bessarion in Italy. The *Epitome* of the *Almagest*, half the length of the original and a model of clarity, was printed in 1496; and as copies spread throughout Europe, Ptolemy's mathematical astronomy at last became truly accessible. There were even prospects that the *Almagest* itself might one day be superseded.

In 1471, after some years spent partly with Bessarion, Regiomontanus settled in the prosperous commercial centre of Nuremberg. The best instrument-makers were on his doorstep, and his home became not only his observatory but also his printing house, where he could produce astronomical books that commercial publishers might find unattractive: Peurbach's *Theoricae novae planetarum* was followed by an *Ephemerides* of nearly 900 pages giving the positions of the heavenly bodies for each day from 1475 to 1506, the first such to get into print. Columbus took a copy of one of Regiomontanus's books with him on his fourth voyage (1502–04) and used its prediction of the lunar eclipse of 29 February 1504 to overawe the hostile natives of Jamaica.

In 1475, however, Regiomontanus left Nuremberg for Rome, apparently to help with the calendar reform, and there he too met an untimely death. He had however found a collaborator in Nuremberg, the merchant Bernard Walther (1430–1504), whose own observations began five days after Regiomontanus left the city and continued for nearly thirty years. Walther achieved a new accuracy in astronomical measurements, and his observations were to be used extensively by Copernicus, Tycho Brahe and Kepler.

Copernicus and the culmination of the Greek programme

Nicolaus Copernicus (1473–1543) was born in Toruń in Poland and studied at the University of Cracow. Although remote from the main cultural centres of Europe, Cracow University had a distinguished tradition in astronomy, including teachers who made no secret of their dissatisfaction with the way that equants violated the principle of uniform circular motion.

In 1496 Copernicus went to Italy to study. Constantinople was no longer in Christian hands, for in 1453 the city where Greek culture had survived for so long had at

last fallen to the Muslims. In the decades preceding this
traumatic event, Greek scholars had begun making their
way west, and especially to Italy, where in Florence a
school had been established in 1462 by Marsilio Ficino in
explicit imitation of Plato's own Academy. Greek was now
widely studied in universities, and the elegant dialogues of
Plato – most of which were unknown in the Middle Ages –
had become available and were proving profoundly influential. Although the Platonic tradition was towards abstraction and away from routine observation of the natural
world, it was imbued with a mathematical view of nature,
and demanded of any theory of the cosmos that it reflect
the cosmic harmony and symmetry.

In Italy Copernicus studied canon law and medicine,
eventually taking his doctorate in canon law at the
University of Ferrara in 1603; but he also learned Greek and
enhanced his interest in astronomy. He made his first
recorded observation in March 1497, and his future disciple
Georg Joachim Rheticus (1514–74) tells us that around 1500
in Rome he lectured on astronomy before a large audience.
Astronomy was in a far from satisfactory state. The calendar – society's basic requirement from astronomers – was
clearly getting out of step with the seasons (and would
remain so until the Gregorian reform of 1582 brought the
spring equinox back to its traditional date and modifed the
sequence of leap years to ensure it stayed there). To all students in Arts, the presence of equants in the Ptolemaic
planetary models was a cause for scandal, so that Rheticus
would describe the equant as 'a relation that nature abhors'.
Those who delved further into the technicalities of
mathematical astronomy were disturbed because the
Ptolemaic model of the Moon's motion implied that the
Moon's apparent size varied greatly, which it clearly did
not. Ptolemy's *Planetary Hypotheses* had been lost, and
with it his exposition of his integrated system of the
planets; in the current Platonist climate, the fact that in the
Almagest Ptolemy treated each planet independently,
without any attempt to develop a coherent cosmic system,
made the work deeply unsatisfactory. Past astronomers,
Copernicus would later write,

> have not been able to discover or to infer the chief point of
> all – the structure of the universe and the true symmetry of
> its parts. But they are just like someone taking from differ-
> ent places hands, feet, head, and the other limbs, no doubt
> depicted very well but not modelled from the same body and
> not matching one another – so that such parts would
> produce a monster rather than a man.

On the other hand, the geometrical models from which planetary tables were calculated were seen as reasonably satisfactory for this purpose: both the level of complexity of the models and the accuracy of the resultant tables were acceptable. Modern legend has it that by the sixteenth century, the Ptolemaic models had been elaborated in a desperate search for greater accuracy, to the point where the number of circles required was beyond all reason. The legend goes on to claim that on the Sun-centred hypothesis the number of circles was greatly reduced, and that the motion of the Earth was considered an acceptable price to pay for this simplification.

In this legend there is no truth: Copernicus's detailed models are every bit as complicated as Ptolemy's. Indeed, it is obvious that the legend cannot be true; for the supposed elaboration of the Ptolemaic models could only have been necessitated by new observations, observations of such precision that mere adjustments in the parameters were not enough. But observations of such accuracy not only were not, but could not be made at this period – the necessary instruments simply did not exist. Sadly, the legend will doubtless persist; which is a shame, because it offers a pedestrian motivation for what is in fact one of the greatest of the intellectual leaps known to the history of science.

For those (like Copernicus) who saw other reasons for a reform of astronomy, there were several clues as to the direction such a reform might take. Some were to be found in Aristotle's own writings. In discussing the stability of the Earth, for example, the Philosopher followed his usual custom and reported the views of his predecessors before demolishing them; in consequence, every student in a medieval or Renaissance university heard of ancient Greeks who believed the Earth to be in motion. To take a single example: Albert of Saxony, who was Rector of Paris University in the 1350s, expounded these ancient opinions in his *Questions* on Aristotle's *On the Heavens and the Earth*, concluding, '*Ita dixerunt solem non moveri circa terram sed magis terram circa solem*' – 'Hence they said that the Sun does not move around the Earth but rather the Earth about the Sun'. It is no surprise to find that Albert concluded in favour of Aristotle; but his *Questions*, with their account of this alternative view, was reprinted no fewer than six times during the lifetime of Copernicus.

Another clue came from the mystery as to why the annual period of the Sun had found its way into the model of each planet. So Peurbach had written: 'It is clear that each of the six planets in its motion shares something with the Sun, and the Sun's motion is, so to speak, the common mirror and measure for their motions.' With hindsight we know the reason: relative to us on Earth, the Sun orbits once a year taking with it all the other planets, so that satisfactory representations of their motions relative to us – which is what Ptolemy attempted – must in every case involve the Sun's period of one year.

In 1503 Copernicus returned to his native Poland, where he had meanwhile become a canon of the cathedral of Frauenburg (Frombork), where his uncle was bishop. As a canon (but not a priest) he was an administrator, and a private astronomer. We know little of the development of his thinking in the years that followed, but before long a manuscript by him later entitled *Commentariolus*, or *Little Commentary*, began to circulate. In it he briefly stated his dissatisfaction with existing planetary astronomy, singling out equants for special mention. He then set out the postulates of a Sun-centred alternative (in which the Earth became a planet with the Moon as its satellite), showed how this approach allowed an unambiguous order to be assigned to the planets (now six in number), and went on to develop equant-free models for the planets and the Moon.

Little more was heard of the Polish canon until Rheticus, then a teacher of mathematics in the University of Wittenberg, decided to visit him in 1539. Copernicus agreed to allow Rheticus to publish a *First Report* of his work, and the slim volume appeared the following year. No violent controversy ensued, and perhaps it was this that persuaded Copernicus to allow Rheticus to take his complete work, *On the Revolutions of the Heavenly Spheres*, for printing in Nuremberg. *De revolutionibus*, as it is always known, was published in 1543, the year of its author's death.

Rheticus's teaching duties had not permitted him personally to oversee the passage of the manuscript through the press. This task fell to a Lutheran clergyman, Andreas Osiander (1498–1552). No doubt with the good intention of shielding the author from controversy, Osiander inserted in the front of the work an introduction that was unsigned (and so apparently by Copernicus himself), declaring that the author was not maintaining that the Earth truly moved around the Sun, only that this was a convenient hypothesis on which to base efficient mathematical models of the planetary motions.

Not surprisingly, this served to obscure Copernicus's message. The author's true position should have been clear to anyone carefully studying the work as a whole; but the readers of *De revolutionibus* were mainly technical

astronomers preoccupied with computing tables of planetary positions and eager to plunder the work for any help it
could give them in this. In 1551, for example, Erasmus
Reinhold (1511–53), rector of the University of Wittenberg,
based his *Prutenic Tables* on *De revolutionibus*; he made
minor adjustments in the parameters in Copernicus's
models, but the models themselves he followed slavishly,
seemingly uninterested in the truth or otherwise of the
Sun-centred hypothesis.

Copernicus's treatise attempted simultaneously to serve
two quite different purposes. In Book I he demonstrated the
attractive consequences of the proposition that the Earth
was an ordinary planet orbiting the Sun: the planets would
form a coherent and integrated system, and many otherwise
puzzling facts of observation would become natural and to
be expected. Book I was a treatise on the cosmos, and
although Copernicus could offer no compelling proof of
what we might term his cosmovision, Book I was concerned
with the fundamental structure of the world in which we
live. The remaining 'books', however, served a quite

Erasmus Reinhold, one of the first great students
of Copernicus's *De revolutionibus*, was especially interested in the various mechanisms that
Copernicus proposed for eliminating Ptolemy's
equant. Here his marginal notes tabulate and distinguish the circles used for the Earth's motion.
In the text Copernicus states that he does not
know which is correct 'but it must be one of
them since they all give the same result'. Royal
Observatory Edinburgh.

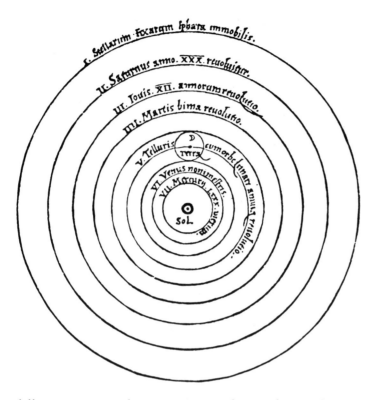

Copernicus's diagram of the solar system, from Book I of *De revolutionibus*, with (very approximate) periods. Note that the Earth ('*Tellus*', here in the genitive '*Telluris*') is anomalous in carrying a satellite around the Sun. From N. Copernicus, *De revolutionibus*, 1543, by permission of the Syndics of Cambridge University Library.

different purpose. They were to provide 'justification by works' for the cosmovision, by showing that adequate planetary tables could indeed be calculated from geometrical models with the Sun at the centre. These books were concerned with quantity and with the close match between prediction and observation, not with truth and falsehood.

In Book I Copernicus showed that one natural consequence of his making the Earth into a planet in orbit about the Sun, was that the other planets would divide themselves into two groups, those whose orbits lay inside the path of the Earth and those whose orbits lay outside. It had been known since long before Ptolemy that Venus and Mercury are to be seen only around dusk or dawn, while Mars, Jupiter and Saturn may be visible at any time of night. But what had been so mysterious a division on the Ptolemaic view, now became a natural and expected consequence of the Sun-centred hypothesis.

In Ptolemaic astronomy the Sun orbited the Earth with a period of one year; and since Venus and Mercury kept company with the Sun, they too inevitably had a period of one year. But if Copernicus was right and it was the observer on Earth who had the period of one year, it was a simple matter to make the appropriate allowance for this and to disentangle the true periods of Venus and Mercury. These turned out to be around seven and a half months and eighty-eight days, respectively. As a result, every planet now had its own individual period, and they could at last be ordered by increasing period in a well-defined sequence: Mercury, Venus, Earth, Mars, Jupiter, Saturn.

But it is not only the period of the Earth that enters into any observations we make of the other planets – so does the Earth's distance from the Sun (the 'astronomical unit'). Once this was recognized, the common distance scale could be applied to each of the geometric models of the planets, and it was then a simple matter to calculate the (relative) distance of each of them from the Sun. This allowed Copernicus to order the planets by a second criterion: their increasing remoteness from the Sun at the centre of the system.

To the delight of his Platonist soul, Copernicus found the two lists identical: 'Therefore in this arrangement we find that the world has a wonderful commensurability, and that there is a sure linking together in harmony of the movement and magnitude of the orbital circles, such as cannot be found in any other way.'

What of the occasional and very puzzling backward ('retrograde') motions that mathematical astronomers since the time of Plato had struggled so hard to reproduce in their geometrical models? Copernicus showed that these too were to be expected on the Sun-centred view, as no more than the natural consequence of our observing the other planets from a moving platform. For example, we on Earth orbit the Sun on the inside of Mars. Most of the time we see Mars moving around the Sun in the same direction as ourselves; but for the relatively short time that we are actually overtaking Mars, Mars appears to us to be moving backwards. The age-old mystery of the retrogressions that earned the planets the name of 'wanderers' had at long last been given a simple explanation.

If one looked a little more closely at the sizes of the orbits and the relative periods, one could easily calculate many of the finer details of the planetary retrogressions – such as where a planet would be when it appeared to come to a halt, for how long it would seem to move backwards, and where it would be when it appeared to resume its onwards movement. Copernicus triumphantly listed many such observed facts, pointing out that 'all these proceed from the same cause, which rests in the movement of the Earth'.

He reserved his greatest display of emotion for the passage where he reflected on how the Sun, symbol of the good in the Platonist tradition, was now in the middle of the universe:

> In the centre of all resides the Sun. For in this most beautiful temple, who would place this lamp in another or better place than that from which it can illuminate the whole at one and the same time? As a matter of fact, not inappropriately do some call it the lantern of the universe; others, its mind; and others still, its ruler. The Thrice-Great Hermes calls it a 'visible god'; Sophocles's Electra, 'that which gazes upon all things'. And thus the Sun, as if seated on a kingly throne, governs the family of planets that wheel around it.

Copernicus could not of course ignore the astronomical objections to the motion of the Earth. If the stars lay at the relatively modest distance traditionally ascribed to them, and if they were observed from an orbiting Earth, then they would appear to move with a cyclic annual movement of as much as 7°; and this they clearly did not. Furthermore, Earth-dwellers would sometimes see more than half the stars, sometimes less. In response, Copernicus, like Aristarchus before him (see page 34), could do no more than banish the stars to a great distance – so great that their movements would be too small for detection by contemporary instruments. Yet, as Tycho Brahe was to insist, this would leave a vast, and unintelligible, gap between the outermost planet and the stars; and the stars would have to be of colossal size, to appear so large when they were so far away.

The proclamation of Copernicus's cosmovision occupied no more than the first half of Book I of *De revolutionibus*. The remaining 95 per cent of the volume was dauntingly mathematical, offering proof that the Sun-centred hypothesis could deliver the goods – that it could be made the basis of an equant-free mathematical astronomy at least as good as Ptolemy's *Almagest* in the computation of planetary tables. After thirteen centuries, Ptolemy had at last been beaten at his own game.

De revolutionibus was a book whose reception was affected by the complexity of its construction. It attempted within the covers of a single volume not only to provide mathematical astronomers with the geometrical machinery for the calculation of planetary tables to a new level of accuracy, but also to canvas support for a revolutionary cosmovision; and even the few pages devoted to the cosmovision were undermined by the anonymous and wholly misleading preface. The cosmovision badly needed a publicist; but for half-a-century, Rheticus apart, no publicist of stature would be forthcoming.

5

From geometry to physics: astronomy transformed

Michael Hoskin

The publication in 1543 of Copernicus's *De revolutionibus* marked the culmination of the campaign that had begun in the time of Plato in the fourth century BC, to develop geometrical models that reproduced the observed motions of the planets. The aim had been to reveal the regularities that underlay the planetary odysseys against the background of the 'fixed' stars, and so to permit predictions of their future positions. Interest was focused on how the planets moved, not what caused these movements.

The essential components of these geometrical models had been uniform circular motions. That is, the planetary motions were analysed into their component cycles; and since cycles endlessly reproduce themselves, the parameters could be improved without the need for observations of great accuracy.

All this was to change in the century following Copernicus's death in 1543. He had been a traditionalist in purpose, methodology and techniques, yet his *De revolutionibus* sowed the seeds of revolution – for how could the stable Earth not only spin, but hurtle through space, its passengers going about their lives in blissful ignorance of what is happening to them?

Indeed, how had the Earth come to be spherical in the first place? To Aristotle, the answer had been easy: any earthy body that had found itself away from its natural place in the centre of the cosmos would have moved naturally towards the centre, and so it was no surprise that the resulting aggregate approximated to a sphere. Copernicus could only suggest that earthy bodies belonged together and had assembled to form the planet Earth, just as venusian bodies belonged together and had assembled to form the planet Venus.

To explain the daily motion of the Earth, Copernicus argued that it is a natural sphere, and that natural spheres naturally spin. He may have thought that the Earth was itself embedded in a vast but invisible sphere, whose

spinning carried the Earth in its annual orbit around the
Sun; but his views on this are unclear. Evidently, he had
advanced the solution of problems in kinematics – how the
planets move – only to create new problems in dynamics –
what are the causes of their movements.

In the transformation brought about in response to these
unanswered questions, crucial roles were played by four
men, of different countries and contrasting talents. Tycho
Brahe of Denmark was an observer, who made accuracy and
completeness his first priority; Kepler was a German mathematician, who transformed astronomy from applied geometry into a branch of dynamical physics; Galileo was an
Italian physicist, who used telescopes to reveal celestial
truths hidden since the Creation, and developed a new
concept of motion to underpin Copernicus's claims; and
Descartes was a French philosopher, who conceived of an
infinite universe, in which no position and no direction was
special, and the Sun became merely our local star.

Tycho Brahe and the quest for accuracy

Copernicus had uncritically used the observations of previous astronomers when these were conveniently to hand; he
himself did only what was necessary, content with the
modest accuracy that the available instrumentation
allowed. In the later decades of the sixteenth century,
Tycho Brahe (1546–1601) was to bring about a revolution in
the attitude of astronomers to observation.

Tycho was born into the Danish nobility, but he was fos-
tered by the academically-inclined family of his father's
brother, and so was able to escape the restrictions of feudal
life. Instead, he wandered from one university to another,
exempt by the privilege of birth from the normal pressures
of career. By 1563, when for the first time for twenty years
Jupiter overtook Saturn as the two planets slowly moved
against the background stars, the sixteen-year-old Tycho
was already sufficiently interested in astronomy to make
simple observations of this 'conjunction'. He found that the
prediction of the date in the thirteenth-century *Alfonsine
Tables* (calculated from Ptolemaic planetary models) was a
month out, while even the modern *Prutenic Tables* based
on Copernicus's models were nearly two days wrong.
This convinced him that there must be a reform of astronomy built on a solid foundation of accurate observations,
and that such accuracy could come only from a combination of improved instrumentation and improved observing
techniques.

A 1531 woodcut showing an Aristotelian interpretation of the appearance that year of a bright comet. The comet (which was in fact Halley's) rounded the Sun on 26 August, having earlier in the month passed under the legs of the Great Bear (the constellation that includes the Plough or Big Dipper). From Regiomontanus, *De cometae magnitudine . . . problemata XVI*, 1531, reproduced with kind permission of Dr Owen Gingerich and the Houghton Library, Harvard University.

The new star of 1572 and the comet of 1577

In the years that followed, Tycho increasingly sought the company of astronomers, both amateur and professional, and meanwhile he took his first serious steps as an observer. Then, in November 1572, Nature treated mankind to an astonishing spectacle: a star-like object, bright enough to be seen in the daytime, appeared in the constellation of Cassiopeia. Was it what it appeared to be, a new star? New heavenly bodies were unheard of, and there were compelling reasons for thinking such a celestial novelty to be impossible: the received Aristotelian cosmology was founded on the dichotomy between the unchanging heavens of the planets and stars, and the changeable central region of the four elements.

Had the 'nova' appeared even a decade earlier, it is possible that no consensus would have emerged from the clamour of contradictory arguments. In 1572 Tycho was still only in his twenties, and the astronomical instrumentation at his disposal was as yet modest in both scale and quality. But he was able to satisfy himself that the object was far above the Moon, and therefore celestial.

More importantly, he recognized the question-mark that the apparition placed against the accepted theory of comets. In his *Meteorology*, Aristotle had discussed the effects of the contact between the uppermost of the four terrestrial elements, namely fire, and the adjacent part of the rotating heavens. He thought of the fire (an unsatisfactory name for which he apologised) as a kind of fuel, 'so whenever the circular motion stirs this stuff up in any way, it bursts into flame at the point where it is most inflammable'. In his view, the whole of the element of fire, and much of the element of air

below it, was carried round by the rotation of the heavens,
and 'in the course of this movement it often ignites wherever
it happens to be of the right consistency', sometimes because
of the motion of a particular star or planet. This resulted in
shooting stars, and in comets of various shapes.

According to Aristotelian doctrine, therefore, comets
'came to be and passed away' in the manner characteristic
of the region of the four elements, and so were atmospheric
rather than celestial. Their study belonged not to astronomy but to 'meteorology', and medieval astronomers had
seen little point in spending time trying to measure the
heights of comets, when the answer was already known.
But Tycho now had proof that changes occurred also in the
heavens. He therefore promised himself that if a comet
appeared one day, he would carefully measure its height and
see whether or not it was indeed terrestrial.

Nature obliged: in 1577 a brilliant comet appeared, and
Tycho's observations showed it was celestial. More precisely, it was located among the planets. Yet, if so, it was
passing effortlessly through the invisible spheres that were
thought to carry the planets around the central Earth. The
implication eventually dawned on Tycho: these spheres did
not exist after all.

The effects of this discovery on Johannes Kepler a
generation later would be profound. Few had seen any great
problem in explaining the spinning of the spheres that
carried the planets: if angelic intelligences were no longer
acceptable, perhaps Copernicus had been right and natural
spheres naturally rotated. But if the spheres did not exist
and the planets were isolated bodies moving in orbit, it
would be altogether more difficult to explain the causes of
their motions. As it turned out, to provide an answer,
astronomy had to migrate from kinematics to dynamics,
from geometry to physics.

Tycho had published his observations of the nova of 1572,
but his tract had had little impact. On the question of the
comet of 1577, he was determined to be heard, and heard
clearly. While a hundred other writers rushed to cash in on
the alarm aroused by the awesome apparition, Tycho prepared
an elaborate analysis of his observations running to over 200
pages, to which he appended an even longer critique of the
results arrived at by other observers. It was a blockbuster of a
book, and when in 1588 it finally appeared, the transfer of
comets from meteorology to astronomy was assured; with the
transfer came recognition that changes in the heavens are by
no means unusual. In a second and this time exhaustive
study on the nova of 1572, Tycho's contemporaries were

again subjected to severe examination, though this work was not to appear until 1602, after Tycho's death.

These two treatises – which were to form the bulk of Tycho's literary output – served notice on other observers that new standards now applied. Tycho's own observations of the nova and the comet had been made with instruments commercially available; but by the time of the comet's appearance, Tycho was committed to a fundamental reform of both instrumentation and observational technique. For this he needed funding on a princely scale.

The observatory on Hven

Astonishingly, the money became available. In 1575, Tycho had visited the Landgrave of Hesse, Wilhelm IV, who was himself an enthusiastic observer, and it may have been on the Landgrave's recommendation that in the following year the King of Denmark granted Tycho the lordship of the island of Hven in the Danish Sound. There Tycho had the space, the time, and the financial resources to establish the first major observatory in Christian Europe.

On his island, between 1576 and 1570, Tycho built for himself a handsome observatory, Uraniborg ('Heavenly Castle'), with everything an observer could wish for. At the very top of the building were eight bedrooms for assistants. On the principal upper floor were four observing rooms, and other chambers that included a summer dining room with a view out to sea. On the ground floor were the library, kitchen and winter dining room, and three spare bedrooms,

The 'great mural quadrant' at Uraniborg was mounted onto a north–south wall. The instrument itself consists of a quarter-circle of brass, measuring over 6 feet in radius and divided at intervals of 10 seconds of arc. A fixed foresight at the centre of the circle is built into the wall at the left. There are two (alternative) movable backsights that slide on the arc. The observer (only partially visible at the extreme right), who is perhaps Tycho himself, adjusts one of the back-sights to the correct position as the star approaches the meridian, calls out the moment when the star is on the meridian, and reads the angle on the scale; the time is announced by an assistant at the clocks (which have only an hour hand), and the angle and the time are written down by a second assistant seated at the desk. Although the two clocks are prominent in the figure, we know that Tycho had little confidence in mechanical timekeeping, and often determined the time from measurements of a star.

As decoration, the wall onto which the quadrant is mounted has a painting of Tycho with his faithful dog. In the background we see, on the top floor of the observatory, other assistants making observations with various instruments; below them, in the library, more students are working near the great celestial globe; and the apparatus in the basement is for Tycho's alchemical experiments.

In a niche in the wall above the painting of Tycho is a brass globe that was fitted with wheels and could represent the daily spin of the Earth and the paths of the Sun and Moon, and much else besides. To either side of the globe are portraits of Tycho's patron, Rudolf II, and his consort, and above the globe is part of Tycho's library. From Tycho Brahe, Astronomiae instauratae mechanica, 1598, reproduced with kind permission of the Institute of Astronomy, Cambridge University.

while the basement contained the alchemical laboratory. Elsewhere on the island was a paper mill, and a printing office was established in a corner of the observatory grounds, so that Tycho could publish his results himself. Later he built machine shops where skilled artisans could construct instruments under his direct supervision.

About 1584, when even the lavish facilities of Uraniborg were proving inadequate, Tycho built a satellite obervatory, Stjerneborg ('Castle of the Stars'), nearby. Constructed entirely of masonry, its rooms were below ground level, so that the instruments were shielded from the wind. Tycho attempted to link Stjerneborg to Uraniborg by an underground passage, but this was never completed. Instead, an observer walked across from Uraniborg and descended to the underground complex down steps that led directly to the heating chamber. From this he could pass to any one of five chambers, each of which contained a single major instrument; observations were made through ground-level windows or through openings in a rotatable roof. Above ground the enclosure contained stable supports onto which other apparatus might be placed when necessary. Having two separate buildings in this way also enabled Tycho to ensure that the measurements made by different teams of assistants really were independent.

Year by year Tycho constructed, tested, modified, and tested again his instruments of many different types, until they were capable of measuring angles to an accuracy of better than one minute (a sixtieth of a degree). Meanwhile, he and – more especially – his assistants maintained an intensive observing programme, to a standard of accuracy, reliability and completeness never before attempted. Following his success in establishing the negligible parallax of the nova of 1572 and the comet of 1577, Tycho embarked on a more ambitious research program to find the parallax of Mars. Because the distance to the Sun was universally underestimated by more than an order of magnitude, Tycho had a reasonable expectation of finding this quantity from the observed differences in Mars's position when it was near the horizon and high overhead. Furthermore, according to the Copernican system, Mars could come twice as close as assumed in the Ptolemaic arrangement, so Tycho hoped to provide the first observational test between the two cosmologies. After nearly a decade of intensive work, with ever more refined instruments, Tycho's plan failed, but it left a legacy of superb observations of Mars that Kepler would exploit.

Tycho also compiled a catalogue of 777 stars, to supersede existing catalogues that derived from Ptolemy's

Almagest. First he and his teams determined the positions of selected 'reference' stars as carefully as possible, and then they measured the positions of other stars relative to appropriate reference stars. The accuracy of the catalogue therefore depended in the first place on the accuracy of the positions of the network of reference stars, and by the late 1580s Tycho had determined these angles to within half a minute or so of the true values. Yet although the places of the brightest of the non-reference stars are mostly correct to around the minute of arc that was his standard, the fainter ones are less accurately located, and there are many errors. It seems that Tycho may have lost enthusiasm for the work once the zodiacal stars – especially important as reference positions for future observations of planets – were completed. The catalogue eventually appeared posthumously, in his second publication on the nova of 1572.

The Tychonic system

Inside every astronomer there lurks a cosmologist, and Tycho was no exception. To us, his reform of observation is his outstanding achievement, for it gave to astronomy the respect for facts that we think of as characterizing modern science. But he himself was probably proudest of the cosmology expressed in the 'Tychonic system', which soon replaced the Ptolemaic as the most popular of the Earth-centred world pictures. Tycho was well able to appreciate the merits of Copernicus's cosmovision, but he was enough of a traditionalist to regard the price as too high. Galileo would later admire Copernicus for 'committing such a rape upon his senses' as to move the Earth; Tycho's feet were solidly on *terra firma*. As a Protestant he saw difficulties for Copernicus in certain passages from the Old Testament. The ancient proof of the immobility of the Earth, based on the behaviour of projectiles – an arrow fired vertically in the air hit the ground at the place from which it had been fired, showing that the ground below had not moved while the arrow was in flight – seemed to him as valid as ever. And, even with the aid of his superb instrumentation, he could not detect the apparent annual movement of stars ('annual parallax') that was to be expected if they were being observed from a moving platform on the orbiting Earth.

There were two possible explanations of his failure to detect annual parallax. Either Copernicus was wrong and the observer on Earth was in fact at rest, or the stars were so far away that their apparent movements were too small to be detected even with Tycho's instruments. Tycho

estimated that, in that case, the stars would have to lie at
700 or more times the distance of the outermost planet.
This would create an unintelligible gap between the planets
and the stars – and the stars would have to be of colossal
size in order to appear so large when they were so far away.
To Tycho, such a universe made no sense at all.

How then to retain the advantages of the Copernican cosmology without becoming entrapped in the absurdity of a
moving Earth? With hindsight the compromise seems
obvious. Any motions that we see a planet perform are
motions *relative to us*, and any geometrical explanation of
what we see is irrelevant to the question of whether it is the
Earth or the Sun that is *absolutely* at rest. Therefore, if
Tycho retained the geometry of relative motions as proposed
by Copernicus, but declared the Earth to be the body that is
absolutely at rest, he would have the best of both worlds.

This step may seem obvious now, but to Tycho it was far
from obvious. By 1578 he had reached a half-way stage, the
system described by Martianus Capella in the fifth century,
in which Venus and Mercury were satellites of the Sun,
while the Sun, along with the Moon, Mars, Jupiter and
Saturn, orbited the central Earth. Six years later he was con-
templating making all five lesser planets – Mercury, Venus,
Mars, Jupiter and Saturn – into satellites of the Sun. But
there was a physical problem: the sphere that carried Mars
would have intersected the sphere that carried the Sun.

At last it dawned on him that the unimpeded passage of
the comet of 1577 through the planetary regions implied
that these spheres did not exist after all, and now there was
nothing to prevent him from making the five lesser planets
into satellites of the Sun. In 1588, in the book on the
comet, he published the resulting system in outline, along
with detailed geometrical models for the motions of the
Sun and Moon.

In the Tychonic system, the Earth was at rest at the
centre. Around it orbited the Moon and the Sun. The other
five planets were satellites of the Sun, and were carried
with the Sun around the Earth. Just beyond the outermost
position attained by any planet was a thin shell of space,
centred on the Earth, and within this space the stars were
located. The Tychonic universe was reassuringly compact,
with a radius equal to some 14,000 Earth-radii. Even
Ptolemy's universe had had a radius half as large again.

The end of a dream
Tycho's 'Heavenly Castle' was always vulnerable, depending for its existence on the continuing favour of a princely

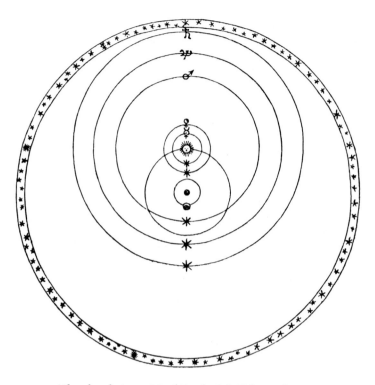

The Tychonic system. The Earth is at rest at the centre, and the Moon and the Sun orbit around it. The other five planets are satellites of the Sun and are carried with it about the Earth, while just beyond the planetary regions lie the stars. The orbit of Mars intersects that of the Sun; in Tycho's opinion, this was phys-ically possible as the free passage of the comet of 1577 had shown that the celestial spheres did not in fact exist. The relative motions are the same as in the Copernican system, so that observations that con-cerned only relative motions – such as the phases of Venus observed by Galileo – were neutral in the debate between the Copernicans and the advo-cates of the Tychonic system. From Tycho Brahe, *De mundi aetherei recen-tioribus phaenomenis*, 1588.

patron. The death in 1588 of Frederick II brought no immediate change, for the regency that followed was domi-nated by Tycho's family and friends. But as the years passed and the young Christian IV began to take an increasing role in government, it became clear that the halcyon days of Hven were over. In 1597 a resentful Tycho quit Uraniborg, and after two difficult years he crossed Europe to Prague, and took service with a more appreciative patron, the Emperor Rudolf II.

Rudolf did his generous best, but by now Tycho had lost his appetite for observations. Four of his instruments were still in place on Hven, and the biggest of the remainder were in store at Magdeburg. His chief concern was for the publication of past research. Tycho now had only months to live; but in those months he renewed an earlier invitation to Johannes Kepler (1571–1630) to come and join him as his assistant; and this time Kepler accepted.

Kepler and the introduction of dynamics into astronomy

Astronomers of the early and middle decades of the seven-teenth century were often at a loss to know what to make of the writings of Johannes Kepler. He was that rarity, a

astronomer who is frank and open about his mistakes:
unlike his modern counterparts, who 'launder' their
accounts before publication and give them an artificial
simplicity, Kepler required his readers to share with him
both triumphs and disappointments on the road to discovery. In the endless pages of calculations, he made mistakes
that sometimes led him to draw wrong conclusions.
Because the mathematical techniques he needed did not yet
exist, he resorted to procedures that we can recognize as
flawed. In dynamics he was a transitional figure, whose
idiosyncratic conception of motion (according to which a
planet would instantly come to a halt unless driven on at
every moment by an outside force) was to be repudiated by
the next generation. And he was motivated by a religious
mission, to penetrate the mind of God the geometer.

Kepler had none of the social advantages of Tycho Brahe.
The son of a quarrelsome father and a mother whom he
would later have to defend from charges of witchcraft, he
was born in 1571 in Weil der Stadt near Stuttgart and
studied at the University of Tübingen. Although he
intended to enter the Lutheran ministry, his studies at
Tübingen began with courses that included astronomy. The
professor was Michael Mästlin (1550–1631), an exceptionally competent mathematical astronomer, who – wherever
his true preference lay – made sure that his students were
aware of the merits of the Copernican hypothesis. Kepler
began his studies of theology in 1591; but in his third year a
mathematics teacher in Graz died, and the Tübingen
authorities, asked to propose a replacement, nominated
Kepler. Under protest, Kepler complied.

The Cosmographic Mystery

Established at Graz, Kepler began to reflect on the universe
created by God whom, like so many in the tradition of
Plato, he saw as a geometer. Copernicus had discovered the
layout of God's universe – but not what had motivated God
to choose this layout rather than another.

The immobile parts – the central Sun, the exterior
sphere, and the space between – were easy to understand:
they mirrored God the Father, God the Son, and God the
Holy Ghost. The moving parts – the planets – were more
difficult. Why were there six (as Copernicus had found),
rather than, say, five or seven; and what had motivated God
to locate a given planet at one distance from the Sun rather
than another, and to give it one speed rather than another?

In retrospect, the answer to the question of numbers
must have seemed obvious. As every student of Euclid

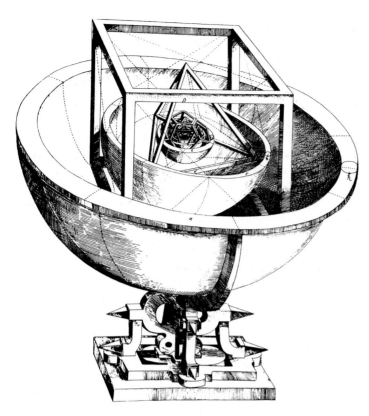

A diagram from Kepler's *Cosmographic Mystery* (1596) in which he reveals the geometrical relationships underlying God's scheme for the planetary orbits. Each orbit is represented by a sphere, and the separation between one sphere and another is defined by one of the five regular solids. The outermost sphere is that of Saturn. Inscribed within it is a cube, and within the cube is inscribed the sphere of Jupiter, so that the cube defines the ratio of the radius of Jupiter's orbit to that of Saturn. Similarly, the tetrahedron defines the ratio of the radius of Mars's orbit to that of Jupiter; and so on.

knew, there were exactly five regular solids (pyramid, cube, octahedron, dodecahedron and icosahedron), neither more nor less. There were therefore five appropriate figures for God the geometer to use to space out the pairs of adjacent planetary spheres: five figures, five spaces, and therefore six planetary spheres to be separated by these spaces.

The idea of a nest of spheres alternating with regular solids opened his way to the explanation of the distances of the planets from the Sun. After a little creative manipulation, Kepler found a particular nest in which the radii of the spheres were in fair agreement with the distances from the Sun derived by Copernicus from observation of planets in the real world.

There still remained the question of the speeds of the planets in their orbits. Copernicus had found with great satisfaction that the further a planet was from the Sun, the longer it took to complete a circuit. This was for him a demonstration of the geometrical harmony in the universe. But Kepler began to take astronomy along a decisively new path when he convinced himself that this harmony was brought about through the physical – that is, dynamical – influence of the Sun at the centre, as it drove the planets

onwards in their orbits. Obviously this solar force would be
less effective at greater distances, which was why the outer
planets travelled more slowly.

Kepler's physical intuition was reinforced by his
appreciation that the Sun was larger than the Earth or
indeed any of the planets: it made sense to have the huge
and therefore powerful Sun at the centre of the system.
When the next generation of astronomers accepted *forces* as
the key to understanding the heavens, and as the enormous
size of the Sun was increasingly recognized, the days of
Earth-centred cosmologies (such as Tycho's) were num-
bered.

Kepler had penetrated the mind of God and uncovered
the geometrical motivation underlying the structure that
God had chosen to give the universe. His proudly-titled
Cosmographic Mystery appeared in 1596, when Kepler was
just twenty-five. Even without its hints of new physical
insights, it was significant as being the first irredeemably
Sun-centred book: there was no way in which the nest of
spheres and solids could be adapted so as to have the Earth
at its centre.

Kepler sent a copy of his book to Tycho. Tycho had by
then developed his detailed models for the motions of the
Sun and Moon, but had made no headway with the planets.
Undeterred by the Copernican stance of Kepler's book,
Tycho recognized it as the work of a very special
mathematical talent, and he pressed Kepler to visit him on
Hven.

The remoteness of the Danish island, however, made
such a visit impracticable, and Kepler had other commit-
ments. But soon conditions for Protestants in Graz took a
turn for the worse, and when Kepler learned that Tycho had
left Hven and was in Prague, he decided to pay him an
exploratory visit.

Warfare with Mars and the New Astronomy
Kepler arrived at Tycho's observatory in February 1600, and
for three months worked on the orbit of Mars. Except for
Mercury, whose proximity to the Sun makes it hard to
observe, Mars is the planet whose orbit differs most from
the circular, and it was therefore the most difficult to
handle in terms of the traditional circular motions. When
the three months were up, Kepler returned to Graz, but by
October he was again knocking on Tycho's door. The Dane
gladly took him in, and Kepler resumed work on Mars. A
year later, Tycho took ill and died. Within forty-eight hours
Kepler was appointed his successor.

Kepler's 'warfare' with Mars, the god of war, was to last for years. His campaign had, he said, a threefold foundation: the Sun-centred vision of Copernicus, the incomparable observational records of Tycho Brahe, and the magnetical philosophy of the Englishman William Gilbert (1544–1603). In 1600, Gilbert, a prominent London physician, had published an experimental treatise *On the Magnet* in which he argued that the Earth itself was one vast spherical magnet with a magnetic axial rotation. The resulting force – which Gilbert used to explain the fall of bodies to the surface of the Earth, the northerly orientation of compasses, and much else – provided Kepler with a suggestion as to how to envisage the Sun's influence on the planets. The Sun, Kepler believed, was rotating and sending out an influence that was pushing the planets round, the nearer planets being influenced more than the distant ones. Since the planetary orbits are not perfect circles with the Sun at their centres, Kepler had also to provide a force to vary the distances of the planets from the Sun. Adapting a concept from magnetism, he supposed that the Sun also exerted on each planet an attraction over one part of its orbit and a repulsion over the remainder.

This physical intuition was to prove decisive. For example, at a crucial moment it led Kepler to realise that he must refer planetary orbits to the real, physical Sun; until then, like Copernicus before him, he had referred planetary orbits instead to the geometric centre of the Earth's orbit. More generally, it caused Kepler to focus his attention on the actual planetary orbit that resulted from the solar forces, whereas his predecessors had been preoccupied with the individual components of the geometrical machinery – epicycles and so forth – that generated the orbit. Accordingly, Kepler would be satisfied with nothing less than a single model that generated the motions of a planet in both longitude and latitude; by contrast, even Copernicus had been content to work with one geometrical model for the motion in longitude and another (incompatible) one for the motion in latitude.

Yet his physical intuition could also prove a stumbling block, most notably when, after analysing the movements of Mars, he returned to the problem of the exact shape of the orbit. For a time he excluded ellipses from consideration even though their geometry had been fully understood since the time of Apollonius in Antiquity: an ellipse had two axes of symmetry, whereas dynamical considerations suggested that the orbit would have only one such axis. Even today it can strike one as bizarre that a planetary orbit

– an ellipse with the Sun at one focus – has geometrical symmetry about the minor axis, whereas dynamically the Sun is to one side of this minor axis and has no physical counterpart at the other, 'empty' focus.

Equally decisive for Kepler's investigations was the legacy of Tycho's observations, even though Kepler's Copernicanism caused him endless diplomatic problems in his dealings with Tycho's heirs. The sheer volume of the observations was to be of the greatest importance to Kepler's assault on Mars. For example, at one stage Kepler realized that any Earth-based observations involved measuring the position of Mars from a moving platform, of whose location in space he was unsure. But he was able to circumvent this problem by using Tycho's ten observations of Mars at 'opposition' (when the Earth lies directly between the Sun and Mars), for then the Earth-based observer sees Mars in exactly the same direction as would a hypothetical observer on the motionless Sun. And when he later needed to reverse the procedure and 'observe' the Earth from a determined position in the orbit of Mars, he found Tycho had left enough observations to make this possible.

As luck would have it, Tycho's observations were also of just the right accuracy to enable Kepler to discover his laws of planetary motion. On the one hand, they were accurate enough to force him to consider non-circular orbits. Kepler had been able to generate the motions of Mars in longitude with a model consisting of a single circle; but when he investigated whether the circle would account for the planet's motion in latitude, he found errors of as much as eight minutes of arc. With any of Tycho's predecessors, such errors would have been acceptable; but Tycho's accuracy was better than eight minutes, and the circle that had promised to be an ideal solution had to be abandoned.

On the other hand, in our post-Newtonian world we know that planets do not move *exactly* in ellipses with the Sun at a focus. Even without the presence of other planets (whose pulls in fact cause disturbances or 'perturbations' of the orbit), a planet would have, at the focus of its orbit, not the Sun itself, but the centre of gravity of the Sun-planet system. Had Tycho's observations been of even greater accuracy, they might have ruled out, not only the circle, but ellipses as well.

Kepler's 'first law', that a planetary orbit is an ellipse with the Sun at one focus, appeared in the book in which he described his warfare with Mars. The law was a strikingly elegant and simple solution to the problem of orbits, and it broke the spell of circularity that had dominated astronomy for two millennia.

His 'second law' (which in fact preceded the first) was, however, confused and confusing. Kepler proposed first a variant of the Ptolemaic 'equant', a position from which the speed of the planet appeared uniform. This position was later to be recognized as the 'empty' focus. He next took the speed v to be inversely proportional to the distance r of the planet from the Sun ($v \propto 1/r$). Finally, he arrived at the rule that we today know as the second law, whereby the line from the Sun to the planet traces out equal areas in equal times. It would be far from clear to other astronomers which (if any) of the three versions was correct and which were approximations, for observations could scarcely distinguish between them. Dynamically, however, the differences were profound, and the situation would become clear only after 1687, when Isaac Newton incorporated motion under the area rule within a unified theory of dynamics. In the meantime, while the elliptical shape of the orbits themselves would be accepted by many, the crucially important question of the ever-changing speed of the planet was left in confusion, and many preferred to work with some form of equant.

Kepler's *New Astronomy* eventually appeared in 1609, a vast book of forbidding aspect, conceptually suspect both in its mathematics and in the underlying physics. The revolution it embodied was proclaimed in the full title: *New Astronomy Based upon Causes, or Celestial Physics, Treated by Means of Commentaries on the Motions of Star Mars*. Kepler had found astronomy a branch of geometry; he had converted it into a branch of physics.

The Harmony of the World

In *New Astronomy* Kepler had investigated how an individual planet behaves as it orbits the Sun, but his long-standing ambition to comprehend the overall structure of God's creation called for knowledge of the relationship of one planetary orbit to another. This was one of the topics he explored in *The Harmony of the World*. The study of cosmic harmony went back to Pythagoras (see page 26), who alone of mortals had been privileged to hear the music of the heavenly spheres. Kepler looked for harmony in the arithmetic ratios to be found in various aspects of geometry and of astronomy, and in Book V he investigated such matters as the speeding up and slowing down of the planets in their orbits, from which he believed he could derive the actual notes of the celestial music. He also

looked for arithmetical patterns that might reveal a relationship between the periods and the sizes of the planetary
orbits; and he somehow hit upon the fact that the square
of the period of a planet is in a fixed ratio to the cube of
the radius of its orbit (Kepler's 'third law'). Copernicus
had been delighted to find that the further a planet is from
the Sun, the longer it takes to complete an orbit; Kepler
now knew the arithmetical rule that ensured that this
was so.

When *The Harmony of the World* appeared in 1619,
Kepler was in process of making his insights and discoveries available to a wider audience, in an *Epitome of
Copernican Astronomy*. Because of the disruption to his
work caused by his mother's trial for witchcraft, and by the
war raging around him, Kepler's *Epitome* appeared piece-
meal, in 1618, 1620 and 1621. It was in fact to be his largest
publication. Cast in the form of question and answer, it
covered the whole range of his astronomical thinking, from
the use of regular solids by the divine geometer, via the
quasi-magnetic forces exerted by the Sun, to the details of
the planetary orbits. Copernicus's name appears in the title;
but nothing better illustrates the revolution brought about
by Kepler's creation of 'celestial physics', than a comparison of the forces of the *Epitome* with the geometrical epicycles and deferents of Copernicus's *De revolutionibus*.
And though the *Epitome* does not of course list 'Kepler's
laws' in the bald manner of a modern textbook, in this
work they were at last made accessible to a wider reader-
ship.

However, Kepler's astronomy had yet to face the traditional test: could it provide the foundation for planetary
tables of improved accuracy? Tycho's own involvement in
astronomy had been triggered by his dissatisfaction with
the *Prutenic Tables* based on Copernican models; and in
1601, when Tycho presented Kepler to Rudolf II, the
emperor had commissioned Kepler to work alongside Tycho
on the new planetary tables that Tycho planned. They were
to be called the *Rudolphine Tables*.

The tables appeared at last in 1627, when Tycho and
Rudolf were both long since dead. Their accuracy was strikingly demonstrated four years later. On 7 November 1631,
when Kepler himself had been dead a year, the French
astronomer Pierre Gassendi became the first observer in
history to see Mercury crossing the face of the Sun, in
fulfilment of a prediction by Kepler. Kepler's tables were in
error by only one-third of the solar diameter, whereas even
the Copernican tables they had replaced were in error by

thirty times that amount. If Kepler's tables were so accurate, surely the planetary laws on which they were based were worthy of serious consideration.

Galileo deploys the telescope in support of Copernicus

Copernicus had signally failed to promote his own vision of the cosmos: the insights in the brief cosmological Book I of *De revolutionibus* were obscured by the sheer scale of the mathematics that dominated the rest of the volume. Not surprisingly, at the end of the sixteenth century publicly convinced Copernicans were still rare; and as yet they did not include the man who would prove his most effective advocate, Galileo Galilei (1564–1642).

Galileo went to school in Florence and then, in 1581, became an unenthusiastic student of medicine at the University of Pisa. After four years he returned to Florence where he privately studied (and taught) mathematics. In 1589 he became professor of mathematics at Pisa, moving in 1592 to a similar but better-paid post at Padua where he was to teach for eighteen years. It was at Padua that he used the daily and annual motion of the Earth to formulate a possible explanation of the puzzling phenomena of the tides; and when Kepler in 1597 sent him a copy of *Cosmographic Mystery*, Galileo hinted at this theory in his reply. But as yet his Copernicanism lacked conviction, and Kepler's appeal for moral support went unanswered. Galileo's telescopic discoveries in 1609–10 removed any remaining doubts he may have had.

Extending the human senses
Aside from exceptional events such as a new star, a typical Renaissance astronomer looked out on very much the same universe as had his predecessor in Antiquity. If the Renaissance astronomer was blessed with a better grasp of astronomy, this was chiefly because he had been in a position to study printed copies of the books written by his predecessors and contemporaries, or the observational records they had compiled: he had had the advantage of reading more, rather than seeing more.

All this was about to change. From now on, astronomers of every generation would have an overwhelming advantage over their predecessors however brilliant, simply because improved instrumentation would allow them to view objects hitherto unseen and unknown, and therefore unstudied. They would then be able to say of their predecessors, as Galileo did of his, 'If they had seen what we see, they would have judged as we judge.'

In the summer of 1609, when Galileo was in Venice, news arrived that in Holland a device consisting of a tube with two pieces of curved glass was being used to make distant objects seem near. Curved glass, like a curved mirror, was well-known as leading to distortion, the very opposite of the truth. It was likely that still greater distortion would result from the use of two pieces of curved glass in combination, and so Galileo took the precaution of verifying the rumour. Only then did he attempt to construct such an instrument for himself.

The task presented little difficulty to one with Galileo's gift for mechanical improvisation, and by August he was again in Venice, where he demonstrated to the authorities a telescope of 8× magnification, 'to the infinite amazement of all'. He was offered a big increase in his Paduan salary. However, the following year the Grand Duke of Tuscany, Cosimo II de' Medici, made him a still better career offer; and so in September 1610 he returned to Florence, this time with a life appointment as mathematician and philosopher to the Duke.

News from the stars

By the time 1609 drew to a close, Galileo had improved his magnification to some 20×. When he used the instrument to look at the stars, he saw many that had been invisible to the unaided eye – stars that had remained hidden since the Creation, awaiting their discovery by Galileo. The mysterious Milky Way he was able to resolve into myriads of tiny stars, so confirming the speculation reported by Aristotle two millennia before. He also found that whereas the planets appeared proportionately enlarged as expected, the same was not true of stars. This was welcome news for Copernicans. Tycho (see page 102) had estimated that to explain away his failure to detect annual parallax, Copernicans would have to banish the stars to at least 700 times the distance of Saturn; and to appear disc-shaped at such a distance the stars would have to be of colossal size. Now, however, their disc-shaped appearance was exposed as illusory.

His most striking discovery concerned the planet Jupiter. When he first examined it on 7 January 1610, he found the planet in the midst of three little stars ranged – curiously – in a straight line. Jupiter was then in westwards ('retrograde') motion, and Galileo therefore expected that the following night Jupiter would be west of the supposed stars;

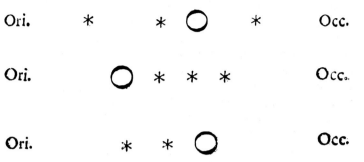

Ori.   *   *   ○   *   Occ.

Ori.   ○   *   *   *   Occ.

Ori.   *   *   ○   Occ.

Observations of moons of Jupiter, from the *Starry Messenger*. The upper sketch shows the supposed stars on 7 January 1610, in the midst of three little stars. The planet was then moving west (to the right). In the middle sketch we see the supposed stars on 8 January, with Jupiter to the east (left) of them, contrary to what Galileo expected. The lower sketch shows the configuration as it was on 10 January. Galileo devoted nearly half his little book to observations of these satellites. From Galileo, *Siderus nuncius*, 1610.

but in fact it was to the east. The next night was cloudy, but on the 10th he found the planet to the west of two stars, with the third star nowhere to be seen. By the 13th the number of stars had increased to four; and by the 15th Galileo had realized that the supposed stars must in fact be satellites, moons circling around Jupiter and carried along by the planet as it orbited the Sun.

This was more good news for the Copernicans. Even in the basic model of the Copernican system presented in Book I of *De revolutionibus*, there had been one serious anomaly: the Earth, while in other respects an ordinary planet, was unique in carrying a satellite Moon around the Sun. The telescope had now revealed that another planet was carrying no fewer than four moons.

Though welcome to Copernicans, the discovery was far from welcome to those who had opted for the system of Tycho, or for another of the compromises then in circulation. In these systems, the Sun orbited the central Earth, and Jupiter orbited the Sun. But now the hierarchy had, implausibly, to be extended a stage further, to accommodate these moons that were orbiting Jupiter.

Our own Moon must have been among the objects in the sky that Galileo had examined with his earlier instrument, but now it was time for a serious examination. For all its mottled appearance, the Moon was in the Aristotelian heavens and therefore perfect; yet Galileo's telescope showed its surface to be irregular, with mountains like those on Earth – mountains whose reality he was to bring home to his readers with his estimates of their actual heights, so that the reader might imagine attempting to climb them.

Galileo moved rapidly into print, any lingering doubts about the truth of the Sun-centred system banished by what he had seen. He was anxious to exploit his discoveries and so further his career. His little *Starry Messenger*, written

The Copernican system in outline as presented by Galileo in his *Dialogue on the Two Great World Systems* (1632). It is similar to Copernicus's figure (see page 91), with the important difference that the Earth is no longer alone in carrying a satellite. Galileo makes no attempt to represent the two mysterious appendages of Saturn, which disappeared two years after their discovery in 1610, only to reappear a few months later.

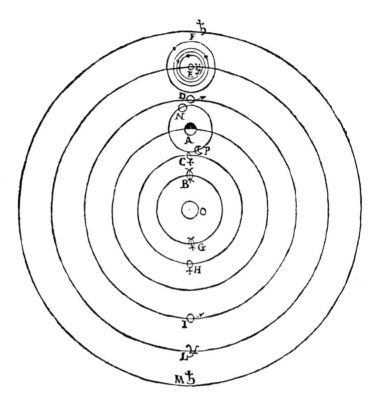

and published in a matter of weeks, proclaimed his astonishing news.

Many were incredulous: peering through a tube with two pieces of curved glass was no way to penetrate the secrets of the cosmos. But decisive support came a year later, when four Jesuit astronomers in Rome signed a statement confirming his discoveries. It would in any case be only a matter of time before telescopes became widely available, and then Galileo's claims could be verified by anyone who had a mind to.

In the months ahead, Galileo was to make three further discoveries, the first of which concerned the Sun. Under suitable conditions, sunspots can be seen with the naked eye, but the first telescopic observation of a sunspot was made in England by Thomas Harriot (c. 1560–1621) late in 1610. Galileo himself discovered them soon afterwards, and the following spring, when he was in Rome, he showed them to other observers. Unknown to him, the Frisian astronomers David Fabricius and his son Johann were by then making regular observations of sunspots, and in June 1611 Johann completed a tract in which he argued that the spots were actually on the surface of the Sun, which was therefore rotating.

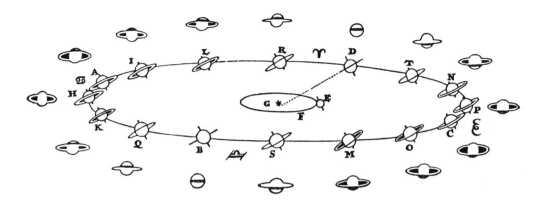

In 1612 Galileo – who as yet had published nothing on the subject – received from Germany a copy of another tract on sunspots whose author proved to be the German Jesuit, Christoph Scheiner (1573–1650). Scheiner, who had first observed sunspots in March 1611, believed that the spots were satellites of the Sun, rather than being on the body of the Sun itself. The resulting dispute – which was enlivened by the underlying controversy over priority – led Galileo to write his *Letters on Sunspots*, in which, like Fabricius, he argued that the spots were on the rotating Sun. By this time he had realized the significance of spots being on the Sun itself: in received thinking the Sun was the symbol of perfection, and a Sun that was 'spotty and impure' represented another blow to the traditional cosmos.

The second of Galileo's new discoveries concerned Saturn. The planet, it seemed, had mysterious appendages, which subsequently disappeared and then reappeared; these would puzzle observers until the Dutch physicist Christiaan Huygens (1629–95) hit upon the extraordinary explanation in the winter of 1655/56, long after Galileo's death: the planet was 'surrounded by a thin flat ring which does not touch him anywhere'.

The third discovery was of more immediate importance. Venus, he found, went through a complete sequence of Moon-like phases, appearing at times with a circular disc like the full moon, and at others having a thin crescent like the new moon. This was totally incompatible with the Ptolemaic geometry for Venus. Ptolemy's model (see page 47) implied that Venus was always in between the Earth and the Sun. As a result, the hemisphere of Venus illuminated by the Sun was always facing away from the observer on Earth, who therefore would never see the planet with the circular disc of a full moon.

Huygens's diagram showing how the ever-changing appearances of Saturn's appendages could be explained by supposing the planet to be surrounded by a ring. Interestingly, the idea occurred to him at a time when the ring was edge-on and invisible. From Christiaan Huygens, *Systema Saturnium*, 1659.

Unfortunately for Galileo's work as a propagandist for Copernicus, the phases of Venus tell us only about the *relative* motions of Earth, Sun and Venus, not which of them is absolutely at rest. And since the relative motions were the same in the Sun-centred Copernican and the Earth-centred Tychonic system, the observed phases of Venus could not decide between them. Galileo had no patience with such logical niceties. To him the Tychonic system was a blatant compromise, and not to be taken seriously; Ptolemy was wrong, and so Copernicus must be right. Others were less dogmatic in their judgement.

The Copernican propagandist
In both the *Starry Messenger* and his *Letters on Sunspots* (1613), Galileo was careful not to make his support for Copernicus appear too strident. Patronage was as important to astronomers and mathematicians outside the universities as it was to painters and poets; and Galileo was already taking a calculated risk by christening the moons of Jupiter the 'Medicean Stars', for what if the moons proved illusory? But in private Galileo made no secret of his Copernicanism; and if he had the gift of making friends, so too he had the gift of making enemies. In the tense theological climate of the post-Reformation era, his enemies saw good prospects in accusing him of denying truths of Scripture, by holding the Sun to be at rest.

The matter became public in December 1614 when he was denounced from a Florence pulpit. A Scripture reading that day had been from the Book of Joshua. Joshua commanded the Sun to stand still – which implied that it would otherwise be moving. 'Ye men of Galilee', the preacher supposedly took as a punning text from Acts of the Apostles, 'why stand you gazing up to heaven?'

Galileo responded by venturing into theology. In an open *Letter to the Grand Duchess Christina* (1615), he recalled the traditional Catholic teaching, that the scriptural authors wrote in the language of the common man and that allowance must be made for this; and he argued that the purpose of Scripture was to teach men how to go to heaven, not how the heavens go. But his enemies continued their agitation, until at last the Holy Office in Rome became involved. The key figure was the Jesuit Cardinal Robert Bellarmine (1542–1621), a saintly and scholarly man prepared to reinterpret Scripture in the light of new discoveries – but only when the discoveries were established beyond question. In Copernicus's *De revolutionibus*, Bellarmine found from the introduction (which was in fact the unauthorized insertion of the proof-reader, see page 89), that even the Polish canon had – or so it seemed – taken the traditional view: the job of the astronomer was to predict accurately, not to propose truths of cosmology.

The upshot was that in 1616 the various branches of the Holy Office were notified that *De revolutionibus* was to be suspended until amended to ensure it conformed to the traditional role of astronomical texts. Bellarmine notified Galileo privately, that he might no longer believe that the Copernican system was true, or defend this belief.

In 1623 Galileo's Roman fortunes were transformed with the election to the papacy of his friend and supporter, Maffeo Barberini. Memories of the 1616 censure of Copernicanism faded, and after an encouraging series of audiences with the newly-elected pope, Galileo finally judged it safe to go ahead with the book on cosmology he had long planned. His *Dialogue on the Two Great World Systems*, published in 1632, was a discussion of the relative merits of Ptolemy and Copernicus. But the title was seriously misleading: by that time the Ptolemaic system had been largely abandoned by believers in a central Earth, and astronomers who could not accept the Sun-centred system of Copernicus – the great majority – were opting for the Tychonic or one of the other Earth-centred compromises on offer.

Written in the vernacular, the *Dialogue* gave a brilliant presentation of the advantages of the Copernican cosmology, and of the telescopic evidence in its favour. So far so good; but its author had to recognize that his readers' sensations were the opposite of what they might have expected as passengers on a spinning sphere hurtling in orbit about the Sun. And many contemporary thinkers agreed with Aristotle: an arrow fired vertically in the air, by falling to ground at the very place from which it had been fired, demonstrated that the ground had not moved while the arrow was in flight.

Galileo's response involved a reappraisal of the very concept of motion. For Aristotle, a natural object, whether living or not, revealed its nature by how it 'moved' (that is, changed – change of place, or 'local motion', being but one of several kinds of motion). Motion, in this general sense, was therefore central to his natural philosophy, which in the thirteenth century had gained added strength from its assimilation into a Christian framework.

For Aristotle, local motions, like all motions, required a cause, and so called for an explanation; rest, on the other hand, did not. Galileo set out to present an alternative view of local motion. According to him, it was not the motion itself but change of motion – acceleration – that called for explanation. Steady motion – of which rest was simply a special case – was a state, and to continue in the same state generated no sensation of movement. This was why Earth-dwellers were oblivious of the speed of the Earth as it orbited the Sun.

The *Dialogue* is presented as a discussion between three friends: Salviati, who speaks for Galileo; Sagredo, the man of good sense who therefore tends to end up agreeing with Salviati; and Simplicio, the Aristotelian. The real Simplicius had been a sixth-century commentator on Aristotle. But the name also had unflattering overtones, and Galileo made the mistake of putting into the mouth of Simplicio the Pope's own view: that, in the last resort, we have to accept that God could – and indeed may – have brought about the observed effects 'in many ways unthink-able to our minds'. Galileo had intended to call his book, *On the Ebbing and Flowing of the Sea*, thus focusing atten-tion on his conviction that the tides provided clear proof of the Earth's motion. He had indeed altered the title at the Pope's insistence; but the argument from tides survived in the text, and even though it was dressed up as being merely hypothetical, Galileo's remarks to this effect were uncon-vincing.

No wonder the Pope was displeased, though this does not seem enough to explain why Rome took the extreme step of summoning Galileo and charging him with dis-obeying the order of 1616. In the end, 'plea-bargaining' took place – Galileo was not of the stuff of which martyrs are made. He abjured his Copernicanism in a scene that was to do incalculable harm to the subsequent reputation of the Catholic Church, even though his 'house arrest' meant a life of comfort, first with his friend the Archbishop of Siena and later in a villa near Florence. Even the symbolic weekly recitation of the penitential psalms was said for him by his daughter.

Had Galileo realized the significance of Kepler's achieve-ment – and possessed the patience to master his writings – how much stronger would have been the arguments at his disposal! But although Kepler had enthusiastically sup-ported Galileo's telescopic claims in *Starry Messenger* by publishing a *Dissertation with the Starry Messenger*, and although we have a letter from a friend to Galileo, written in 1612 and speaking of Kepler's ellipses as common knowl-edge, Galileo never appreciated the intellectual weapons that Kepler had made available to Copernicans.

Galileo no doubt recoiled from the theological geometry and mystical harmonics that permeated Kepler's thinking, and he would certainly have been repelled by the endless pages of calculation in *New Astronomy*. Another factor was Galileo's lifelong inability to escape from the lure of circles. For him, it was not a ball on a plane that would roll and roll indefinitely (as we have been taught to imagine), but a ball on a smooth horizontal surface; and a horizontal surface was spherical, curving with the curvature of the Earth.

Newton was to synthesize the very different achievements of these two great contemporaries. But first, to make this possible, conceptual abstraction and clarification were needed. The Galileian cosmos still retained privileged places, such as the Earth's centre round which balls could roll, or the Sun round which the planets orbited. These last vestiges of spatial inequality had to be outlawed, and a universe conceived whose space was infinite and perfectly uniform – the universe of Cartesian physics which, as its author explained, was nothing but geometry.

Descartes and the geometers' universe

René Descartes (1596–1650) was composing his major works at much the same time as Galileo, but he was a generation younger, and the contrast in style is striking. In the medieval universities, Aristotle had been simply 'the Philosopher', and even in Galileo's time, the Philosopher's influence was still dominant. Galileo accordingly saw him as the arch-enemy, and could rarely write a page without making some criticism of him. Descartes, a man of independent means who lived his life outside the university system, saw the battle against Aristotle as already won, and seldom so much as mentions him; to Descartes, the central task was not destruction but reconstruction – what to put in place of the rejected reasoning of Aristotle. His own ruthless methodology led him to an extreme position: that the universe was infinite and uniform throughout, its space completely filled with undifferentiated matter moving under the laws of impact.

Descartes was born at La Haye, between Tours and Poitiers, and was fortunate to study at the Jesuit college of La Flèche. The Jesuits saw (and see) no contradiction in being both a priest and an astronomer or mathematician, and the students at La Flèche were introduced to Galileo's telescopic discoveries within a few months of their first announcement.

The quest for certainty

Descartes's teachers instilled in him a love of mathematics, and still more so a love of the certainty that mathematics made possible. After leaving school and seeing military service as a gentleman volunteer, in the early 1620s Descartes increasingly reflected on the problem of the method to be used in reasoning if one was to attain a similar certainty outside of mathematics; and he gradually came to the conclusion that the method of the geometers was itself the key. But how to apply this method to the study of the natural world?

In 1628 Cardinal de Bérulle heard Descartes speak in Paris on the gulf between certainty and mere probability, and he put it to Descartes that he had a moral duty to devote himself to the development of his insights into the problem of 'the manner of philosophizing'. Descartes thereupon withdrew to Holland, so as to distance himself from the distractions of Parisian life. There he lived until 1649, when he allowed himself to be tempted to Stockholm by an invitation to be philosopher to Queen Christina. He succumbed to the rigours of the Swedish winter in February 1650.

The difficulty with the wider application of the method of the geometers was that geometrical reasoning began from established truths, whereas in everyday life people took for true what may in fact have been mistaken assumptions that they had adopted in childhood and had unthinkingly accepted ever since. How to get rid of these errors? Descartes's solution – or so he claimed – was to attempt to doubt any and all alleged truths, because those that were immune to such drastic treatment must be genuine. His doubting knew no bounds: in particular, he asked himself whether the sensory world, of whose existence he had hitherto felt so sure, might not be as illusory as a dream.

But in the midst of all this Descartes found something whose truth he could not doubt: his own existence, for how could he think about such a question without existing in order to do the thinking? Furthermore, he could conceive of a perfect being (God); and such a concept of perfection could not originate in Descartes's own imperfect mind but must come from outside, from such a being. Furthermore, this perfect being would not play tricks on Descartes, by equipping him with an intellect that was essentially and irredeemably flawed.

The Cartesian universe

Confident therefore of the power of human reason rightly used, Descartes began to develop an insight into the uni-

verse, based on the 'clear and distinct' ideas in his mind –
those ideas that were still in the condition in which they
had come to him from God. They included mathematical
concepts of space and motion, and Descartes tells us that
the ancient geometers had understood these concepts with
the same admirable insight they had displayed in their
method of reasoning. The *space* of the geometers – totally
undifferentiated and extending infinitely in all directions –
was the space of the real world. The *motion* of the geometers, likewise, was the motion of the real world. The
Aristotelian concept of motion Descartes dismissed out of
hand as incomprehensible (though it had been compre-
hended well enough for two millennia!). Motion, he tells
us, was now revealed as simply movement from one location to another, as when a point moved to define a curve or
a curve to define a surface.

Descartes also analysed his concept of matter, and he
compared it with his concept of space. Any given piece of
matter could have been of any colour, blue, or green, or
indeed colourless: colour therefore was not part of the
actual concept of matter. Nor was taste, or smell, or any
other quality: an isolated piece of matter, as such, had
shape and size and location, nothing more. But the same
was true of an isolated region of space. And so Descartes
drew the astonishing conclusion that – aside from the
crucial consideration that matter could move *in* space –
matter and space were identical.

Important consequences followed. Since there was but
one kind of space, there was but one kind of matter; space
did not vary in density, so neither could matter; and space
without matter – a vacuum – was conceptually absurd. The
result was that, had it not been for the motions in the universe, there would have been no difference whatsoever
between one region and another: the entire universe would
have been totally uniform.

That there were differences was the result of motion. To
understand what Descartes had in mind, it may be helpful
to imagine a vast tank of motionless water into which a
spherical lump of ice is placed. The lump of ice is made to
spin; and as it spins it melts. Suppose that the spinning
then continues within this spherical space, occupied by
water that was formerly ice. We now have a tank wholly of
water (symbolizing Descartes's undifferentiated matter),
within which however a sphere can be identified because it
is distinguished from its surroundings *by its motions*.

In the Aristotelian cosmos, individual bodies existed,
and moved in accordance with their natures: a body's move-

ments reflected its nature. It was very different in the Cartesian universe; there, bodies were defined – one might almost say, created – by motion. What more could be said about this motion? God was above all changeless, and this characteristic was reflected in his creation, by what we today might term 'laws of conservation': space/matter was conserved, and (more interestingly) motion was conserved – both the total amount of motion in the universe (speed times amount of matter summed throughout the universe) and, in tendency at least, the motion of individual bodies.

The motion of a body tended to be conserved as it was at that moment, for God lived in the eternal present. The body was then moving with a certain speed in a certain direction, and so it was this motion – straight-line motion – that tended to be conserved.

But in practice this could not happen, for there was no space without matter, and so no body was isolated: nothing could move without compatible movements in the matter surrounding it, both in front and behind. As a result, motions that tended in principle to be straight-line, in practice took the form of a circulation of matter, in whirlpools or 'vortices'.

The solar system was one such vortex, mostly made up of matter whose motions made it invisible. It was this invisible matter (rather than the discredited solar magnetism of Kepler, of whose achievements Descartes was seemingly unaware) that carried around the visible planets. Meanwhile, by a kind of centrifuge effect, little pieces of agitated matter that the human eye perceived as luminous were being forced to the centre of the vortex, where they collected to form the Sun. The same happened in other large vortices, each of which had a sun (that is, star) at its centre: the Sun, therefore, was a typical star.

This conception Descartes believed he had derived from his insights into the immutable nature of God the Creator. Indeed, in his first exposition of his position – withheld from publication on news of the condemnation of Galileo – he made a significant point, that this immutability of God must inevitably be reflected in any world that He created, or indeed could create. This meant that when Descartes showed that a feature of our world was implied by the immutability of God, this feature didn't simply happen to be true of our world, it had to be true. Things could not be otherwise, for God had to be faithful to his own nature.

However, there was a limit to how far Descartes could reason on this basis, not only in practice but even in theory. He explained that when God created the infinite matter/space that constitutes our world, he imposed motion upon it and left it to develop in accordance with the laws of motion. But God was at liberty to impose any one of many patterns of motion, and it was not possible to infer from first principles which of these patterns he had in fact selected; and hence it was not possible to infer any of the detailed consequences that had followed from his particular choice. Similarly, features of the world about us are localized in space or time – for example, we happen to inhabit a planetary system which currently has a given number of planets; but this could have been otherwise, and so to learn about our particular system we have to use observation.

Despite its author's attempt to arrive at certainty, therefore, the Cartesian programme in science depended upon observation and experiment, with all their attendant uncertainties. And since Descartes himself had set out at length the metaphysical foundation for his cosmology, his immediate disciples felt able to skip most of this metaphysics when writing their textbooks. To persuade their readers, they relied instead on the intuitive appeal of using matter in motion to explain the universe. This approach to nature was known as the 'mechanical philosophy', and in Descartes's *Principles of Philosophy* it received its most radical expression.

The mechanical philosohy had been inherited from the atomists of Antiquity. It was especially plausible in a period when machinery (of which the elaborate clock of the cathedral at Strasbourg was the outstanding example) offered a persuasive model of the universe – persuasive because clockwork illustrated how immensely complex were the effects that could result from mechanisms, the individual components of which nevertheless had explanations that were reassuringly clear and intelligible.

Descartes was no astronomer; but so radical was his world picture that whereas all previous cosmologists had looked back (approvingly or otherwise) to Aristotle, now all would look back to Descartes. From the publication of his *Principles of Philosophy* in 1644, it was increasingly accepted that the Sun was one of innumerable stars in homogeneous and boundless space, and that the planets circulated about the Sun in paths resulting from rectilinear inertia (whereby an unimpeded body moves steadily in a straight line) modified by mechanical forces of impact. Newton was to spend his early maturity under the spell of Descartes; and when at last he abandoned the Cartesian plenum for an almost-empty universe bound together by a mysterious 'attraction', he was to scandalize many of his contemporaries.

Astronomy transformed

In the century that separated the publication in 1543 of
Copernicus's *De revolutionibus* from the publication in
1644 of Descartes's *Principles of Philosophy*, both the
science of astronomy and the universe that astronomers
studied were transformed. For all its novelties, Copernicus's
book was solidly in the tradition of the astronomy of
circles. For him, the task of the astronomer was what it had
always been, to devise geometrical models that replicated
the movements of the planets. Only a few of his contemporaries appreciated that, in carrying out this task,
Copernicus had found himself forced to modify a fundamental aspect of the closed, hierarchical cosmos of
Aristotle: its opposition between the central earth and the
encompassing heavens.

Within decades, astronomers had learned the fundamental importance of repeated and accurate observations,
made with instruments designed and refined to meet the
goal of precision. They had discovered how to devise apparatus to extend the human senses, and to see celestial
bodies hidden since the Creation. And they had extended
their goals, to include the study not only of how celestial
bodies moved but why – the investigation of the forces to
which the planetary movements were responses.

In the process the universe itself had changed, and
changed rapidly. Descartes's *Principles* marked with final
break with the Aristotelian cosmos with its natural places,
substituting instead the totally uniform and undifferentiated space of the geometers, in which a totally uniform and
undifferentiated matter endlessly redistributed itself.

'My physics is nothing but geometry', Descartes wrote in
a letter. But the mathematical tools needed to study the
Cartesian universe had not yet been invented, and so he and
his followers were forced to use words as substitutes for
symbols. In the short term this was an advantage: the liter-
ate but innumerate gentility who frequented the Paris
salons could understand Cartesian physics, as could the
undergraduates of Cambridge where Cartesianism became
all the rage. But although the Cartesians could explain the
celestial bodies plausibly enough, they could rarely forecast
how they would behave in the future. The predictive power
of quantitative Newtonian physics would in time prove
irresistible, and then Descartes would follow Aristotle into
history.

The telescope in the seventeenth century
J.A. Bennett

The refracting telescope

When, in 1609, Galileo turned a telescope of his own construction to the heavens, he was making imaginative use of an instrument that had been known for a decade or so as an optical device of possible military importance. Galileo's instruments had convex 'objective glasses' (which bent or *refracted* the rays so that they came to a focus) and concave eyepieces, and magnified up to 20 times. The lenses were available to him because they were manufactured for spectacles, but for telescopic work their quality was poor, and consequently the resolution of his instrument was primitive.

Gaileo is the most celebrated telescopic observer at this time, but was not the only one. In England Thomas Harriot (c. 1560–1621) was similarly engaged, but being in the service of the Earl of Northumberland, was less free, and had less need, to publish his results. After Galileo's *Starry Messenger* appeared, however, a telescope became a desirable tool for any participant in the cosmological debate, so that the demand for the instrument grew, along with interest in improving its performance.

Kepler responded to Galileo's initiative by designing a different construction of telescope, one with a convex eyepiece, and his scheme was published in 1611. His instrument had a larger field of view and could be used to project an image onto a screen, useful when observing sunspots. A feature that became important later in the century was that a Keplerian telescope could be fitted with an eyepiece micrometer (see below) or be adapted as a telescopic sight on an instrument for measurement. The image was inverted, but this was of little matter in astronomy, where the construction – to be known as the 'astronomical telescope' – became the standard one.

Its resolution, however, was limited by a problem that Kepler himself had indicated: lenses with spherical curvature – the only type that could be made at the time – did not bring all parallel rays of light to a unique focus. In other words, what became known as 'spherical aberration' resulted in a blurred image. All attempts to grind and polish aspherical lenses in the seventeenth century failed. In any case, there was a second problem, 'chromatic aberration', which was not clearly distinguished from spherical: even if a lens could be formed capable of focusing monochromatic rays, the different refractive characteristics of differently coloured rays would continue to deny the astronomer a sharp image.

The only practical measures that could be taken were to use lens of slight curvatures, that is, of long focal lengths, and to introduce apertures that restricted incident light to their central portions. Long focal lengths meant long telescopes, and by the mid-century a good instrument for astronomy would be 30 ft or more in length, requiring masts and

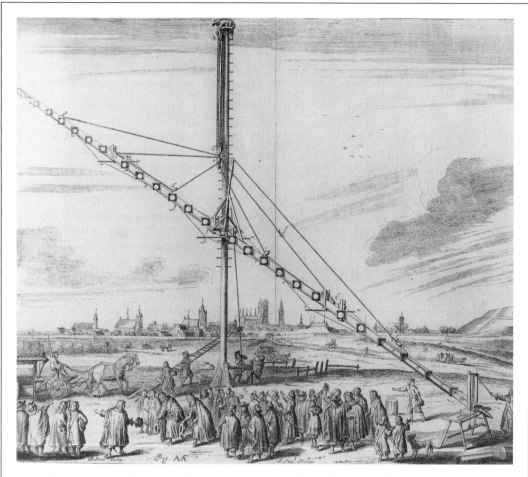

One of the telescopes of enormous length (and with only the skeleton of a tube, to reduce weight and the effects of wind) used by Johannes Hevelius to limit the effects of chromatic and spherical aberration. From J. Hevelius, *Machinae coelestis*, 1673.

systems of pulleys to mount and manage it. The leading telescopic observer of the century, the wealthy Danzig brewer Johannes Hevelius (1611–87), had telescopes of extraordinary lengths, up to 150 ft, but an instrument of such a size was quite unmanageable. Christiaan Huygens even resorted to mountings where there was no physical connection between objective and eyepiece, other than a line operated by the observer.

The reflecting telescope

The practical difficulty in giving the objective glass of a refractor the ideal geometrical shape, and the excessive tube lengths required to circumvent this problem, in 1663 led the Scots mathematician James Gregory (1638–75) to propose an alternative design. In it the light passed down the tube to a concave mirror at the bottom, which reflected it back up the tube. A second, small

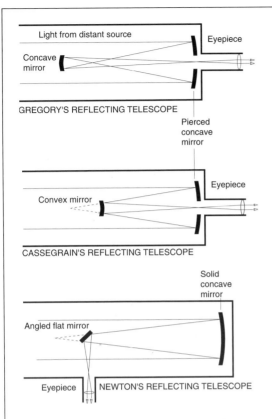

GREGORY'S REFLECTING TELESCOPE

CASSEGRAIN'S REFLECTING TELESCOPE

NEWTON'S REFLECTING TELESCOPE

The three forms of reflector invented in the 1660s and 1670s. In the Newtonian design (below), it was not necessary to pierce the primary mirror, and the secondary mirror was flat rather than curved. This simplified the task of the builder, and so this form became popular in the eighteenth and nineteenth centuries, especially for major instruments with massive mirrors.

concave mirror, centrally placed, reflected the light back again down the tube. The light then passed through a hole in the primary mirror, and so entered the eyepiece.

Gregory took the design to a London firm of opticians, but their rudimentary trial did no more than confirm the feasibility of the design.

A similar design, but with a convex secondary mirror, was proposed in 1672 by a Frenchman called Cassegrain, of whom almost nothing is known. A century later it was shown that the convex secondary mirror helped to avoid the 'spherical aberration' introduced by departures of Gregory's concave mirrors from their ideal shapes, and in modern telescopes the Cassegrain design is widely used.

Meanwhile, Isaac Newton in Cambridge had made a discovery of fundamental importance (see page 224). It had long been believed that white light was simple and that colours were 'modifications' of white light, but Newton showed that – on the contrary – white light was compounded of the colours, each of which was refracted by a given objective lens through a slightly different angle. The blurring caused by this 'chromatic aberration' was, he thought, inherent in refracting telescopes. The mirror that Gregory used in place of a lens avoided this problem; but unfortunately Gregory's design called for a second curved mirror, and the primary mirror had to be pierced.

In 1668, therefore, Newton built himself a little reflector to a design that avoided these drawbacks. A mirror at the foot of the tube again reflected the incoming light, but now a small, flat mirror inclined at 45° to the axis of the tube reflected the light sideways to where the eyepiece was located.

This reflector became known only to a handful of Newton's Cambridge acquaintances, but in 1671 he made a similar instrument, which he presented to the Royal Society. Neither of these reflectors survives, but parts of a third instrument, which he made in the winter of 1671/72, are believed to be incorporated in a reflector that was presented in 1766 to the Royal Society.

Because it used only a single curved mirror, the Newtonian design was widely used in the eighteenth and nineteenth centuries, especially for the study of faint objects for which a mirror with large 'light-gathering power' was required.

Newton was overly pessimistic in thinking the problem of chromatic aberration insoluble. In 1729 a London barrister, Chester Moor Hall (1703–71), devised an 'achromatic' lens by combining two glasses of different refractive properties, a concave of flint and a convex of crown glass. But Hall did not pursue the idea commercially, and this allowed John Dollond (1706–61), a leading London instrument maker, to revive the idea in a paper to the Royal Society in 1758, and to patent it. Thereafter 'Dolland achromatics' became much sought-after among observatories and amateurs alike.

Measuring with the telescope

The telescope allowed observers to see new things, and to see familiar things enlarged. By contrast, the astronomer's traditional instruments were for the measurement of angles. It was not immediately obvious that the two could be combined.

In the telescope as designed by Kepler, the image was brought to a focus and examined with the eyepiece, which served as a microscope. About 1640, an English amateur astronomer named William Gascoigne found that a spider had spun a web in the focal plane of his telescope, and this web appeared superimposed on the astronomical image. He realized that cross-hairs could be inserted in this same plane, to locate the exact centre of the field of view and so enable

the telescope to be aligned accurately on the target object. Alternatively, a measuring device or 'micrometer' could be similarly placed, and used to measure either the width of an object such as a planet, or the angle between two nearby objects, such as two mountains on the Moon.

Unfortunately Gascoigne was killed in 1644 in a battle in the Civil War, but his techniques were applied in Oxford in the 1650s, notably by Christopher Wren, who used micrometers in a survey of the Moon. In 1663 Wren also demonstrated to the Royal Society an astronomical instrument that incorporated telescopic sights.

Meanwhile Huygens announced his (independent) discovery of a form of eyepiece micrometer in 1659 in his *Systema Saturnium*. Other forms of micrometer

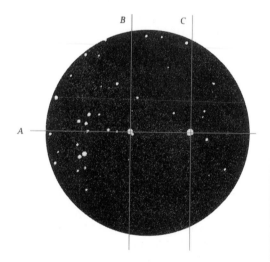

The principle of the wire micrometer. To measure the angle separating two stars, the observer aligns the micrometer so that the fixed wire A passes through the two stars. He then adjusts the moveable wires B and C so that each passes through one of the stars. The amount of the adjustment required is then a measure of the angle separating the stars. From J. J. le F. de Lalande, *Traité d'astronomie*, 1771–81.

were developed in Paris in the 1660s, by Adrien Auzout, Pierre Petit, and Jean Picard. Soon, the transformation of the telescope into an instrument of measurement was complete.

The incorporation of a telescope into traditional measuring instruments, to improve the resolution of the unaided eye, was slower in coming. Resistance came most notably from Hevelius, who was a highly respected observer. Hevelius had built for himself the finest observatory of the day (though it would be destroyed by fire in 1679), and while he of course used telescopes for observation, he insisted on making measurements with the naked-eye. Hooke and other innovators saw this as setting an unfortunate example to would-be observers. A public squabble ensued, with Hevelius refusing to budge; even a visit from the honey-tongued Halley failed to move him. But by the time of his death Hevelius was almost alone in denying the value of telescopic sights.

The foundations of astronomy

The equivalent of terrestrial longitude is known as right ascension (RA), and is measured from the point on the celestial equator where the Sun crosses the equator at the spring equinox. As the Earth rotates once a day, RA can be expressed in degrees or in time (with one hour equivalent to $15°$), and the fundamental measurement is of the moment when the celestial object crosses (or

'transits') the meridian of the observatory. The first precision-mounted instrument dedicated to the measurement of transits was built by Ole Römer after he returned from Paris in 1681 to take charge of the Copenhagen Observatory. The location of an observatory's transit instrument came to define the longitude of the observatory, and since 1884 the transit instrument at Greenwich has defined the world's zero of longitude.

Eighteenth- and nineteenth-century observatories came to see it their duty to record the positions of the stars and other celestial objects with ever increasing precision and completeness. To measure the altitude of the celestial body at the moment of transit, and so obtain its declination (the celestial equivalent of latitude), observatories followed the example of Tycho Brahe and used quadrants or other sectors of a circle mounted against a wall precisely aligned north–south – but of course with telescopic sights. In the nineteenth century, leading English makers such as Edward Troughton developed a mural circle, which (like its predecessors) was to be used in conjunction with the transit instrument. This however was expensive, not least in manpower, since two observers were required. Continental makers responded to demand by building transit instruments that incorporated circles of sufficient accuracy to render a mural arc (or circle) unnecessary, and English makers were soon forced to follow suit.

6

Newton and Newtonianism
Michael Hoskin

We can all too easily imagine Kepler's three 'laws' of planetary motion to have been straightforward generalizations from observational data, and therefore trustworthy and uncontroversial. The truth is very different.

What we know as the first law, that planets move in ellipses with the Sun at a focus, was proposed in 1609 in Kepler's *New Astronomy* after complex and confused calculations guided by a very dubious dynamics. It became more widely known when repeated a few years later, clearly and simply, in his *Epitome of Copernican Astronomy*. The law was an astronomer's dream come true: each planetary orbit was described by one single geometrical curve, of a familiar type whose properties had been established in Antiquity. It was no wonder that Kepler's ellipses proved attractive to mid-century astronomers.

Yet we who live after Newton know that the law is not strictly true: in the real solar system, planets are 'perturbed' (disturbed) out of pure elliptical orbits by the gravitational pulls of the other planets. Mid-seventeenth-century observations were not sufficiently accurate to reveal these complications; but neither were they accurate enough to choose between an ellipse and various ovals that approximated closely to an ellipse.

The 'second law', again proposed in *New Astronomy* and repeated in the *Epitome*, states that the line from the Sun to a planet sweeps out equal areas in equal times. Not only do perturbations again make this law no more than an approximation, but it was completely beyond the competence of mid-seventeenth-century astronomers to verify such a sophisticated mathematical relationship.

Alternative formulations were however to hand. Because the planetary orbits are nearly circular, numerous 'equant' theories (involving locations from which the planet's motion would *appear* uniform) could be devised that were observationally indistinguishable from the law as we know it. These had the great merit of being familiar in form and mathematically tractable.

Nor were these the only candidates in the field. The area rule in fact implies that the speed v of the planet is inversely proportional to the distance r of the planet from the Sun ($v \propto 1/r$) on just two occasions in each orbit, namely, when the planet is at one or other end of the major axis of the ellipse. However, if the ellipse is nearly circular, then the speed is approximately inversely proportional to the solar distance throughout its orbit; and it was this relationship that some astronomers took to be the true law, to which the area law was itself an approximation.

The 'third law', announced in 1619 in the *The Harmony of the World*, declared that the square of the period of each planet's orbit was in a fixed ratio to the cube of its mean distance from the Sun. One might have expected this law to be easily verified, for in dealing with distances and periods astronomers were on familiar ground. But again there were complications: Mercury, and even Venus, being close to the Sun, are difficult to observe; perturbations have their effects; and, as Newton was to show, the law itself needs modification.

The genesis of Newton's *Principia*

Nevertheless, whatever their limitations, Kepler's 'laws' gave astronomers some grip on how the planets moved. How they moved, but not why: the laws were silent on the forces that caused these movements. Kepler believed that a planet was lazy and continued its onward motion only because the rotating Sun somehow pushed it round – that otherwise the planet would immediately come to a halt. Descartes, on the other hand, taught that a planet would – in principle – continue to move in a straight line at uniform speed, unless subjected to outside influence. His universe, however, was completely filled with matter, and so the planets in it were permanently subject to such influences. In practice, therefore, his planets were carried round in the solar vortex, jostled on all sides by matter whose pressure bent the planetary orbits from straight lines into nearly circular paths.

The magnetical philosophy
However, the Fellows of the London-based Royal Society (founded in 1660) were heirs to a quite different tradition, one that stemmed from William Gilbert (1544–1603), who in 1600 had argued in *On the Magnet* that the Earth is a huge magnet (see page 107). His basic experimental

apparatus consisted of a piece of natural loadstone, turned into a sphere and often floated on water. The sphere, which Gilbert called a *terrella* ('little Earth'), was designed to prove by analogy the Earth's magnetic properties. He hoped that if sufficient data on the difference between magnetic and true north, and that between the 'dip' of a compass needle and the horizontal, could be assembled, the information might be of use to navigators.

Gilbert was a physician, and typical of the practical men who frequented Gresham College, recently founded by the London merchant Sir Thomas Gresham. Gresham College was an institution of higher learning quite unlike the universities of Oxford and Cambridge. At Gresham the professors lectured mainly in English, and approached their subjects in the manner appropriate for audiences that included instrument makers, seafarers, physicians, and others of practical bent.

Throughout the early and mid-seventeenth century, the Gresham circle maintained its own tradition of experimental science in which the Gilbertian 'magnetic philosophy' held an honoured place. In the late 1640s, in the aftermath of the English Civil War that was to result in the execution of King Charles in 1649, the group was weakened as the victors imposed some of its members on Royalist Oxford, where they found themselves in key positions. But the tradition survived and, with the approach of the restoration of the monarchy in 1660, Gresham College became once more the focus of scientific life in the capital. Indeed, the informal meeting in November of that year that led to the foundation of the Royal Society took place after an astronomy lecture at Gresham College.

A leading Greshamite of the mid-century was John Wilkins (1614–72). In 1640 Wilkins published a second edition of his *Discovery of a World in the Moone*, and in it he discussed the theoretical possibility of a journey to the Moon. In his view such a possibility existed; for just as Gilbert had believed a spherical magnet to be surrounded by a strictly finite sphere of attraction, so Wilkins thought that the space traveller would be able to escape the magnetic influence of the Earth, once he was perhaps twenty miles above the surface. However, as Wilkins reflected further on this sphere of influence, he decided that it was unreasonable to expect there to be a sharply-defined boundary beyond which gravity had zero effect: 'it is probable, that this magneticall vigor dos remit of its degrees proportionally to its distance from the earth, which is the cause of it.'

Robert Hooke and the early Royal Society

In 1648 Wilkins was imposed on Wadham College, Oxford, as its head, and it was at Oxford that two of the brilliant minds of the next generation came under his influence: Christopher Wren (1632–1723) and Robert Hooke (1635–1703). Wren is remembered as the architect who, in collaboration with Hooke, rebuilt London after the Great Fire of 1666; but he held the post of professor of astronomy, first at Gresham from 1657 to 1661, and then at Oxford. Hooke became Curator of Experiments to the infant Royal Society in 1662, and three years later added to this the post of Gresham Professor of Geometry. He was widely talented, but unfortunately his multifarious duties as Curator of Experiments encouraged his innate tendency to throw out ideas without following them up.

How the planets moved in their orbits, what paths were followed by comets, and how the magnetic pull of the Earth diminished with increasing distance, were problems repeatedly discussed by the two friends and their circle of acquaintances. In 1662 and again in 1664, Hooke attempted to confirm by experiment that the pull of the Earth varied with height. Perched high up in Westminster Abbey or in old St Paul's Cathedral, he measured the weights of bodies, first when alongside him and then when suspended near to ground level – but of course without managing to detect any change.

In 1666, to help illustrate his developing ideas concerning the interaction of the Sun with the Earth and Moon, Hooke showed to the Royal Society a pendulum whose string was divided near its end into two strings, one with a large weight (representing the Earth) and the other with a small one (the Moon). These two weights were set rotating about each other, together forming a system which was then rotated as a whole about the 'Sun'.

Thereafter Hooke's progress in understanding was rapid. In 1674 he took the opportunity to summarize the position he had reached, in one of the most remarkable passages in the history of astronomy ever published. It was in the form of three 'Suppositions' appended to his *Attempt to Prove the Motion of the Earth*:

> First, That all Coelestial Bodies whatsoever, have an attraction or gravitating power towards their own Centers, whereby they attract not only their own parts, and keep them from flying from them, as we may observe the Earth to do, but that they do also attract all the other Coelestial Bodies that are within the sphere of their activity . . .

Hooke held that the Earth both attracted and was attracted
by, not only the Sun and Moon, but all the other planets as
well – and that this interplanetary force was the same as
the gravity that held the parts of the Earth together.
However, his mention of 'the sphere of their activity' shows
that he did not think of these attractive powers as universal.

> The second supposition is this, That all bodies whatsoever
> that are put into a direct and simple motion, will so con-
> tinue to move forward in a streight line, till they are by
> some other effectual powers deflected and bent into a
> Motion, describing a Circle, Ellipsis, or some other more
> compounded Curve Line.

This was the first correct published statement of the
dynamics of motion in an orbit.

> The third supposition is, That these attractive powers are so
> much the more powerful in operating, by how much the
> nearer the body wrought upon is to their own Centers.

The attractive powers, Hooke held, diminished with
increasing distance. But did they vary inversely with the
distance itself ($f \propto 1/r$), or with the distance squared ($f \propto$
$1/r^2$), or what? Hooke could not say; but he regarded the
answer as relatively unimportant, merely one of the loose
ends:

> Now what these several degrees are I have not yet experi-
> mentally verified . . . He that understands the nature of the
> Circular Pendulum and Circular Motion, will easily under-
> stand the whole ground of this Principle.

Of the possible laws, the obvious candidate was that the
forces varied inversely with the distance squared (the
'inverse-square' law). After all, the brightness of the Sun or
a planet reduces in just this way. But there was another
reason for suspecting the inverse-square law, which had to
do with the attempts of the Cartesians to analyse the forces
at work in orbital motion. When a slinger began swinging
the stone around him, he felt the stone pulling forcefully
away from him as he stood braced against it. Prior to the
release of the stone, the circular path in which the stone
travelled at the end of the rope seemed the result of a
balance between the restraining pull of the slinger directly
towards the centre and the outwards pull of the stone
directly away from the centre. (Today we can see that this
is a misleading analysis, the result of working with a rotating frame of reference: the analysis involved consideration
of the pulls, inwards and outwards, in the direction of the

rope – which was rotating all the while.) In 1673, in a book by the Dutch physicist Christiaan Huygens (1629–95), the outwards pull of the shot was shown to be proportional to the speed squared divided by the radius (v^2/r).

Now Kepler's third law stated that the square of the period of a planet was in a fixed ratio (say, k) to the cube of the radius of its orbit. Suppose the planets had all been travelling in circles with constant speeds, rather than in nearly-circular ellipses with varying speeds. Then since the period was equal to the circumference ($2\pi r$) divided by the speed v, the law would have stated that for each planet, $(2\pi r/v)^2 = kr^3$; that is, v^2/r – Huygens's pull – would be in a fixed ratio to $1/r^2$.

The inverse-square law was therefore the obvious candidate. However, the real planetary orbits were not circular with constant speed, but elliptical with varying speed; and so the crucial question was: would an inverse-square law of attraction to the Sun result in elliptical orbits? In January 1684, in discussion at the Royal Society with Wren and the young Edmond Halley (c. 1656–1742), a future professor of geometry at Oxford and Astronomer Royal, Hooke claimed to have a demonstration that this was so, though he would not reveal it. Wren, who knew Hooke well enough to take the claim with a pinch of salt, offered a book prize to him or to anyone else who actually produced a proof within two months. The prize went unclaimed.

Later that year, Halley visited Isaac Newton (1642–1727) in Cambridge, where he was professor, and asked him what would be the orbit of a planet moving under an inverse-square law of attraction to the Sun. Newton gave Halley the answer he had hardly dared hope for: the orbit would be an ellipse.

The dynamics of elliptical orbits

The Cambridge that Newton had entered in 1661 still respected the primacy of Aristotle in its statutory teaching, but the semi-independence of college tutors had permitted the unofficial introduction of Cartesian physics, and with it the mechanical philosophy – the doctrine that matter in motion is the explanation of all things and that one body can affect another only by direct contact.

In the universe of Descartes, who thought the very idea of a vacuum to be a self-contradiction, the Earth and the other planets were carried round in the huge vortex that was the solar system, while the Earth in turn was the centre of a subordinate vortex that carried the Moon. The motion of the Moon therefore involved both the solar and

the terrestrial vortices, and this presented the investigator
with a ill-defined situation.

By comparison, the analysis of the abstract notion of
motion in a circle was straightforward, and by the mid-
1660s the young Newton had independently derived for the
outwards pull the same formula as Huygens. But thereafter
his progress was slow and uncertain. One obstacle lay in
the variants of Kepler's second law in circulation: although
observationally these variants were almost indistinguishable from each other, conceptually the differences were profound. Newton himself, between 1665 and 1679, worked
with a number of equant versions of the law. Eventually,
however, he came across a published statement of the law
in its area form.

By 1679, Newton had been professor of mathematics at
Cambridge for a decade. He was still a Cartesian in planetary dynamics, still struggling to make sense of the
complex interactions within the solar vortex, and still confused over inwards and outwards pulls. That November he
received a letter from Hooke, with whom he had earlier
clashed in a dispute over the nature of light. Hooke was
now Secretary of the Royal Society and his purpose in
writing to Newton was to involve this temperamental
genius in the Society's activities. In his letter he invited
Newton to consider the consequences of 'compounding the
celestiall motions of the planets of a direct motion by the
tangent [inertial motion] and an attractive motion towards
the centrall body'. Where Newton had thought of orbital
motion as a tussle between opposing forces, Hooke was
treating it as one in which a body that tended to continue
in a straight line was instead pulled into a curved path.

Hooke also told Newton that a claim he had made in his
Attempt to Prove the Motion of the Earth (1674), to have
detected an apparent annual movement in the star Gamma
Draconis that reflected the movement of the Earth-based
observer (pages 180–2), had now been confirmed by John
Flamsteed (1646–1719), the recently-appointed Astronomer
Royal at Greenwich (though both were in fact mistaken):
and it was on the question of the motion of the Earth that
Newton touched in his reply. The traditional proof that the
Earth was at rest had involved thought experiments with
projectiles: an arrow fired vertically fell to ground in the
same place, proving that the Earth had not moved while the
arrow was in flight. Along the same lines, it used to be
argued that a stone released from the top of a tower, and
thereafter able to move freely, would take the opportunity
to fall in a straight line to the centre of the Earth. If the

Earth were to be spinning (it was claimed), then the tower
would move to the east while the stone was falling, and the
stone would therefore hit the ground to the west of the
tower. It was believed that such a stone would hit the
ground at the foot of the tower, and this was taken as proof
that the tower, and the Earth, had *not* moved during the
time of fall.

Newton however had a more sophisticated understanding of such a test. He pointed out that the top of the
tower was further from the centre of the Earth than was the
foot of the tower. If therefore the Earth was indeed spinning, a stone at the top of the tower would be travelling
horizontally with a velocity greater than that of the base.
Since the stone in falling would retain its horizontal velocity, it would actually hit the ground ahead (that is, to the
east) of the base of the tower. Perhaps, Newton suggested,
with care the difference would be measurable.

Newton illustrated his letter with a drawing to show the
path of fall of the stone, from the top of the tower to the
ground. Extending the path into the interior of an imaginary
Earth that he pictured as offering no obstacle to the passage
of the stone, he went on to suggest that if the force pulling
the stone towards the centre of Earth was a constant irrespective of distance, then the stone as it continued to fall
would spiral in towards the centre. By analysing the motion
of the stone in this way, Newton converted Hooke's
problem from one of free fall into one of orbital motion.

Hooke in reply could not resist pointing out that, with
his spiral, Newton had made an elementary blunder; and
nothing concentrated Newton's mind more than being
proved wrong. Yet Hooke was soon to tire of pursuing the
consequences of this constant pull that was little more than
a mathematician's game, for he was very much the physicist: 'my supposition is that the Attraction always is in a
duplicate proportion to the Distance from the Center
Reciprocall', in other words, the inverse-square law.

By now, however, all Newton's defence mechanisms
were fully deployed, and he withdrew into his shell. But,
unknown to Hooke, he had solved the problem Hooke had
set him: the imagined planet would move in an ellipse with
the central body at a focus, just as the real planets did
according to Kepler's first law.

Universal gravitation

Newton indeed had gone further. Provided Hooke's
approach to the analysis of planetary orbits was valid, he
now had reason to think that an inverse-square pull exerted

by the Sun would result in planetary orbits that obeyed both the first and the second of Kepler's laws, and doubtless the third as well.

Hooke believed that attraction operated between the Sun, the planets, and sometimes – but not invariably – the comets. Indeed, in his recent *Cometa* he had suggested that the matter forming comets could lose its attracting power, owing perhaps to a 'jumbling' of its internal parts. Cometary orbits had long presented mathematicians with a challenge, for to derive an orbit in three dimensions from observations made from Earth was very difficult – so difficult that Newton was later to struggle with the problem for months while composing the *Principia*. However, comets were seen to follow paths across the sky that were roughly arcs of 'great circles' (circles centred on the human observer), and the general opinion was that these arcs were projections onto the heavenly sphere of orbits that were nearly straight-line. It was thought that when a comet appeared, heading for the solar system, it approached along a straight line, accelerating as it came; then, after a small swerve as it passed the Sun, it continued straight on and out of the solar system, decelerating as it went.

Flamsteed, however, was one of several astronomers who took a different view. A comet had appeared in November 1680 heading towards the Sun, and another comet was first seen on 10 December heading away from the Sun in pretty much the opposite direction. Flamsteed believed that these were not two distinct comets, but one and the same. To explain the reversal of direction, he argued that as the comet had been moving inwards in response to the Sun's attraction, it had been thrown off course by the solar vortex. Eventually it had found itself in a new magnetical relationship to the Sun, and been repelled. As a result, it had been driven back before it ever reached the Sun.

Flamsteed went out of his way to inform Newton of his theory and to provide him with the observational data on which it was based. In the exchange that followed, Newton floated the alternative suggestion that the comet had 'fetched a compass round the Sun' – that it had passed around the back of the Sun. He may well have been wondering whether the comet was attracted by the Sun in accordance with the inverse-square law; should this prove to be the case, cometary orbits would be no different in principle from those of the planets. Eventually, however, he concluded that the evidence told otherwise.

We know little else of how Newton's ideas developed between the exchange with Hooke in the winter of 1679/80 and the arrival of Halley at his door in 1684 – a visit that revived his interest in the dynamics of planetary orbits. Halley asked him 'what he thought the Curve would be that would be described by the Planets supposing the force of attraction towards the Sun to be reciprocal to the square of their distance from it'. Newton revealed to the young and sympathetic Halley what he had concealed from the combative Hooke: the curve would be an ellipse.

Halley was 'struck with joy and amazement'. When he could speak, which we may suppose was not immediately, he begged to see the calculation 'without any further delay'. Newton was unable to lay hands on it, but he promised to send it to him.

The writing of the Principia

What arrived was a draft treatise of only nine pages, much of it devoted to bodies moving in empty space. Hypothesis 2 is a statement of the law of inertia: 'Every body, unless impeded by something extrinsic, by its innate force alone proceeds uniformly into infinity, along a straight line.' It is expressed in terms of infinite straight lines, lines that had been inconceivable in closed, finite universes such as those of Kepler and Galileo.

Newton proved – or believed he had proved – a series of propositions relating to Kepler's laws or to generalizations of them. If a body moved under a pull to a 'centre', it would obey the area law. If the pull was inverse-square, then the orbit would be conical (possibly, but not necessarily, elliptical); and conversely. If bodies moved in elliptical orbits and the pull was directed towards their foci, then the orbits would obey Kepler's third law; and conversely.

These abstract results in mechanics could be applied to the solar system, raising to the status of true laws what had until now been viewed as Keplerian generalizations of questionable reliability: 'The major planets orbit, therefore, in ellipses having a focus at the centre of the Sun [Kepler's first law], and with their radii drawn to the Sun describe areas proportional to the times [the second law], exactly as Kepler supposed.'

The focus of a planetary orbit, we note, was 'at the centre of the Sun'. Newton was still thinking of the Sun as doing the pulling and the planet as being pulled. He had yet to match the insight that Hooke had expressed, however imperfectly, in the first of his 'suppositions' of 1674: that the attraction between celestial bodies is mutual.

This is a little surprising, in that his draft stated that Kepler's third law applied also to the moons of Jupiter dis-

covered by Galileo, and to the moons of Saturn (Huygens
had come across Titan in 1655, and in Paris Gian Domenico
Cassini had recently discovered four more). These moons
were therefore subject to the inverse-square pull of the
parent planet. If Saturn pulled Titan, why should it not also
pull the Sun?

By the time Newton composed a revised draft a few
weeks later, he had taken the crucial step: celestial bodies
did pull each other. He was now committed to the path that
would lead him to universal attraction. Conceptually,
confusion was giving place to clarity on a scale and at a
pace unique in the history of the physical sciences. But
even Newton was aghast at the implications for the unfortunate investigator. With so many celestial bodies, each
pulling all the others, how could anyone hope to cope with
the mathematics involved: 'But to consider simultaneously
all these causes of motion and to define these motions by
exact laws admitting of easy calculation exceeds, if I am not
mistaken, the force of any human mind.'

The transformation that had recently taken place in
Newton's world picture could hardly have been more dramatic. When the 1680s began, his world was crammed full
of matter, a confused Cartesian plenum in which bodies
endlessly impacted on their immediate neighbours. By the
middle of the decade his world was almost empty; in it, isolated bodies were diverted from their onward paths by the
attractive pulls of other isolated bodies, reaching out to
them across seemingly empty space.

Halley had no trouble in realizing the revolutionary significance of all this, and he began to coax Newton towards
publication. He knew that Newton was what was termed a
'nice' person, one who had to be handled with great delicacy. Newton was also an inveterate drafter and polisher, an
author who rarely considered his work to be fit for publication in its present state. But he became increasingly
obsessed with the task in hand, and the proposed treatise
grew and grew.

One prize within his grasp was a theory of comets – even
before writing the first draft he had revised his opinion of
the comet of 1680 and now believed it had indeed 'fetched a
compass' about the Sun. Comets, he was convinced, obeyed
the inverse-square law. Their paths were as lawlike as those
of the planets, and after the mathematical problems had
been overcome it would be possible 'to ascertain . . .
whether the same comet returns with some frequency to
us'. Newton's investigations were to result in a Proposition
in the *Principia* declaring that comets moved in conic sec-

tions – possibly, but not necessarily, ellipses – with the Sun
at a focus (Kepler's first law, generalized), and that the line
to the Sun swept out equal areas in equal times (the second
law).

In his first 'supposition' of 1674, Hooke had identified
the pull that the Earth exerted on other celestial bodies at
great distances with the pull it exerted on falling stones a
few yards away. The same idea now occurred to Newton.
Could it be that the inverse-square law of attraction applied
to matter large and small? In particular, did the Earth exert
the same attractive force on a stone as on the Moon?

Before Newton could attempt an answer to this question,
his faith in the law – like that of his future readers – had to
survive a test. Could one seriously maintain that two equal
lumps of matter at the same distance from a stone (or the
Moon), one lump deep in the bowels of the Earth and the
other lying on the surface, would attract the stone (or the
Moon) with exactly equal force? Surely in the case of the
lump inside the Earth, the screen provided by the surrounding rocks must reduce its pull? So strong did this objection
seem to some of Newton's later disciples that they declared
that, since attraction depended on distance alone, and inter-
vening obstacles made no difference whatsoever, attraction
must be a law imposed directly by God.

Granted, then, that attraction varied precisely with the
inverse-square of the distance, Newton had to overcome a
mathematical hurdle before he could make the Moon test.
The total pull of the Earth on a falling stone was the
combination of innumerable individual pulls, some over
distances of a few yards, others over distances of thousands
of miles. How to sum up the combined effects of all these
pulls, operating in different directions and over such dispar-
ate distances? Newton succeeded in proving a very remarkable theorem, that in a uniform spherical body (as the Earth
approximately is), all such pulls are together equivalent to
the pull of the combined matter imagined as concentrated
at the very centre of the sphere: the falling stone was, in
effect, exactly one Earth's radius away from the attractive
pull of the entire Earth.

This done, Newton was able to compare the Earth's pull
on the stone (at a distance of one Earth radius) with the pull
on the Moon (at some sixty times the distance). The pull on
the stone was the cause of its acceleration in free fall; the
pull on the Moon caused it too to 'fall', out of straight-line
motion and into an orbit around the Earth. His calculations
showed that the pulls were indeed in about the ratio of
$1:60^2$. It was the final blow to Aristotle's dichotomy

between terrestrial and celestial: Kepler's planets and the falling stones studied by Galileo were moving under the same law of force.

As Newton's treatise grew, so too did the number of observed facts for which attraction now emerged as the cause. For example, he gave the first satisfactory explanation of the tides, as resulting from differences in how the solid Earth and the flowing water of the seas were pulled by the Moon and the Sun. And he pointed out that, as a result of its spinning, the Earth would bulge slightly at the equator and be slightly flattened at the poles; because of this non-sphericity, the pulls of the Moon and Sun would cause the Earth to wobble like a child's top, resulting in the precession of the equinoxes that had been discovered by Hipparchus in Antiquity (see page 39).

Because the Moon is so near to the Earth-based observer, its complex movements can be studied in unique detail; and because the Moon's mass is significantly large in comparison with that of its parent planet (unlike, say, those of the moons of Jupiter), its dynamical interaction with the Earth is exceptionally complicated. Indeed, in the eighteenth century the Moon was to become a testing ground for Newtonian dynamics: could the law of attraction explain, not only qualitatively but quantitatively, the many departures of the Moon from a simple orbit?

Tycho, for example, had found that, when other factors had been taken into account, there was a residual variation in the speed of the Moon's movement around the sky (that is, in longitude), such that the resulting displacements reached alternating positive and negative maxima in the four 'octants' (the positions that lie midway between one of the quarters and either full or new moon). He had also found that the Moon's position in longitude displayed a further annual irregularity, amounting to as much as 11 minutes of arc.

In his *Principia*, Newton was able to explain why these effects came about. To a first approximation the centre of gravity of the Earth–Moon system moved around the Sun in an ellipse, while the Earth and Moon moved in ellipses around this centre of gravity, the Earth in a smaller ellipse and the Moon in a larger one, each opposite the other. But the Sun interfered with this simple picture: it pulled the two bodies with different pulls if their distances from the Sun were different, and it pulled them in different directions when their positions formed an angle at the Sun. These different effects had to be combined with the Earth's pull on the Moon.

Twice a month, around new moon and full moon, the
effect was to subtract from the Earth's pull on the Moon,
and twice a month, around the quarters, it was to add to the
Earth's pull. As the Moon passed from the quarters to new
or full moon, the Moon's linear speed was increased; as it
passed from new and full moons to the quarters its linear
speed was diminished. As a result of these fluctuations, the
Moon's angular position underwent deviations having
extremes four times each month, in the octants – just as
Tycho had found.

As the Earth–Moon system orbited the Sun each year in
its elliptical path, it of course varied in its distance from the
Sun. When the system was near the Sun, the Sun's force
was more powerful than usual and this tended to diminish
the Earth's influence on the Moon, while when the system
was further from the Sun the effect was reversed. This
caused the further, annual irregularity that Tycho had
noticed.

Newton's qualitative derivations of these and other lunar
irregularities (including Ptolemy's 'evection', see page 44)
are to be found in Book I of the *Principia*. But his attempts
to obtain quantitative results in Book III and elsewhere
were only partially successful. The *Principia* was indeed the
end of one era in the history of astronomy, but it was also
the beginning of another. It set the agenda for 'celestial
mechanicians' in the decades to come.

The law of attraction gave Newton other insights into
the planetary system. From the inward accelerations of the
moons of those planets known to be endowed with them –
Earth, Jupiter and Saturn – Newton could calculate their
causes, the masses of the parent planets. The Earth, it transpired, was small compared to Jupiter and Saturn. There was
reason to think that the three remaining planets were likewise small, in which case the major example of attractive
interaction between planets would be the pull between the
massive Jupiter and Saturn. Astronomers were indeed
having difficulty getting Keplerian-style theories to work
for Jupiter and Saturn over the long term: there was trouble
here, though its exact nature would not be clarified for
nearly another century.

Newton believed that the foresight of Providence in constructing a stable, clock-like universe was demonstrated by
the layout of the solar system: the planetary orbits were in
planes that were nearly parallel, all the planets moved in the
same direction, and the two massive (and potentially dis-
ruptive) planets had been banished to the outer recesses. Yet
even this degree of planning (he believed) would not ward

off forever the collapse of the solar system. To prevent this disaster, Providence must have arranged to intervene from time to time, to repair the disruption resulting from the perturbations. To those able to read the Book of Nature, this would be one illustration of how much God cared for his creation. We shall meet another when we come to discuss Newton's model of the universe of stars (see page 191).

Meanwhile, as Newton's treatise took shape in response to Halley's anxious encouragement, it occurred to Hooke that important elements of the conceptual jigsaw had been announced earlier in his own writings, as indeed they had; and he asked to have this recognized in the forthcoming publication.

It fell to Halley to break the news to Newton. Newton allowed himself few favourites and Hooke was not one of them. In his view Hooke was a poor mathematician and a worse philosopher of science, with no appreciation of the difference between unsupported speculations and a fully-articulated mathematical theory in which hypotheses are used to explain a wide range of phenomena. The long-suffering Halley poured copious oil on the troubled waters, and the crisis passed. When Newton, however, came to redraft the section he had intended to call *System of the World*, he chose instead to replace it with material that was pitched mathematically at a level well beyond anything that Hooke could write, or even understand.

The impact of the *Principia*

Newton's *Principia* appeared in 1687. The title – in unabbreviated form, *Mathematical Principles of Natural Philosophy* – was a clear challenge to the generally accepted view of nature, as embodied in Descartes's systematic textbook, *Principles of Philosophy*. The two books were themselves in sharp contrast. Where Descartes had offered the ambiguity of language, Newton offered the precision of mathematics; where Descartes had been content to offer explanation, Newton's geometry opened the way to quantitative prediction.

But Descartes's world view had the great merit of intelligibility: in the Cartesian universe one lump of matter could affect another only if they were in direct contact, like the cogs of a clock. Descartes had taught natural philosophers to reject 'tendencies', 'appetites', 'attractions' and the like, as words coined to disguise ignorance. To Cartesians, therefore, Newton's 'attraction' was retrograde beyond belief. Newton himself several times tried to identify a physical cause for gravity, but with little success. He could only insist that the force must be real because its consequences were evident. It was a subtle methodological point that not all would accept.

Another problem for Newtonians stemmed from the antiquated style of geometry that Newton had chosen to employ. The mathematics of the *Principia* made its argument incomprehensible to all but a very few. By the turn of the century Newton himself was no longer in Cambridge – he was a man of many interests, 'science' as we know it being no more important to him than the study of alchemy and biblical chronology, and after the publication of the *Principia* he had begun to look for work outside academic life. In 1696 was appointed Warden (and later Master) of the Mint in London, where he supervised the great recoinage that was then just beginning. He finally resigned his Cambridge chair in 1701, but meanwhile his deputy, William Whiston (1667–1752), was approached by a young man with a moral dilemma: his college tutor, a Cartesian 'zealot', had asked him to prepare a new Latin edition of the classic classroom text of Cartesian physics, the *Treatise of Physics* of Jacques Rohault. Was it morally acceptable to perpetuate in this way a physics that was false? Whiston replied to the effect that as Newton's work was at present unintelligible, and as Cartesian physics was a big improvement on the Aristotelianism that had preceded it, the student – Samuel Clarke (1675–1729) – might go ahead with clear conscience.

In the event, in the later editions of his translation Clarke eased his scruples by using editorial footnotes to wage a running battle with the Cartesian argument of the main text. He did not hesitate to dismiss Rohault's teaching with '*Hoc falsum est*', and to offer in extended notes the alternative, Newtonian explanation. And so, in this roundabout way, students became familiar with the basics of Newtonianism.

To his followers, Newton was unique among mortals, the one man in history privileged to reveal to the rest of the human race the fundamental truths about the universe: in the couplet by the English poet Alexander Pope,

> Nature and nature's laws lay hid in night.
> God said, Let Newton be!, and all was light.

Once the Newtonian framework was accepted, the agenda was clear: to demonstrate mathematically that *all* the observed movements of the celestial bodies were the result of attraction.

But even among those prepared to countenance attraction
as a valid concept in physics, a choice between the two
philosophies of nature was by no means clear, nor was it
evident that a compromise was ruled out. A direct choice
was the more difficult in that Cartesianism did not lend itself
to quantitative predictions, so that issues in which experiment might decide between contradictory forecasts were
few. One such, however, concerned the shape of the Earth.

Newtonians versus Cartesians on the shape of the Earth

Since Antiquity, every educated person had known that the
Earth was roughly spherical (see page 26): Christopher
Columbus understood perfectly well that the Earth had no
edge for his ship to fall off, and the approximate size of the
Earth had been known since the work of Eratosthenes in
the third century BC (see page 26). But no one person would
be able to determine with precision the shape of the Earth;
this called for an organization with sufficient funds to
attract the necessary talent and to meet the considerable
expenses involved.

Such an organization had come into being with the
formation of the Paris Academy of Sciences in 1666 by Jean
Baptiste Colbert, the finance minister of Louis XIV. One
immediate result was the foundation of the Paris
Observatory, and the recruitment of astronomers from all
over Europe, to be led by the Italian Gian Domenico
Cassini (1625–1712, the founder of a dynasty of
astronomers and who is therefore known as Cassini I).

Perhaps the most eminent recruit to the Academy,
acquired at great expense, was Christiaan Huygens from
Holland. Another was the Frenchman Adrien Auzout
(1622–91), who had played a central role in the development
of the wire micrometer; this enabled the astronomer to
measure a tiny angle by adjusting two hairs or wires in the
field of view until they coincided with the angle to be measured (see page 128). He in turn promoted the recruitment
of Jean Picard (1620–82), who brought to the problem of the
size of the Earth the skills of a dedicated observer.

Huygens had assumed that the time a pendulum of given
length took to swing was the same everywhere on Earth, in
which case the length of a pendulum that beat so that each
swing took one second was a natural unit; and in 1661 he
had proposed to the Royal Society that this length be taken
as a universal standard. However, different investigators
began to obtain significantly different lengths. Matters
came to a head after the Academy sent Jean Richer in 1671

to the island of Cayenne off the Atlantic coast of South
America, where he was to make observations, especially of
the close approach of Mars to Earth the following year. He
took with him a seconds pendulum; or, more exactly, a pendulum that in Paris had beaten seconds. But in Cayenne
Richer found that it would beat seconds only if shortened
by a tenth of an inch.

Newton in his *Principia* offered an explanation: the
Earth bulged at the equator. Huygens, working in the
Cartesian tradition, came to a similar conclusion, but dif-
fered as to the amount of the bulge.

A lengthy debate ensued as to who was right. Eventually
the Academy decided to use its financial muscle to settle
the matter, by funding expeditions to measure the length of
a degree of latitude in two very different locations: Lapland
and Peru. The conclusion was that the Earth did indeed
bulge at the equator; but by exactly how much was still
uncertain. Other measurements were made, but still there
was no definite conclusion.

This attempt to decide between Newton and Descartes
had ended in confusion. But the issue was soon to be
settled, among astronomers and the general public alike, by
the appearance of a comet in 1759. This would lead to the
apotheosis of Newton, for the comet had been forecast two
generations earlier by Edmond Halley: a Newtonian had
predicted the unpredictable.

The return of Halley's Comet

Newton had shown in the *Principia* that comets were
lawlike and moved through the solar system in conic sec-
tions, though their paths were very different from the
nearly-circular ellipses of the orbits of the planets. So elongated were the cometary orbits that near the Sun they
approximated to a parabola.

However, unless its velocity was sufficiently large, the
departing comet would be unable to escape from the
domination of the Sun. Instead, it would leave along an
elliptical orbit, destined one day to be hauled back again to
the solar system – though on a path modified compared to
that of its previous visit, if it had then passed close to a
major planet. But even with such modifications, the
comet's orbit would have similar characteristics on each
appearance, and these appearances would be separated in
time by similar intervals. Did historical records provide any
examples of this?

Halley undertook the necessary enquiry, and quickly
found a prime candidate in the comet of 1682: its orbit was

retrograde (that is, opposite in sense to the orbits of the planets), and so had been the orbits of the comets of 1531 and 1607. Further enquiry showed that the orbits had much else in common, and in 1695 Halley told Newton he felt sure these were reappearances of the same comet.

The problem was that the intervals between these appearances were similar but not identical. Halley realized that this was because the comet's orbit had been modified as the comet had experienced the pull of one or more planets, and he predicted the comet would return 'about the end of the year 1758, or the beginning of the next'.

In the popular mind comets had always been portents of disaster, while even to astronomers the nature of comets and their role in the cosmic order were still shrouded in mystery. As 1758 approached, therefore, interest mounted. Some feared that the returning comet might collide with the Earth and put an end to the human race.

In his prediction, Halley had taken into account the acceleration the comet had experienced from the pull of Jupiter as it approached its 1682 passage round the Sun, but he had not allowed for the opposite consequences of the pull of Jupiter as the comet left the solar system. To remedy this called for detailed and laborious calculations, and in Paris, in June 1757, Alexis-Claude Clairaut (1713–65) embarked on the task with the help of two associates. They first analysed the orbit of the comet as it left the solar system after the 1531 encounter, and on this basis 'predicted' its return in 1607 which they compared with what actually occurred. They then analysed the comet's orbit as it left in 1607, and similarly 'predicted' its return in 1682. Learning from these quasi-experiments, they similarly predicted the comet's forthcoming return: it would swing round the Sun in mid-April 1759, give or take a month.

It was a good forecast. The comet was first seen on Christmas Day 1758 by an amateur astronomer living near Dresden, while the first professional to observe it was the Parisian comet-hunter, Charles Messier (1730–1817), four weeks later. On 13 March the comet rounded the Sun.

The characteristics of the comet's orbit were very similar to those of the comets of 1531, 1607 and 1682. Clearly, the comets were one and the same: Halley had been proved correct, and Newtonianism had won its most public triumph.

The Newtonian programme

While these very public trials of Newtonian attraction were taking place, a tiny band of the mathematical elite were

working at their desks, exploring the consequences of the
attractive pulls between the Sun, the planets, and their
moons. To implement this Newtonian programme,
Providence supplied an extraordinary cluster of talent in the
middle decades of the eighteenth century: Jean le Rond
d'Alembert (1717–83), who took his name from the Paris
church where he had been found abandoned, and who later
devoted his life to mathematics while living in Paris on an
annuity from his natural father; the Swiss Leonhard Euler
(1707–83), lured by Catherine the Great to St Petersburg,
and then by Frederick the Great to Berlin, and who finally
returned to St Petersburg where he died; the precocious
Parisian Alexis-Claude Clairaut, whose life-long association
with the Paris Academy of Sciences began when he was
only twelve; and Joseph Louis Lagrange (1736–1813) of
Turin, who was persuaded first to Berlin and then to Paris.
The last decades of the eighteenth century and the early
decades of the nineteenth would be dominated by Pierre
Simon de Laplace (1749–1827), who survived the French
Revolution and ended his life as a marquis.

Often in competition with each other, these mathematicians tackled one problem of 'physical astronomy' after
another, developing the necessary mathematical tools in
the process, as Newton's cumbersome geometrical approach
to trigonometry was replaced by techniques involving infinite series that could be treated in an abstract manner. Even
so, investigators were forced to proceed by approximation,
and in handling infinite series they often had to make a
judgement as to which terms in a series were significant
and which might safely be neglected.

One example of the difficulties this entailed is given by
Clairaut's investigation of the motion of the lunar apogee
(the point in the Moon's orbit where it is furthest from the
Earth). In 1747 he announced that attraction accounted for
only half of the observed motion (a conclusion independently reached by Euler and d'Alembert), and he suggested adding a fourth-power term to the inverse-square
law. But a year later, having carried the approximation
further, he found that in fact the law as stated by Newton
did account for the whole of the observed motion.

The Moon was of special interest, not only because
observational data concerning the many irregularities in its
movements were exceptionally complete, but because there
were hopes that a mastery of these irregularities would
allow mathematicians to solve the urgent problem that
faced every mariner who sailed far from land: how to determine his longitude. The 'finding' of longitude was also of

great importance to geographers and astronomers: geographers needed it in order to prepare accurate maps, and astronomers in order to correlate observations made at different observatories.

The finding of longitude

Latitude north of the equator is, by definition, equal to the altitude of the celestial North Pole, the point in the sky about which the stars revolve, and so on land the determination of latitude presented no problem. Even at sea, the Renaissance mariner could derive his latitude from measurements of the altitude of the Pole Star or of the Sun at noon (see page 80).

Longitude – the angle by which a town, or ship, is east or west of a standard location such as Greenwich – was altogether more difficult to determine. Differences in local times correspond to differences in longitude, one hour being equivalent to $\frac{1}{24}$ of a circle or $15°$, and to determine local time in a given place by the Sun or stars was straightforward – for example, if the Sun was due south, it was midday. But to determine the difference in local time between two places, it was necessary to determine their local times *simultaneously*. How was this to be done?

In Antiquity, Hipparchus had pointed out that an eclipse of the Moon was in effect a time-signal that would be simultaneously visible from wherever on Earth the Moon could be seen; and on occasion down the ages, at each of two places on land, a lunar eclipse had been observed and the local times recorded, and from the difference in the times the corresponding difference in longitude was later calculated. For example, an eclipse was due to take place in 1631, and Henry Gellibrand (1597–1636), professor of astronomy at Gresham College in London, arranged with a sea captain setting out to find the Northwest Passage, that both men would observe the eclipse and note the times when it occurred. In an appendix to *The Strange and Dangerous Voyage of Captain Thomas James*, published in 1633, Gellibrand concluded that the difference in longitude between London and Charlton Island in James Bay, Canada, was $79°30'$ (a value some $15'$ in error). But eclipses of the Moon were much too rare to be of use to navigators on the high seas, and even observers on land found the moment of eclipse poorly defined.

The discovery of four moons of Jupiter in 1610 by Galileo opened up better prospects. Their eclipses were much more clearly defined than those of the Moon, and much more frequent. Unfortunately, although these

eclipses would prove invaluable for use on land, they were
not frequent enough to answer the needs of mariners.
Tables that set out the changing configuration of Jupiter's
moons, on the other hand, would provide a useful (though
less precise) alternative.

In 1598, the King of Spain, whose ships were sailing the
oceans of the world, had offered a handsome reward to
whoever could 'find longitude'. Galileo, never slow to turn
his discoveries to his financial advantage, entered into
negotiations over the use of Jupiter's moons for this
purpose. But from the deck of a rolling ship it rarely proved
possible to see Jupiter through a telescope, let alone its
moons; and despite attempts to rig for the navigator a
swinging platform that would reduce the movement, the
practical difficulties proved too great for the method to be
adopted at sea.

For use on land, where time was on the side of the
observer and accuracy the main criterion, the Jupiter
method called for reliable tables of eclipses of the moons;
but it was not until 1668, through the work of Cassini I,
that such tables became available. French astronomers in
particular were quick to exploit the new technique. For
example, before the observations made in the previous
century by Tycho Brahe on the Baltic island of Hven could
be correlated with observations in progress at Paris, the
difference in longitude between the two sites had to be
determined with the greatest possible care. In 1671 Jean
Picard was entrusted with the task by the Paris Academy of
Sciences and, in company with the Danes Erasmus
Bartholin and Ole Römer, he spent eight months on Hven
making a series of observations of eclipses of the first satellite of Jupiter. Meanwhile, Cassini did the same in Paris,
and on Picard's return, a comparison of their data gave the
required difference in longitude.

Quite different methods also came under consideration.
There were periods in the sixteenth and seventeenth centuries when it seemed that the mariner equipped with suitable charts might be able to use the difference between
magnetic north and true north, or between the angle of dip
of the magnetic needle and the horizontal, as a coordinate
taking the place of longitude; but these hopes came to
nothing. Seafarers meanwhile continued to rely on 'dead
reckoning' – in other words, 'guestimates'.

The public were constantly reminded of the perils of
navigation out of sight of land, as regular toll was taken of
warships and merchantmen alike. For Britain the worst
such disaster occurred in September 1707, when a fleet *en*

route from Gibraltar, and sailing east in the belief that it was in the English Channel, had four of its ships wrecked on the Scilly Isles with the loss of two thousand men.

In 1713 William Whiston and a schoolteacher named Humphrey Dutton announced they had a solution to the problem of longitude at sea which they would disclose if suitably rewarded. The method was fantastic – it involved mooring ships at fixed positions across the oceans – but their claim stimulated the British Parliament to establish a Board of Longitude, with authority to award the huge prize of £10,000 to whoever produced a practical method of determining longitude at sea to one degree, or twice the amount for twice the accuracy.

There were two methods seriously in contention. The first was a question of technology: how to develop precision clockwork capable of withstanding the motion of a ship buffeted by a gale. This would provide the navigator with a standard time, which he could compare with his local time. Christiaan Huygens had devised a pendulum-driven clock that he hoped would keep a regular beat even in a rough sea; but tests in 1668–70 on voyages between France and the Mediterranean, and even as far as North America, showed that the problem had not yet been solved.

The other strongly-backed candidate in the race was of considerable scientific sophistication: the method of lunar distances. The Moon travels rapidly, moving every hour among the background stars by the equivalent of its own diameter. Just as the hour hand of a conventional clock tells the time as it moves among the numbers engraved on the clockface, so – it was hoped – the Moon might be used to tell the time as it moved among the stars.

To read the time by the lunar clock, the mariner would require three things: (i) an accurate and convenient instrument for measuring the angle between the Moon in its current position and a suitable star, (ii) an accurate catalogue of star positions, and (iii) accurate tables of the Moon's motion. All three were lacking at the end of the seventeenth century. No suitable instrument existed; those available, like the cross-staff, required the user to look at both objects simultaneously, and it was difficult to do this at sea with the any accuracy. As to star positions, Tycho Brahe's catalogue of 777 stars had been compiled before the invention of the telescope, while Johannes Hevelius (1611–87) had spurned the use of telescopic sights when compiling the catalogue of over 1,500 stars that was published posthumously in 1690. Yet it was the lack of reliable tables of the Moon's motion that posed the greatest

problem. Even after Newton had revealed the dynamics
underlying the Moon's complex manoeuvres, the errors
resulting from his lunar theory as set out in the 1713
edition of the *Principia* still came to several minutes of arc,
and this alone could easily have resulted in an uncertainty
of a hundred miles in a ship's position.

On two of the three counts, matters improved early in
the eighteenth century. The invention in 1731 of a double-
reflection quadrant, the ancestor of today's sextant, provided a method of measuring angles between heavenly
bodies that was both accurate and suitable for use at sea.
The positions of 3,000 stars were to be found in the 'British
Catalogue' of John Flamsteed published posthumously in
1725, in fulfilment of the royal warrant issued half a
century before, requiring the Astronomer Royal 'forthwith
to apply himself with the most exact care and diligence to
the rectifying . . . the places of the fixed stars, so as to find
out the so much-desired longitude of places for the per-
fecting the art of navigation'.

It was now up to the mathematicians to get a firm grip
on the motion of the Moon and to use this to calculate
tables of lunar positions. Around the middle of the century
both the Paris Academy of Sciences and the St Petersburg
Academy focused attention on questions of lunar theory by
means of challenge prizes; and it was by using equations
developed in this connection by Leonhard Euler, that the
Göttingen professor Tobias Mayer (1723–62) – an
astronomer of practical bent – produced tables of the Sun
and Moon which he sent to London in 1755. Trials of the
tables were interrupted by the Seven Years War, and Mayer
prepared improved tables shortly before his death. These
eventually earned his widow £3,000 from the British
Parliament (and a surprised Euler £300), and they enabled
the Astronomer Royal, Nevil Maskelyne (1732–1811), to
publish in 1766 the first of the annual volumes of *The
Nautical Almanac*, designed to make it convenient for navigators to determine longitude by the method of lunar distances.

But meanwhile advances in technology were offering a
more straightforward alternative. In 1735, the English
clockmaker John Harrison (1693–1776) produced his first
marine timepiece (known today as H1), and this was taken
in trials to Lisbon and back the following year. The Board of
Longitude then granted Harrison £250 towards an improved
clock; and things continued in this way, with successive
improvements and successive grants, for nearly thirty years.

In 1764 Harrison and H4 sailed to Barbados in a warship.

H4 did all that was asked of it, but Parliament decided that Harrison should have £10,000 then, and the second £10,000 when copies had been made and tested. A copy was taken by Captain James Cook on his voyage to the tropics and the Antarctic (1772–75), and 'our never failing guide, the Watch' performed triumphantly. As soon as suitable chronometers could be constructed, they became the preferred solution to the problem of longitude. Astronomical observatories were established alongside major ports, and astronomers found a new role, making accurate observations of noon-time (or at 1 p.m.) and immediately dropping timeballs as a signal by which the ships' chronometers could be checked.

On land, however, the roughness of the roads tended to disturb the running of chronometers as they were transported from one site to another. Yet pairs of observatories were increasingly anxious to establish their differences in longitude in order to correlate their observations. Sometimes artificial means – the firing of rockets from an intermediate hilltop – were used to generate a signal to be seen simultaneously from two different places. Greenwich and Paris, for example, were linked in this way in 1825 by a succession of such signals, the astronomers being supplied with large parties of troops for the purpose. But before long the introduction of the electric telegraph solved the problem of longitude on land, just as the chronometer had solved the problem at sea.

The past and future of the solar system

By the mid-eighteenth century, 'celestial mechanics' based on Newtonian principles had registered many successes. In particular, the new solar and lunar tables took account of perturbations, and were far more accurate than the tables they replaced. But two puzzling anomalies remained: an apparent acceleration of Jupiter and deceleration of Saturn, evident since the time of Tycho Brahe, and an apparent acceleration of the Moon, shown by Halley to have been going on since Antiquity. These trends had profound implications: if they continued indefinitely, with Jupiter spiralling into the Sun, Saturn receding, and the Moon falling into the Earth, the solar system was doomed to change drastically, and perhaps to perish.

In the work of Euler (who was one of those prepared to accept the possibility of such a catastrophe), Lagrange, and Laplace, a distinction came to be made between two kinds of variations, the 'periodic' and the 'secular'. Periodic variations were seen as pendulum-like oscillations, in longitude,

latitude, and distance from the central body, that were
restored in a relatively short time. The secular variations
were long-term, and Euler at first conceived them as operating always in one direction. They affected the shape and
orientation of the ellipse in which the body orbited: its
eccentricity, the position of the axis of the ellipse, the
inclination of the orbital plane to the ecliptic (the plane of
the Earth's orbit), the position of the line of the nodes
(where the ellipse cut the ecliptic), and perhaps even the
average distance to the Sun.

In 1772 Laplace, who was then only twenty-three years
old, proved that, to a high degree of approximation, the
attractive pulls between two planets could not cause
permanent, one-directional change in the average
Sun–planet distance. In the light of Kepler's third law (as
corrected by Newton), this implied that these attractive
pulls could not cause a one-directional change in the
periods of the planets. Laplace drew the conclusion that the
apparent acceleration of Jupiter and deceleration of Saturn
could not be due to their mutual gravitational interaction –
he thought it likely that these effects were due to interactions with comets.

In 1774 Lagrange showed that the secular variations in
the inclinations of the planetary orbits to the ecliptic and in
the position of the line of nodes were, to a first approximation, oscillatory and periodic, with periods measured in
thousands of years. On reading Lagrange's demonstration,
Laplace immediately applied the same type of analysis to
other aspects of the planetary orbits. It now appeared that
all long-term changes in the planetary orbits resulting from
their mutual attractive pulls were oscillatory and periodic.

In 1785, Laplace discovered the cause of the long-contin-
ued acceleration of Jupiter and deceleration of Saturn: the
variation was not one-directional but periodic, with a
period of about 900 years. It differed from the still longer-
term secular changes because it depended on the positional
relationship between the two interacting planets (Jupiter
and Saturn) and the Sun. To find this effect, Laplace had to
pursue the series approximations to the order of the third
power of the eccentricities and inclinations – an under-
taking that called for finesse but also involved a great deal
of sheer calculative drudgery.

In 1787 Laplace identified a cause for the secular
acceleration of the Moon. It was a secondary effect, arising
from the secular reduction currently taking place in the
eccentricity of the elliptic orbit of the Earth, which led to a
net reduction in the Sun's action on the Moon. This effect

would be reversed when, in its long-term oscillation, the eccentricity stopped diminishing and began increasing again. (In the 1850s, it would be found that Laplace's explanation accounted for only about half the Moon's apparent acceleration; the other half came to be attributed to the slowing down of the Earth's rotation, caused by the friction of the Moon-induced tides in shallow seas.)

These discoveries led to a picture of a solar system whose internal motions and geometrical parameters were subject only to minor oscillations about their average values. Laplace in fact believed that he had proved the solar system to be a stable, self-regulatory system, similar in this respect to the self-regulation evident in living Nature (though, as we shall see on page 166, in the late twentieth century it would become clear that Laplace's insight here was only part of the truth).

Laplace described this picture of a stable solar system in a brilliant work of popularization, his *Exposition du système du monde*, which appeared in 1796. But he did something else as well: he attempted to explain how such a system could have come about. There could have been, in the first beginnings of the solar system, a giant nebula or vortex whirling about the Sun, and the planets and their satellites could have condensed out of this whirling matter as the myriad particles attracted each other. This supposition would account for the fact that all the planets and then known satellites circulated about the Sun in the same direction, from west to east, and in nearly the same plane. Granted this explanation, the nicely balanced oscillations in the motions of the solar system would be the result of spatial relationships that had survived from the system's chaotic origins.

The *Exposition du système du monde* was translated into many languages, and widely read. Laplace's 'nebular hypothesis' of the origins of the solar system harmonized well with William Herschel's contemporary theorizing about the origins of systems of stars, which we discuss in the next chapter, and helped prepare the way for Darwin's promulgation in 1859 of his theory of organic evolution. Today the nebular hypothesis is the accepted account of the beginnings of the solar system, and one of the main pre-occupations of solar-system theorists is to understand how the solar system has evolved, by means of tidal and other gravitational interactions.

Laplace's magisterial five-volume synthesis, *Traité de mécanique céleste* (1799–1825) – especially in the English translation by the American amateur Nathaniel Bowditch

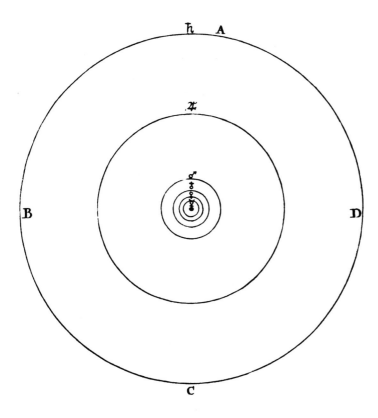

A diagram of the solar system, from David Gregory's *Astronomiae elementa* (1702). The orbits of Mercury, Venus, Earth and Mars are clustered at the centre, and there is then what seems a disproportionately large gap between the orbits of Mars and Jupiter.

(1773–1838), which contained many helpful explanatory notes – became the bedside reading of astronomers and celestial mechanicians, describing the methods and setting the problems for future research.

'Missing' planets

Early on in his attempt to discover what had motivated God the geometer in his choice of the layout of the solar system, Johannes Kepler had been disturbed at what he saw as the disproportionately large gap between the fourth planet, Mars, and the fifth, Jupiter; and he had toyed with the idea that the gap was occupied by an undiscovered planet. A century later, Newton saw the gap as evidence of how Providence had contrived to limit the disruption caused to the structure of the solar system by the attraction of the two massive planets, Jupiter and Saturn, by banishing them to the outer regions of the system.

Other, more prosaic minds offered physical explanations of how the gap might have been generated. The Alsatian polymath, J. H. Lambert (1728–77), thought that Jupiter and Saturn may have 'plundered' planets that were once in the

The Sun ('Sol') and the
planets drawn to scale,
from Christiaan Huygens's
Cosmotheoros (1698).
Jupiter and Saturn are huge
by comparison with
Mercury, Venus, Earth
('Tellus'), and Mars.

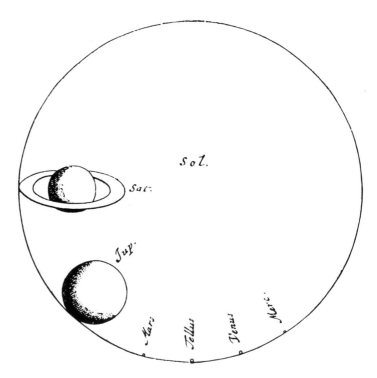

gap, while the eccentric English amateur Thomas Wright of
Durham (1711–86), in an unpublished speculation, won-
dered if a planet that once existed in the gap had been
destroyed by collision with a comet.

Bode's 'Law'

The suggestion that the gap contained (or had contained) a
'missing' planet was encouraged by an arithmetical relationship that emerged during the eighteenth century. In his
Astronomiae elementa of 1702, the Oxford professor David
Gregory (1659–1708) noted that the radii of the planetary
orbits were roughly proportional to the numbers 4, 7, 10,
15, 52, 95. Christian Wolff, a German philosophical popularizer, republished these figures in a work that came to the
attention of Johann Daniel Titius (1729–96), professor of
physics at Wittenberg University. In 1766 Titius produced a
German translation of the *Contemplation de la nature* by
the distinguished French naturalist Charles Bonnet; and
into Bonnet's text he interpolated a paragraph in which, by
altering Gregory's 15 into 16 and 95 into 100, he made the
numbers equal respectively to 4, 4+3, 4+6, 4+12, 4+48 and
4+96.

None of the known planets corresponded to the missing
term in the sequence, 4+24. Bonnet commented: 'But

should the Lord Architect have left this space empty? Never!' His own method of filling the gap was absurd – undiscovered satellites of Mars – but in 1772 a second edition of his translation came to the eyes of a young German astronomer, Johann Elert Bode (1747–1826), who was then putting the finishing touches to a new edition of his highly successful introduction to astronomy. Bode dismissed the nonsense about the satellites of Mars, but he was fascinated by the relationship and agreed about the gap: 'Can one believe that the Creator of the Universe has left this position empty? Certainly not!' Bode became convinced that a primary planet lay undiscovered in the Mars–Jupiter gap, at some 4+24 units from the Sun (the Sun–Earth distance being 4+6 units).

The discovery of Uranus

Inclusion in Bode's book guaranteed the relationship wide publicity. And in 1781, the relationship received a sensational and wholly unexpected boost, when William Herschel (1738–1822), a Hanoverian-born organist living in the English spa resort of Bath, became the first person in history to discover a planet.

When he came across Uranus, Herschel had not been looking for a planet – indeed, as a self-taught amateur astronomer, he was unaware of speculations about 'missing' planets. His interests lay in the stars, not the solar system, and at the time he was using a home-made reflecting telescope to familiarize himself with the brighter stars, which he was examining one by one. On 13 March 1781, his systematic 'review' of the sky led him to the constellation Gemini. There he came across an object that professionals had earlier mistaken for a star. But so good was Herschel's telescope, and so skilled was he as an observer, that he was able instantly to recognize the anomalous nature of the object. It was, he noted in his journal, 'a curious either Nebulous Star or perhaps a Comet'.

If it belonged to the solar system, it might well be moving perceptibly against the background stars. Herschel therefore returned to the object four days later, and found that his suspicions had been justified, for it had already moved. Since it was not one of the known planets, he assumed it must be a comet. A friend with scientific contacts informed the Astronomer Royal, Nevil Maskelyne, and the astronomy professor at Oxford, Thomas Hornsby. Though observing with professionally-made instruments, neither could see any unusual object in that region of the sky, and it was some time before they could identify

Herschel's 'comet'. It proved instead to be a planet, the first
to be discovered since the dawn of history.

The sensational discovery gave Herschel's allies the
opportunity to secure for him a royal pension, which
allowed him to give up music and devote himself to astronomy. Herschel wished to show his gratitude by naming the
planet Georgium Sidus in honour of his royal patron, but
Continental astronomers preferred the more conventional
Uranus, and their wishes prevailed.

Uranus – astonishingly – was found to have an orbit that
corresponded well to the next term in the sequence,
$4+192=196$. The court astronomer at Gotha, Baron Franz
Xaver von Zach (1754–1832), now became a convinced
believer in the arithmetical relationship, and in 1787 he
undertook a search for a planet in the Mars–Jupiter gap –
but without success. In 1799 Zach visited a number of
German colleagues, and from their discussions the idea of
an organized attack on the problem began to evolve. On 21
September 1800 Zach met with five other astronomers in
Lilienthal, home of the prominent amateur J. H. Schröter
(1745–1816). They decided to enlist the cooperation of
leading observers throughout Europe, to form a team of
twenty-four 'celestial police', each of whom would be
assigned a share of the zodiac, with the duty to keep a
lookout for strangers at large in his district.

The discovery of the asteroids

Their plans were overtaken by events. One of the co-opted
members was to have been Giuseppe Piazzi (1746–1826) of
Palermo, the southernmost of the European observatories.
In the 1780s, by dint of residing in London and breathing
down the neck of the renowned instrument-maker, Jesse
Ramsden, while work progressed, Piazzi had coaxed
Ramsden into completing a superb 5-ft vertical circle of
unique design. The turn of the century found Piazzi using
this circle to assemble a star catalogue of greater accuracy
than any of its predecessors.

On New Year's Day, 1801, unaware of the task for which
he had been 'volunteered' by the Lilienthal astronomers,
Piazzi was at work as usual on his catalogue. He measured
the position of no. 87 of N.-L. de Lacaille's catalogue of
zodiacal stars, and took the opportunity of measuring a
(supposed) star of about the eighth magnitude that preceded
it. His careful method of working involved him in remea-
suring his positions on a subsequent night; and when he did
this, he found that the eighth-magnitude 'star' appeared to
have moved, a movement that he confirmed on the nights

that followed. It was therefore no star, but a member of the
solar system.

Throughout January 1801 Piazzi kept watch on his
newly-discovered member of the solar system whenever
weather conditions allowed, and on the 24th he wrote
letters to Bode and others to announce his discovery. In his
letter to Bode he claimed no more than the finding of a
comet; but to his friend Barnaba Oriani of Milan he admitted a suspicion that 'it might be something better than a
comet'.

By mid-February the object was too close to the Sun to
be seen, and so Piazzi began to investigate its orbit on the
basis of the twenty-four observations he had been able to
make. He tried to fit parabolas, first to one triad of observations and then to another, because a parabola was known to
approximate well to the elongated orbit of a comet when
the comet was near the Sun. But no parabola came near to
accounting for all the measured positions. Piazzi then tried
circles, and found that radii of around 2.7 times the
Earth–Sun distance – close to the $(4+24)/(4+6)$ implied by
the unassigned term in Bode's 'Law' – gave promising
results. The object, it would seem, must be a planet, in an
elliptical (but nearly circular) orbit. The problem was, how
to determine its orbit with enough precision for
astronomers to recover it when it emerged from the glare of
the Sun – for it had been only briefly observed, while
moving over a very short arc.

Illness prevented Piazzi from making any further
progress with the mathematical analysis of his observations, and so in April he sent his data to Oriani in Milan, to
Bode in Berlin, and to J.-J. L. de Lalande in Paris, thereby
handing over the problem to the astronomical community
of Europe. By good fortune a brilliant new mathematical
talent had emerged in the person of Carl Friedrich Gauss
(1777–1855); and by November, Gauss had devised a
method that allowed him to calculate the characteristics of
the orbit. On the basis of this information, Zach began to
search for the lost planet; and on the last night of the year
he found it, in just the position forecast by Gauss.

The object, named Ceres by Piazzi in honour of the
patron goddess of Sicily, turned out to lie at about the distance predicted by Bode's 'Law', and at first there seemed no
reason to doubt that it was, like Uranus, a major planet. But
Herschel found to his surprise that, even with his large telescope, he could scarcely detect a planetary disc. It was, he
thought, even smaller than our Moon. Worse still, in March
Olbers found another moving body, which he named Pallas.

Herschel measured its diameter also, and thought this to be
well below 200 miles. Clearly, Pallas was no planet. As a
descriptive term for this new species of celestial body,
Herschel proposed 'asteroid'.

To rescue the 'Law', Olbers suggested that the two asteroids were fragments of a full-sized planet that had once
occupied the gap. If so, other fragments were waiting to be
discovered, and so it proved: by 1807 Juno and Vesta had
also been found.

For a long time thereafter no further asteroids were identified, and astronomers soon wearied of the search. It was
revived by the German K. L. Hencke, an ex-postmaster.
Hencke was a man of extraordinary dedication, for he began
work in 1830 and laboured for fifteen years before enjoying
his first success. He found another asteroid in 1847, and at
this leading observers began to take up the hunt.

By 1891, more than 300 asteroids had been discovered,
and the pace of discovery then greatly increased with the
application of photography. Max Wolf at Heidelberg would
photograph a large star field with an exposure of several
hours, the camera carefully tracking the rotation of the sky.
On the resulting photograph, the stars appeared as point-
like images, but the asteroids betrayed themselves by the
small streaks caused by their movements during the time
the photograph was exposed.

If Olbers had been right, and asteroids were indeed the
fragments of an exploded planet that had obeyed Bode's
'Law', then (initially at least) their orbits would all have
intersected, in the place of the explosion and again on the
opposite side of the Sun. But as increasing numbers of asteroids became known in the second half of the nineteenth
century, it was realized that this was far from being the
case. Modern astronomers favour a contrary theory: that
the asteroids, whose combined mass is thought to be only a
fraction of that of the Moon, are the survivors of a myriad
of small objects that failed to coalesce into a planet,
because of the attractive pull of the newly-formed Jupiter.

The discovery of Neptune

Herschel had come across the planet Uranus, and Piazzi the
asteroid Ceres, quite unexpectedly, when they were engaged
in studies of the stars. But in the mid-nineteenth century
another planet was to be found, this time by observers
searching in a region determined by astronomer-mathematicians who had been armed with nothing more than pen
and paper and a knowledge of Newtonian dynamics.

Soon after Uranus was discovered in 1781, Bode found

that the planet's position had been recorded by Tobias
Mayer in 1756, and by John Flamsteed as long ago as 1690;
both had taken it for a star. These additional observations
allowed his friend Placidus Fixlmillner and others to determine the characteristics of its elliptical orbit and to calculate tables of its future positions.

The planet, however, soon began to deviate from its predicted orbit. Matters improved in 1790 when the Paris
mathematician J.-B. J. Delambre (1749–1822) published
tables that seemed to match the observations well enough,
but in the 1820s and 1830s the theory of Uranus's orbit was
again in trouble.

Various explanations were floated. Some were quickly
rejected – that the planet was being impeded by a cosmic
fluid, that it had an unseen but massive satellite, that it had
been struck by a comet about the time of discovery – but
two others were more worthy of consideration: perhaps the
law of attraction departed noticeably from the inverse-
square at great distances, or perhaps Uranus was being
pulled out of its orbit by an outer planet as yet undiscovered.

Modifications in the law of attraction had been considered from time to time in the mid-eighteenth century, but
by now the law was firmly established. A consensus therefore emerged, that the anomalous behaviour of Uranus was
the result of perturbations by an undiscovered planet.

By 1845 Uranus's movements were under the scrutiny of
a proven master of the techniques of Newtonian mechanics, Urbain Jean Joseph Le Verrier (1811–77) of the Paris
Observatory. He presented his first paper on the topic to the
Paris Academy of Sciences in November of that year, and a
copy soon reached the Astronomer Royal at Greenwich,
George Biddell Airy (1801–92). The following June Le
Verrier presented his second paper. In it he made the
assumption that the undiscovered planet occupied the next
place in the sequence embodied in Bode's 'Law', and after a
lengthy analysis he concluded that its current longitude as
seen from the Sun must be within a few degrees of 325°.

Unbeknown to Le Verrier, a young Cambridge graduate,
John Couch Adams (1819–92), was at work on the same
problem. He too had assumed the planet obeyed Bode's
'Law', and he had arrived at an approximate solution for the
position of the planet by October 1843. Distracted by teaching duties, he did not derive a more precise result until
September 1845; the heliocentric longitude of the planet on
1 October he reckoned to be 323°34'.

Armed with a letter of introduction from James Challis

(1803–82), the Professor of Astronomy at Cambridge, Adams called on Airy to present his analysis, but owing to a chapter of accidents failed to speak with him. He did however leave him a summary of his results.

The arrival next summer of Le Verrier's paper, with its nearly-identical prediction of heliocentric longitude, stirred Airy to action. In his opinion, a hunt for an undiscovered planet was not the business of the publicly-funded Royal Observatory, but he persuaded Challis to make a search at Cambridge, and this Challis eventually did. Unfortunately, Challis did not possess accurate charts of that region of the sky. This meant that the only way he could identify a planet – a temporary visitor to the area – was to re-examine the region at a later date, to see whether any of the 'stars' had moved in the interval. This was a chore that Challis undertook with no sense of urgency.

The delay cost Adams priority, for Le Verrier had meanwhile prevailed on astronomers at Berlin Observatory to make a search, and they were fortunate enough to have the relevant sheet – not yet distributed – of the Berlin Academy's new Star Atlas. Within a few minutes of starting their search on 23 September 1846, the Berlin observers came across a 'star' that was not on the sheet. It was the missing planet.

The discovery of Neptune in 1846 was the ultimate triumph of Newtonian dynamics: two astronomer-mathematicians, sitting at their desks, had calculated from the effects – the deviations of Uranus from its predicted orbit – to the cause, and had pinpointed the whereabouts of the culprit, a major planet whose very existence had until then been unsuspected.

The failure to discover Vulcan

In 1846 celestial mechanics appeared to be the queen of the sciences, the most successfully mathematicized, and the most exact in its predictions. Fate, however, had a trick in store.

Like Uranus, Mercury had an unexplained feature in its orbit: its point of nearest approach to the Sun ('perihelion') was advancing in longitude faster than expected, by about half a minute of arc per century.

That such a tiny discrepancy – a degree every hundred centuries or so – should arouse concern is astonishing testimony to the success of mathematicians in explaining the planetary movements on Newtonian principles. Leverrier naturally considered whether the cause might again be an undiscovered planet, this time orbiting inside Mercury, and in September 1859 he announced the results of his calculations. A body of the same size as Mercury but at half the distance from the Sun would produce just such an advance; so would a similarly placed ring of asteroids. Alternatively, the Sun might not be perfectly spherical in shape.

As luck would have it, earlier that year, in the French town of Orgères, an unknown physician named Lescarbault had seen, crossing the Sun, what he believed to be just such an intra-Mercurial planet. He said nothing about this at the time, but in December he learned of Le Verrier's prediction, and he wrote to him. Le Verrier hurried to Orgères, satisfied himself of the physician's *bona fides*, and named the planet Vulcan. Its period he calculated to be just under 20 days.

In 1876 Vulcan was supposedly seen once more crossing the disc of the Sun; but on further investigation the object proved to be a sunspot. By this time some twenty possible observations of the supposed planet had been assembled, five of which Leverrier was prepared to accept as authentic. On the basis of these he calculated the planet would cross the Sun in March 1877, and again in October 1882; but in the event, nothing was seen.

Searches during solar eclipses revealed nothing, although evidence of the advance of the perihelion of Mercury became ever stronger. By the end of the century Vulcan was accepted as spurious, a 'pseudo-planet', and the advance reverted to being an unsolved problem of celestial mechanics, a rare exception to the seemingly-endless stream of successes for Newtonian dynamics. But by a strange stroke of fate, what had been expected to lead to another Newtonian triumph was to prove the very opposite. As we shall see in Chapter 8, Albert Einstein showed in 1915 that his General Theory of Relativity, based on a radical reappraisal of Newton's basic concepts, implied that Mercury's perihelion would advance by almost precisely the observed amount.

The puzzling movement of the planet had at last been explained, but not on Newtonian principles. Pope's couplet in praise of Newton required amendment:

> It did not last: the Devil howling 'Ho,
> Let Einstein be,' restored the status quo.

Celestial mechanics in the twentieth century

But despite the Einsteinian revolution, most work in celestial mechanics – the plotting, for instance, of orbits of artificial satellites – has continued to be based on the Newtonian equations. Such calculations are inevitably no more than approximations, but for many purposes the Newtonian theory gives results that are entirely adequate.

In the accurate measurement of time, however, this is not the case. Until the mid-twentieth century the mean solar day was taken as the unit of time; but recognition of the role of tidal friction in slowing the Earth's rotation – to say nothing of changes in the rotational rate due to changes in the Earth's shape – made this standard obsolete. Atomic clocks now replace the Earth's rotation in supplying a standard of time. However, like all clocks on Earth, atomic clocks exist in an accelerated frame of reference, and are therefore subject to alterations in rate predicted by Einstein's General Theory of Relativity. Celestial mechanicians must therefore undertake to incorporate these relativistic effects in the computer programs whereby they generate predictions of planetary and satellite positions.

In the late twentieth century, celestial mechanics has had to face up to still another challenge, perhaps even more disconcerting than that posed by Einstein. The view of celestial mechanics promulgated by Laplace, and accepted until recently, was that its predictions could be made as precise as the observations. As Laplace put it, a demon that knew for any instant the exact positions and velocities of all the bodies in the universe, and had a mind capable of performing the necessary calculations, would be able to know the whole future and past of the universe. Human astronomy was but a 'pale image' of such a knowledge, but an image it was. If observations gave the positions of the planets to a certain precision, then their future positions could be predicted to similar accuracy.

In the 1890s the French mathematician Henri Poincaré (1854–1912) introduced a new way of thinking about the problems of celestial mechanics. Instead of focusing on problems that appeared to be soluble by the traditional method of approximations by equations with an infinite series of terms, Poincaré used quasi-geometric methods to test the assumption that small changes in the parameters of the equations would result in only small differences in the numbers forming the solution. He found, instead, that for many dynamic systems, small differences in initial conditions could lead to drastic differences in outcome.

With the advent in recent decades of electronic computers of ever increasing power, it has become feasible to explore such systems quantitatively. Take, for instance, a system consisting of a simple pendulum whose point of suspension is set in oscillation; if there is a near match between the pendulum's natural frequency and the frequency of the imposed oscillation, the sensitivity of the behaviour to the actual initial conditions turns out to be indefinitely large. Since there is a limit to the precision with which the initial conditions can be measured, the motion of the system becomes unpredictable – even though in principle it is completely determined. The connection between determinism and predictability that Laplace had assumed is here broken, and we have a situation of the type now known as 'dynamical chaos'.

It has recently been shown that the inner planets Mercury, Venus, Earth, and Mars, as well as the outer planet Pluto, provide instances of dynamical chaos. Certain perturbations of the Earth's motion, for instance, are in 'near resonance' with its annual motion, in close analogy with the case of the simple pendulum mentioned above. From observation we can never know exactly – with unlimited precision – where any planet is; and for the inner planets and Pluto the initial range of uncertainty increases with time by leaps and bounds. Calculations with computers have shown that the initial uncertainties for the Earth increase by a factor of 3 every five million years, so that an initial error of 10 metres produces an error of a million kilometres after 100 million years.

One implication of the discovery of 'chaos' among the planets is that earlier attempts to prove the stability of the solar system are fundamentally flawed. Systems subject to dynamical chaos are sensitive to chance perturbations: we live in a less certain world than Laplace imagined.

Celestial mechanicians, their dream of deterministic certainty gone, have nevertheless an exciting prospect for future investigation. Their tasks include exploring the detailed implications of planetary 'chaos', refining predictions with the limitations now revealed, and working out plausible scenarios for the evolution of the solar system and its sub-systems.

The astronomy of the universe of stars

Michael Hoskin

In 1833 the English astronomer John Herschel published *A Treatise on Astronomy*, an introductory text for the interested amateur. Although Herschel was the world authority on the universe that lay beyond the narrow confines of the solar system, he allotted just one of his thirteen chapters to the study of the stars: astronomy was still preoccupied with the movements of the Sun's family of planets and comets. Yet nearly two centuries had passed since Descartes had portrayed an infinite universe in which there was no longer a centre and no longer an outside boundary; one in which the Sun was merely our local star, and the stars were suns whose brightness was diminished by their immense distances.

During these two centuries, the stars had continued to play their traditional humble role, as reference points in the sky against which to measure the current positions of members of the solar system. The lack of interest in them as bodies free to move in space is not surprising. Because they are so far away, their movements as seen from Earth are minute – so tiny that even at the end of the seventeenth century, not a single star was known to have altered its position since the first catalogues were compiled in Antiquity. The stars were too remote and appeared too static to arouse the curiosity of most astronomers.

Stellar astronomy became important only with the development of astrophysics in the second half of the nineteenth century. Today, the pendulum has swung, and it is the solar system that is of minority interest. As a result, the investigations pioneered by the handful of earlier astronomers – many of them amateurs – who had the courage to tackle the distant stars have proved to be of immense significance for the future development of astronomy.

The Sun's neighbours in the universe of stars

Depending on their temperament, these men adopted one of
two possible approaches. Some bold spirits – often mixing
astronomy with theology – sought to understand the
cosmos as a whole. Others preferred to proceed step by step,
outwards from the solar system to the Sun's nearest neighbours among the stars. It is with work of these latter, more
cautious souls, that we begin.

Stars that change in brightness

It may seem surprising that astronomers of Antiquity and
the Middle Ages believed the stars to be unchanging in
brightness as well as in position. There are stars whose
light varies regularly, and some of these variations are not
difficult to detect with the naked eye – once you know
where to look and, more fundamentally, once you know
that such variations exist to be observed. These changes
escaped detection for so long because Nature so arranged
the universe that none of the brightest stars varied sufficiently to force such changes on the attention of early
observers.

The dramatic new star of 1572 ('Tycho's nova', see page
96) compelled astronomers to recognize that changes in
brightness could and did occur. Remarkably, Nature forced
home the lesson only a generation later, in 1604, when a
second such star ('Kepler's nova') blazed forth.

It appeared at a time of the utmost astrological significance. Jupiter took some twelve years and Saturn thirty to
complete one circuit of the zodiac, and so their 'conjunctions', when Jupiter caught up with Saturn, occurred at
intervals of some twenty years. Because of their rarity,
these conjunctions were thought of astrologically as 'great'.
In the twenty years betweeen great conjunctions, Saturn
moved ⅓rds the way around the zodiac, in the next twenty
years another ⅓rds, and after a further twenty years another
⅓. By this time it had completed two full circuits and had
returned to near the place of the first conjunction.

The positions of three consecutive great conjunctions
therefore formed (roughly) the vertices of an equilateral triangle around the sky, and the signs of the zodiac within
which they fell would normally be either the first, fifth and
ninth of the twelve signs; or the second, sixth and tenth;
and so on. These groups of three signs were known as
'trigons' (that is, triangles), and each trigon was associated
with one of the four elements, the fiery trigon being the
most portentous.

The conjunctions in practice drifted from one trigon to the next every two centuries or thereabouts; and so after every eight centuries, the conjunctions returned to their original trigon. It happened that around the birth of Christ, the conjunctions entered the fiery trigon (and some use this fact in interpreting the Star of Bethlehem); and in 1603, after two such cycles of eight centuries, the same profoundly ominous event was happening again.

By the autumn of the following year, Jupiter and Saturn had been joined there by the third-slowest planet, Mars. Suddenly, in the midst of the three planets, and on this occasion of the very highest astrological significance, a new star shone out.

Some thought the end of the world had come. Others expected the overthrow of the Turkish kingdom, or the appearance of a great new monarch: 'nova stella, novus rex'. Johannes Kepler, as Imperial Mathematician, rushed out a short tract on the nova, and then, two years later, he published the definitive treatise, On the New Star, in which he ranged over all the issues involved, astronomical and astrological.

Tycho, Kepler and their contemporaries were privileged: astronomers have since been waiting nearly four centuries for the appearance in our Galaxy of the next 'supernova', as these celestial fireworks are now classified.

There had already been talk of another newcomer among the stars: in 1596 the Frisian astronomer David Fabricius (1564–1617) claimed that a nova had appeared in the constellation of the Whale (Cetus). Then, in 1638, another Frisian observer, Johannes Phocylides Holwarda, noticed a star, also in the Whale, that was missing from the star catalogues of Ptolemy and Tycho, and presumably also a nova. The pages of Holwarda's account of the discovery had already been printed off when he was astonished to find the star had reappeared.

Eventually it was realized that Fabricius's nova and Holwarda's nova were one and the same. But the reappearances of 'mira stella' ('wonderful star') seemed capricious. In 1662, Johannes Hevelius gathered together all the known observations, including those that he had himself accumulated in three years of careful scrutiny. By bad luck, three years was not quite long enough to reveal that, although Mira varied in brightness not only within each cycle but even in the corresponding stages of one cycle and the next, it did observe one significant regularity in its behaviour. It attained a maximum brightness every eleven months; and to this extent at least, Mira was predictable.

It was left to the French priest-astronomer Ismael
Boulliau (1605–94) to announce this in a work published in
1667. Boulliau also offered a physical explanation of 'variable stars', one that was to endure into the nineteenth
century. The Sun, he reminded readers, was a star that
rotated. Furthermore, it had spots on its surface, and these
spots varied. A variable star, he believed, likewise rotated,
and it too had dark regions on its surface, though these
regions were much more extensive than the spots on the
Sun. Any regular variation in the star was due to its rotation, and any irregular variation to alterations in the dark
regions.

It was an explanation that could account readily –
perhaps too readily – for almost any change in the stars.
Meanwhile, the hunt was on. Reputations could be made
overnight through the discovery of (supposedly) variable
stars. In their haste, observers were deceived by changes in
seeing conditions, or perhaps by wishful thinking; and sceptics had no way of proving them mistaken. Yet the discoveries – even if genuine – seemed to be leading astronomy
nowhere, and by the end of the seventeenth century the
whole subject had fallen into disfavour.

Part of the problem lay in the lack of a sufficiently delicate technique for monitoring the apparent brightness of a
star. Stars were simply grouped according to the crude
classification inherited from Antiquity, whereby the brightest stars were first magnitude and the faintest, sixth. The
mid-nineteenth century would see the invention of new
instruments to give an objective measure of the brightness
of stars, and a new definition of magnitude (see page 258).
But before then, in the closing years of the eighteenth
century, astronomers were at last provided with a simple
method of determining whether a star had in fact altered in
brightness, when John Herschel's father, William Herschel,
at his observatory near Windsor Castle, prepared a series of
'Catalogues of the Comparative Brightness of the Stars'. In
these catalogues, each of the major stars in a given
constellation was compared with neighbouring stars that
had been selected because they were almost equal in brightness to the listed star; any future change in the brightness
of a listed star would reveal itself by disturbing these published comparisons.

Herschel's technique of comparative brightnesses was a
refinement of a simple method using sequences of stars
arranged in order of brightness, that had been developed in
the 1780s by two amateur observers in York in the north of
England: Edward Pigott (1753–1825), whose father had

recently established a private observatory in the city, and his neighbour, the young deaf-mute John Goodricke (1764–86). In 1781 Pigott decided to dedicate himself to the investigation of new and variable stars, and the following summer the two friends re-examined the known variables, notably Mira, but also Algol (Beta Persei), which a century earlier had on two occasions been seen as fourth magnitude instead of the usual second.

On 7 November 1782, Algol was second magnitude as usual, but five nights later an astonished Goodricke found it reduced to fourth magnitude. The following night it was back to second again, a change so rapid as to be utterly without parallel in the literature.

In the weeks that followed, the two friends kept independent watch on the star, and were rewarded on 28 December, when they both saw it start the evening at third or fourth magnitude and then brighten to second magnitude before their very eyes. Pigott at once suspected that 'the alteration of Algol's brightness, was maybe occasioned, by a Planet, of about half his size, revolving round him, and therefore does sometimes eclipse him partially'; and he even calculated possible orbital periods for the planet.

More of these supposed eclipses followed, and the longest orbital period that would account for every one of them grew ever shorter, until by April it had reduced to a mere sixty-nine hours! Pigott generously allowed Goodricke to appear as the sole author of the paper to the Royal Society announcing their results; but the young man concentrated on observational facts, and mentioned the eclipse hypothesis merely as an alternative to the usual explanation in terms of dark patches.

Whereas dark patches could explain almost anything, and so could scarcely be either proved or disproved, the eclipse hypothesis made exact predictions: that the brightness at minimum would always be the same, that the length of time occupied by one cycle of variation would always be the same, and that on either side of minimum the light curve would be symmetric. Interestingly, in the years to come the two friends abandoned this correct explanation for Algol. Perhaps they were deceived by changes in seeing conditions into thinking that the predictions had not been fulfilled. Or perhaps they were influenced by the impossibility of using the eclipse hypothesis to explain the three other short-period variables they had discovered: one of these, Beta Lyrae, consists of a pair of companion stars which are thought to be very close together and ellipsoidal in shape as a result of tidal distor-

tion, while the other two, Delta Cephei and Eta Aquilae,
are individual stars (of a type now known as Cepheids) that
pulsate, climbing rapidly to maximum brightness and then
slowly declining. Whatever the reason, in their brief years
of activity the two friends had enriched the study of variables with a new and wholly unsuspected class, those
whose periods were measured in days rather than months
or years.

   Their collaboration came to an end in 1786, when the
Pigotts left York and Goodricke met an untimely death
through illness. Thereafter the study of variable stars was to
stagnate, until the development of astrophysics offered the
possibility of insights into the underlying physical processes at work.

   The direction in which the solar system is travelling
In the Newtonian universe, the stars were isolated bodies,
free to move in any direction; yet in Newton's own mind
the stars were effectively as 'fixed' and motionless as ever.
The first announcement of the discovery of motions of individual stars (their 'proper' motions) appeared in
Philosophical Transactions in 1718. In Oxford, Edmond
Halley (*c.* 1656–1743) had been comparing contemporary
positions of stars with those recorded for the same stars in
Antiquity. He found that he could make sense of the latitudes ascribed by Ptolemy to three of the brightest stars
only on the assumption that the stars had moved. After all,
'these Stars being the most conspicuous in Heaven, are in
all probability the nearest to the Earth, and if they have any
particular Motion of their own, it is most likely to be perceived in them'.

   One might have expected that thereafter the number of
known proper motions would rapidly increase, as other
astronomers compared contemporary positions with those
recorded by earlier observers. But such was not the case.
Changes resulting from proper motions accumulated
century by century, and so knowledge of these motions
depended on two factors: the accuracy of the positions
recorded in the past, and the length of the time interval
between past and present observations. Unfortunately, the
search for accuracy in astronomy had been inaugurated by
Tycho Brahe little more than a century earlier – and that
was before the invention of the telescope and the use of
telescopic sights, and before there was a proper understanding of atmospheric refraction.

   The observed position of a celestial body is the direction
from which light from this body is travelling at the very

moment it reaches the observer. Unless the body in question happens to be directly overhead, this light will have entered the atmosphere obliquely, and in the last few miles of its journey its path will have been bent by refraction.

Astronomers in Antiquity had been aware of this problem. Cleomedes, who lived somewhere around the time of Christ, spoke of seeing the Moon eclipsed when both Sun and Moon were above the horizon – even though an eclipse occurs when Sun, Earth and Moon are in a straight line. He (correctly) explained this by saying that the light from the Sun and the light from the Moon were both raised during their passage through the Earth's atmosphere. In fact the Sun on the horizon is raised by refraction by an amount roughly equal to its own size, so that at the moment when an observer on an atmosphere-free Earth would have seen the last rays of the Sun disappear from sight, an observer on the real Earth can still see the whole of the Sun's disc.

The effect of atmospheric refraction on a body anywhere near the horizon is therefore very considerable; but to establish by exactly how much the observed position of a body differs from what the position 'ought' to be is not easy. Tycho Brahe attempted to draw up tables of refraction; but he believed that refraction had different effects on the Sun, the Moon, and other bodies, and he also took refraction to be effectively zero above a certain angle from the horizon – 45° for the Sun, but only 20° for stars. The accuracy of his star catalogue suffered correspondingly.

To make matters worse for the student of proper motions, a large and unforeseen source of error in earlier observations was identified in 1728 by James Bradley (1693–1762), Savilian professor of astronomy at Oxford and future Astronomer Royal, through his discovery of 'the aberration of light'. This is the name given to the effect whereby the observed position of any given star is constantly altered by the ever-changing velocity of the Earth-based observer as he is carried round the Sun, much as the direction of rainfall on a train window is affected by the movement of the train. A further problem was revealed by Bradley's later discovery that the Earth's axis nods or 'nutates' (mainly as a result of the gravitational pull of the Moon on the non-spherical Earth), and this affects the very co-ordinate system we use to measure the positions of stars.

How it was that Bradley came to discover aberration and nutation is a fascinating story that belongs on a later page (see page 182). His achievement meant that future astronomers would know they had to correct their raw

observations to allow for these effects, and this ushered in
an era of greatly-improved accuracy. Bradley himself later
took the lead in this, assembling a treasury of stellar positions which later observers would use to determine proper
motions. As Astronomer Royal at Greenwich from 1742, he
first re-equipped the observatory, and then, from 1750 until
his health began to fail, carried out a massive observing programme during which he carefully recorded all the circum-
stantial information necessary to amend the raw data and
so derive stellar positions of unrivalled accuracy.

Bradley himself did not live to make these 'reductions',
but his unreduced observations were published at the end of
the century. The German astronomer and mathematician F.
W. Bessel (1784–1846) later undertook the necessary reductions, and in 1818 he published a catalogue of over three
thousand stellar positions for 1755, a convenient mid-date
in Bradley's observing programme. Bessel's volume bore the
proud title, *Fundamenta astronomiae*, and its appearance
allowed students of proper motion to use the year 1755 as
the starting point in time from which to determine future
changes in the positions of stars.

Meanwhile, attempts were being made to determine
proper motions, however provisional, and to make sense of
them. This was no easy task. In 1748 Bradley himself re-
emphasized in *Philosophical Transactions* that such apparent motions were relative, and could arise either from
movements in the stars themselves, or from the motion of
the solar system, or a combination of the two.

Tobias Mayer (1723–62) of Göttingen explained in 1760
how these two causes might be disentangled. A *pattern* of
the appropriate kind in the observed movements was to be
ascribed to a single cause, the motion of the Earth-based
observer through space. The *residual movements*, however,
were to be ascribed to motions of the individual stars themselves.

But what kind of pattern would it be that resulted from
the motion of the observer as he is carried through space
with the solar system? Mayer pointed out that someone
walking in the forest saw the trees ahead appear to move
aside at his approach. In the same way, if the solar system
was moving towards a particular point on the heavenly
sphere (the 'solar apex'), the stars would appear to move
aside – that is, away from this apex, each star moving along
a 'great circle' towards the opposite point of the sky (the
'antapex').

Mayer himself could see no such pattern in the (often
unreliable) proper motions known to him, and he concluded

that the Sun is at rest. 'Perhaps the true and genuine reason for these movements', he wrote, 'will still remain unknown for many centuries'.

Mayer could not have foreseen the inventiveness of William Herschel. In a most uncharacteristic investigation, because it was carried out at his desk and simply using the limited data available to everyone, Herschel managed in 1783 to find indications that the Sun was travelling in the direction of the constellation Hercules. Nevil Maskelyne (1732–1811), the Astronomer Royal, had published the proper motions of seven bright stars, and the directions of all seven components in Right Ascension could be accounted for by the assumption of an apex anywhere within a certain zone of the celestial sphere. The same zone would account for the RA component of most of the twelve of Mayer's motions that had been republished by the French astronomer J-J. L. de Lalande (1732–1807) because he thought them convincing, as they showed (over half-a-century or so) a change of at least 18 seconds of arc in at least one of the two coordinates. The star Aldebaran was among these, though in RA its recorded change was a mere 3 seconds, far below the level of instrumental error.

Herschel, however, insensitive to such considerations, gave full weight to the RA of Aldebaran, and so was able greatly to reduce the zone he had derived from the Maskelyne data. After analysing the changes in declination to reduce the zone still further, he proposed an apex near the star Lambda Herculis.

For reasons that have never been fully explained, Bessel a generation later was able to find no such pattern in the much more extensive and reliable data contained in his *Fundamenta astronomiae*. The patterns supposedly discovered by Herschel and others had, it seemed, been illusory. But with every year that passed, the quantity and reliability of the data improved, and by 1837 F. W. A. Argelander (1799–1875), professor of astronomy at Bonn, published the results of a long investigation into no fewer than 390 proper motions, a number large enough for him to divide the motions by size into three groups, which he treated independently. Each group yielded a solar apex not far from that proposed by Herschel.

Similar analyses by other astronomers quickly followed, with similar results; but all the analyses depended on Bradley's observations, and all were confined to stars visible from Europe. Fortunately, records of positions of southern stars in 1750 were available in the catalogue that had resulted from an expedition of Nicholas-Louis de Lacaille to the Cape of Good Hope (see page 214); and reliable modern positions were available from the new Royal Observatory at the Cape and from the East India Company's observatory on the island of St Helena.

In 1847, Thomas Galloway (1796–1851), a London-based actuary, analysed the proper motions of 81 of these, and although this number was relatively small, the apex he obtained from these completely independent data matched the positions recently derived from the northern stars – a striking confirmation of the reality of the phenomenon.

Since then, improvements in the data, and in the sophistication of the mathematics used to analyse them, have permitted regular improvements in the position of the solar apex. But the reality of the solar motion, and its general direction, have not been in doubt.

The distances that separate us from the nearest stars
The stars appear to us as points of light, whose positions we measure on the two-dimensional surface of the heavenly sphere. To investigate the distribution of the stars in three-dimensional space, and so discover the structure of the star system to which the Sun belongs, we must somehow deter-mine the third dimension, that of their distances from us.

According to Copernicus in the sixteenth century, the Earth was orbiting the Sun. If so, two measurements of the position of a given star that were separated by six months in time had been made from widely-separate locations – the opposite ends of a diameter of the Earth's orbit. His opponents, reasonably enough, demanded to know why in that case the position of the star always appeared the same – why the star did not show 'annual parallax'?

Galileo recognized that no evidence in favour of Copernicus would be more convincing than the detection of one or more examples of annual parallax. Accordingly, in 1632, in his *Dialogue on the Two Great World Systems*, he suggested a way of carrying out such delicate measure-ments. Suppose that two stars *appeared* to be close together in the sky, but that this came about purely by accident: the two stars happened to lie in almost the same direction from Earth, though in fact one of them was, say, six times further away than the other. As the Earth-based observer was carried round the Sun, the two stars would each appear to him to move in orbits that were exactly similar in shape; but the orbit of the nearer star would be a whole six times bigger in scale than that of the more distant.

If, therefore, he measured the position of the nearer star *relative* to the more distant – treating the distant star as a

Galileo's method for
detecting annual parallax.
To do this directly, by
tracking the tiny changes
in the co-ordinates of the
star over the year, was
impracticable: atmospheric
refraction varied, and the
measuring instrument
itself would alter shape in
response to changes in
temperature and humidity.
By adopting a background
star as a quasi-fixed reference point these problems
could be avoided, the positions of the two stars being
affected equally by refraction, and other factors.

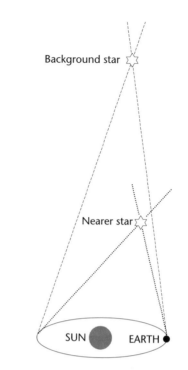

quasi-fixed reference point provided by a helpful Nature –
then the changes he was measuring would in fact be five-
sixths of the true parallax of the nearer star. This slight
reduction in the quantity to be measured was an acceptable
price to pay for the great convenience of measuring small
relative angles rather than large absolute ones.

A still greater advantage of this method was that the
positions of the two stars would be equally affected by
atmospheric refraction (and by aberration and nutation,
factors unknown to Galileo), so that by measuring the position of one relative to the other, such unwelcome complications would be avoided entirely. Typically, however, Galileo
made no effort to follow up this brilliant suggestion, though
in the nineteenth century measurements of annual parallax
by reference to background stars would become routine.

Meanwhile, as the seventeenth century wore on, the
context of the search for annual parallax altered. As the
study of the forces at work in the heavens became an
accepted part of astronomy, the advantages of the
Copernican hypothesis, whereby the planetary system was
centred on the massive Sun rather than on the relatively-
small Earth, became ever more persuasive. The search for
annual parallax then became less a matter of polemics, and
more an investigation into the inverse problem, that of the

distances of stars – for the nearer a star, the more it would appear to move. Successful measurement of the annual parallax of a star would yield its distance (in multiples of the radius of the Earth's orbit, or 'astronomical units').

While observers pondered the challenge of how to track tiny angular changes over a twelve-month timescale with instruments that were themselves affected by climatic changes, Descartes's claim that the Sun is simply the star closest to Earth was seen as offering an approach that might reveal at least the scale of the distance of one star from another.

Let us suppose that the stars (including the Sun) are not merely similar in their physical nature, but virtually identical. Let us further suppose that starlight reaches Earth without being reduced *en route*, by obscuration or some other unwelcome complication. Then, since light falls off with the square of the distance (double the distance of a candle and it appears one-quarter as bright), we can derive the relative distances of the Sun and a given star – say Sirius – if we can somehow contrive to measure their relative brightnesses. If the Sun proves to be one million times brighter than Sirius, for example, then Sirius will be one thousand times more distant than the Sun (and be 1,000 'astronomical units' from us).

The Dutch physicist Christiaan Huygens (1629–95) attempted to carry out the measurement by putting a screen between himself and the Sun, and making a tiny hole in the screen of such a size that the Sun, viewed through the hole, appeared as bright as Sirius. Measurement of the fraction of the Sun's total surface that could be seen through the hole would then yield the desired result.

Unfortunately the Sun was so bright that it proved impossible to make a measurable hole sufficiently small in size. Huygens reduced the sunlight further by placing a lens in front of the hole, but he could only estimate the effect of this. In the end he concluded that Sirius lay at 27,664 astronomical units.

This result made the scale of the stellar universe astonishingly great. Yet, unknown to Huygens and most of his contemporaries, Newton was working along similar lines and with a more reliable technique, and arriving at a still larger value.

In a book published in 1668, the Scots mathematician James Gregory (1638–75) had suggested a brilliant solution to the problem of comparing the brightness of the Sun with that of Sirius. It involved using a planet as an intermediary: one waited until the planet was equal in brightness to

Sirius, and from then on, one disregarded Sirius and focused on the planet instead. The problem thereby reduced to that of comparing the brightness of the Sun – that is, the light of the Sun that reaches us directly – with the brightness of the planet – that is, the light of the Sun that reaches us via the planet.

For the calculations, one needed to know the distances separating members of the solar system; one also had to estimate the fraction of the incoming light that was reflected by the planet; and one had to assume that the planetary regions were free from obscuring matter. Granted all this, the arithmetic was straightforward.

Gregory arrived at a value of 83,190 astronomical units for the distance of Sirius. But he expressly pointed out that he had worked with a scale of the solar system that needed upwards revision, and that with an improved value the result would be much larger. Newton used such a value in 1685 when drafting his *The System of the World*, and he employed Gregory's method to arrive at a staggering one million astronomical units for the distance of Sirius.

Newton had intended the *The System of the World* to be part of his *Principia*, which was published in 1687; but he decided instead to replace it with more technical material, and the scale of interstellar distances was one of the topics he omitted in the process. The result was that Newton's insight into the immensity of the distances separating the stars from each other was known only to members of his immediate circle, until his *The System of the World* at last appeared posthumously in 1728. Meantime, Huygens's estimate (published in 1698, and much inferior, in both senses of the word) held the field.

The methods used by Huygens, Gregory and Newton had revealed, for the first time, the inconceivable immensity of the universe of stars. But their methods were based on the questionable hypothesis of the physical uniformity of the stars; and no one could regard their conclusions as a substitute for the measurement of the annual parallax of specific stars. The first to mount a well-considered instrumental attack on this problem was Robert Hooke (1635–1703). He was alert to the dangers arising from the uncertain effects of atmospheric refraction; but as luck would have it, Nature had provided him with a solution to the problem. A bright star, Gamma Draconis, passed directly overhead his lodgings at Gresham College in London. When the star was overhead, its light entered the atmosphere at right angles, and so would be unaffected by refraction.

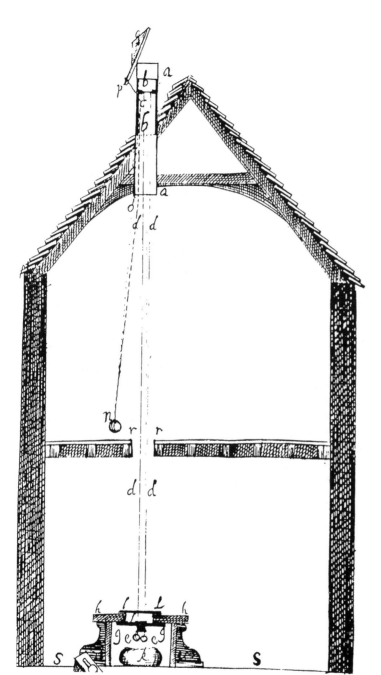

Hooke's zenith telescope, a stable instrument designed to detect annual parallax through measurement of changes in the position of a star passing overhead the observer. Zenith measurements avoided the complications caused by atmospheric refraction. From Robert Hooke, *An Attempt to Prove the Motion of the Earth*, 1674.

But there was another difficulty: his telescope would have to remain motionless, and undisturbed by climatic changes, throughout the months of observation. Hooke therefore decided to incorporate a purpose-built instrument into the very fabric of his own house.

Telescopic astronomy was still in its infancy, and so it is remarkable to find a research programme so focused that it called for the construction of a special form of telescope, designed to measure the position of one single star, and that only when the star was near the zenith. Unfortunately the brilliant conception was ruined by poor execution: in 1669 Hooke was able to make only four measurements before illness, and an accident to the telescope lens, put an end to his investigation. Never one to undervalue his own achievements, he nevertheless declared himself satisfied that his 'Archimedean engine that was to move the Earth' (by proving the Copernican hypothesis) had done just that; but few were convinced.

Half a century later Samuel Molyneux (1689–1728), a prosperous amateur observer living near London, decided to make another attempt to measure the parallax of Gamma Draconis. He enlisted the help of James Bradley, and commissioned from the instrument-maker George Graham a telescope specially designed to measure positions of stars almost directly overhead. Late in 1725 this 'zenith sector' was mounted against a chimney stack within Molyneux's house, so that it could be moved very slightly to either side of the zenith in a north–south direction. As Gamma Draconis passed overhead, the telescope was tilted a little so that the star passed through the centre of the field of view, and the angle of tilt from the vertical was then measured.

Like Hooke before them, Molyneux and Bradley were measuring the position of Gamma Draconis in only one coordinate, the north–south direction. A simple calculation showed that annual parallax would cause this particular star to reach an extreme southerly position on 18 December, around which date its movement from day to day would be vanishingly small. It was therefore with great surprise that on 21 December Bradley found it passing further south than it had a few days earlier. This continued in the weeks that followed, until by March the star was some 20 seconds of arc south of its December position – although annual parallax should by this time have been causing it to move north. The star then stopped, and began to move north, passing through its December position in June and reaching an extreme northerly position in September.

Their first thought was that the axis of the Earth was itself changing direction, in which case the vertical plumbline by which Molyneux's telescope was set was also changing direction. The observed changes would then be, not in the star, but in the coordinate system by which the star's position was being measured. Fortunately another star, on the opposite side of the North Pole from Gamma Draconis, also passed overhead, and although this star was not bright enough to be observed in the daytime, it could be followed sufficiently to show that its changes did not match those of Gamma Draconis in the simple pattern required by their hypothesis.

At the same time the two friends were considering an alternative explanation. They had followed Hooke in observing stars vertically overhead because they believed that by doing so they had circumvented the complication of atmospheric refraction. But this had been on the assumption that the atmosphere was a strictly spherical envelope surrounding the Earth, so that light from stars vertically overhead met the envelope at right angles and was not bent by it. But suppose that the Earth was travelling through a resisting medium that distorted the atmosphere into a non-spherical shape. If so, even the light of a star in the zenith might be bent by refraction, by amounts that would vary with the annual cycle of the Earth's movement around the Sun. This ingenious hypothesis encouraged the observers to look for particular patterns in the movements they were observing, but without success. The extraordinary movements of Gamma Draconis seemed inexplicable.

Bradley decided they needed to bring more stars within the scope of their investigation, and for this he commissioned another zenith sector from Graham, similar in construction but shorter and with a wider field of view. Soon the pattern of movements became clear: stars reached their extreme positions when they passed overhead at six o'clock, morning or evening; and they moved southward while they passed in the day, and northward while they passed in the night. But why?

According to a likely story, the physical explanation occurred to Bradley when he was on a pleasure boat on the River Thames. He noticed that the weather vane altered direction whenever the boat put about – not of course because the wind then blew from a different quarter, but because the direction in which the weather vane pointed depended not only on the velocity of the wind but also on that of the boat.

Back in the 1670s, the Danish astronomer Ole Römer (1644–1710) had shown that the speed of light, though very great, was finite: eclipses of Jupiter's moons were observed on Earth ahead of schedule if the planet was nearer to Earth than usual, but behind schedule if the planet was unusually distant. Bradley realized that since the speed of light was finite, the position of a star – the direction from which the light arrived at the observer – depended, by analogy with the weather vane, not only on the velocity of its light, but also on that of the Earth. He had been looking for annual parallax, an effect that results from the Earth-based observer being located at the outer end of the radius of the Earth's orbit, rather than at the Sun itself. He had found instead 'the aberration of light', caused by the Earth-based observer's velocity, which is tangential to the Earth's orbit. Radius and tangent are at right angles, which was why aberration and annual parallax were out of phase by three months.

The implications of Bradley's discovery were profound. First, he had identified an unsuspected error involved in previous measurements of stellar coordinates, including those in John Flamsteed's great 'British Catalogue' published as recently as 1725. As a result of aberration the apparent position of a star might change by as much as 40 seconds of arc over a six-month period, and Bradley's discovery – in conjunction with his subsequent demonstration that the Earth's axis nutates – ushered in the era of exact positional astronomy.

Second, it was proof – though in a wholly unexpected form – of the motion of the Earth around the Sun.

Third, since all the stars involved were similarly affected, their light must be reaching Earth at the same speed, irrespective of the distance it had travelled – and, as shown by analysis of the timing of eclipses of Jupiter's moons, irrespective of whether the light was direct or reflected. The speed of light was a constant of nature, and Bradley calculated that light took 8 minutes and 12 seconds to reach Earth from the Sun, within 8 seconds of the modern value.

Fourth, Bradley's failure to measure annual parallax implied that this parallax was too small for detection even with his precision instrumentation: it must be less than one second of arc! A simple calculation showed that the stars under scrutiny must therefore be *at least* some 400,000 times the distance of the Sun.

Bradley's results were read to the Royal Society in January 1729, a few months after Newton's *The System of the World* had at last been published. Two major contributions to the problem of the distances of stars had now appeared in quick succession: Newton's argument, based on the hypothesis that the stars were physically uniform, that the nearest (and brightest) stars lay at a distance of a million or so astronomical units; and Bradley's observational proof that certain stars were *at least* 400,000 astronomical units from us. The convergence of these two investigations established once and for all that interstellar distances were to be measured in millions of astronomical units. It also showed that the measurement of annual parallax was a technical challenge of extraordinary delicacy: that of measuring, over an extended period of months, a movement that amounted to no more than the width of a coin several miles away.

Not surprisingly, this led the astronomical community to hesitate long before resuming a direct assault on the measurement of annual parallax. Even Galileo's method of double stars was undermined in 1767, when the Cambridge geologist and astronomer, John Michell (c. 1724–93), used a pioneering probability argument to show that double stars were so numerous that they could not all result from mere accidents whereby the two stars chanced to lie in the same direction from Earth: most must in fact be physically-connected pairs of companions ('binary stars'), at the same distance from Earth and therefore useless for the application of Galileo's method.

This was bad news for the measurement of annual parallax, but good news for Newtonian theorists. For if most double stars – and most star clusters like the Pleiades – were companions bound together by attractive forces, then this was evidence that attractive forces (presumably, but not certainly, Newtonian gravitation) operated beyond the confines of the solar system. Newton himself had declared the force of gravitational attraction to be universal, but had offered no evidence for distances beyond the cometary orbits.

In 1779, unaware of Michell's paper, William Herschel began collecting double stars with a view to their use in the measurement of annual parallax, and three years later published his first catalogue. Michell therefore published a second paper in 1784, repeating his conclusions, and adding: 'it is not improbable, that a few years may inform us, that some of the great number of double, triple, stars, etc., which have been observed by Mr. Herschel, are systems of bodies revolving about each other'.

In 1802, Herschel began to re-examine double stars that he had discovered two decades before. He found that in several, the two component stars had altered position relative to each other, in a way that showed they were indeed companions bound together by an attractive force. The detailed evidence necessary to show that this force was gravitational attraction would not be available for another generation. At last, in 1827, the Paris astronomer Félix Savary was able to confirm that the two stars of Xi Ursae Majoris moved in elliptical orbits about their common centre of gravity, as required by Newtonian theory.

Meanwhile, the precision of astronomical instruments was improving with each new generation of instrument-makers, and by the early nineteenth century astronomers were once more giving thought to the measurement of annual parallax. It was of course crucially important to select for scrutiny the stars that were nearest to Earth, and therefore showed the greatest parallax. The obvious criterion for this had been brightness: other things being equal, the brightest stars would be the nearest. But evidence was accumulating – not least from the differences between companion stars in binaries – to suggest that stars varied enormously in 'luminosity' (the star's 'absolute magnitude'); and astronomers were increasingly coming round to the view that large proper motion was more reliable as a guide to nearness.

Of all the proper motions then known, the largest was, surprisingly, that of a relatively faint (fifth magnitude) star in the constellation of the Swan, 61 Cygni, which was moving across the sky at over 5 seconds of arc per annum. Its motion had been noticed by Giuseppe Piazzi as long ago as 1804, but he was isolated at Palermo, and 'the flying star' did not receive the attention it deserved until 1812, when F. W. Bessel independently published a notice of its large motion. Even so, several attempts to measure its parallax proved fruitless – as did efforts with a variety of other stars.

It was time to regroup and consider strategy. What were the criteria by which astronomers might identify the nearest stars? There were three, according to the German-born Wilhelm Struve (1793–1864) in a paper published in 1837 when he was professor at Dorpat (now Tartu in Estonia): is the star one of the brightest, does it have a rapid proper motion, and, if it happens to be a binary, do the two component stars seem widely separated in view of the time they take to orbit each other? Struve listed the stars that satisfied each criterion, and drew attention to those that satisfied more than one; and if Struve's paper is checked against a modern list of the nearest stars, it becomes evident that astronomers were now selecting the most suitable objects for their parallax measurements.

Equally important was the quality of instrumentation available. At Dorpat Struve possessed a magnificent refractor of 24-cm (9½-inch) aperture, by the German craftsman

and theoretician, Joseph Fraunhofer (1787–1826). Not only
was the object glass of exceptional size and quality, but the
mounting was a mature example of a modern 'equatorial'.
In engineering terms, the simplest mounting for a telescope, whether reflector or refractor, is 'altazimuth': such
an instrument can be rotated horizontally (in azimuth) and
vertically (in altitude). However, the sky spins about the
celestial poles, which means that an observer tracking an
object with an altazimuth telescope must constantly make
adjustment in both directions. By contrast, a telescope
mounted 'equatorially', with one axis pointing to the celestial pole, needs adjustment only in one direction. The
advantages of such mountings had been well recognized
since the time of Tycho Brahe, but it was not until the
eighteenth century that construction techniques had
improved to the point where such mountings became
popular, if only for the portable instruments beloved of
amateur observers. During the nineteenth century such
mountings became standard for refractors and reflectors
alike, and would remain so until the advent of modern
computer-controlled guidance systems provided an alternative solution to the problem of the smooth tracking of a
celestial body.

Meanwhile, at Königsberg in Germany, Bessel had a
16-cm (6½-inch) Fraunhofer refractor of the type known as a
'heliometer'. If an object glass is cut along a diameter into
two semicircles, each semicircle will form a complete
image of the object viewed, though at half the previous
brightness. If the two halves are displaced sideways with
respect to each other, the images are correspondingly displaced. This can be used to measure very small angles, such
as the angle between two stars that are so close together as
to be considered a 'double star'. At first the observer simply
sees the two stars in the normal way; but as the two half-
lenses are displaced, there appear two images of each star.
The amount of displacement required to make one image of
the first star coincide with the other image of the second
star is a measure of the angle separating them. In 1753 the
London instrument-maker John Dollond (1706–61) made
this the basis of a divided object-glass micrometer, which
soon became an accessory sold with many reflectors;
because they were well suited to measurements of the Sun's
apparent diameter at different times of the year, such
instruments were often termed 'heliometers'. Early in the
nineteenth century, particularly through the work of
Fraunhofer and his successors in business at Munich, the
technique was applied to the object glasses of refractors,

though it required the coolness of a diamond-cutter to undertake the halving of a large achromatic lens (see page 128) of high quality.

Struve and Bessel, then, each decided to measure the position of their target star by reference to neighbouring stars that they had good reason to believe were very distant – an adaptation of Galileo's method. Bessel's heliometer was designed for just such measures, while Struve was the leading expert in the measurement of the small angle separating two stars: he had spent two years with the Dorpat refractor examining 120,000 stars to determine which were double, and was now measuring the doubles that he had found, a campaign that would result in the publication in 1837 of data on over 3,000 such stars.

In 1835 Struve selected Vega as the subject for his investigation. The star was exceptionally bright and had a large proper motion, so that it amply fulfilled two of his three criteria of nearness. In 1837 he announced the results of seventeen observations, from which he inferred a parallax of one-eighth of a second of arc (close to the modern value). But he promised to continue his measurements, and in 1840 gave the results of nearly 100 observations, from which he now inferred a parallax that was twice as great. Given the long history of fallacious claims to the measurement of parallax, astronomers were sceptical of Struve's results.

But meanwhile Bessel was directing his heliometer at 61 Cygni. He began observations in 1834, but was soon distracted by the arrival of Halley's Comet, and it was not until 1837 that he returned to the task. Encouraged by the preliminary results from Struve's examination of Vega, for over a year he subjected 61 Cygni to intensive scrutiny, commonly repeating his observations an astonishing sixteen times every night, and still more when the 'seeing' was especially good. By the end of 1838, he announced that the parallax of the star was about one-third of a second of arc. What carried conviction was the way in which the pattern of his many observations matched the prediction from theory. John Herschel, as President of the Royal Astronomical Society, congratulated the Fellows that they had lived to see the day when the sounding line in the universe of stars had at last touched bottom. It was, he said, 'the greatest and most glorious triumph which practical astronomy has ever witnessed'.

It was only a few weeks later that the Scottish astronomer Thomas Henderson (1798–1844), who had been royal astronomer at the Cape of Good Hope, announced a parallax of just over one second of arc for the southern star, Alpha Centauri. This very bright star has a large proper motion, and is a binary star whose components have a wide angular separation. It therefore fulfilled all three of Struve's criteria for nearness. So far as is known, this star and its faint companion, Proxima Centauri, are the stars nearest to the solar system.

Black holes

The study of attractive forces between stars was not the only topic initiated by Michell to which Herschel would make a contribution. Michell was also the first to consider the possibility of what we now term a 'black hole' – a celestial body that was invisible, because its mass was so great that the resultant attractive pull prevented light from leaving it. In his 1784 paper, Michell estimated that if a star of the same density as the Sun had a radius 500 times greater, 'all light emitted from such a body would be made to return towards it, by its own proper gravity'. However, such a body might betray its presence by the effects its pull was having on neighbouring bodies: '. . . yet, if any other luminous bodies should happen to revolve about them we might still perhaps from the motions of these resolving bodies infer the existence of the central ones with some degree of probability, as this might afford a clue to some of the apparent irregularities of the revolving bodies, which would not be easily explicable on any other hypothesis . . .'

Taking up this theme and applying it on the cosmological scale, William Herschel in 1791 wrote of 'the great counteraction of the united attractive force of whole sidereal systems, which must be continually exerting their power upon the particles [of light] while they are endeavouring to fly off'. The idea received wide publicity in 1796, when Pierre Simon de Laplace (1749–1827) included an estimate similar to Michell's in his *Exposition du système du monde*. But Laplace dropped the subject from the 1808 edition of his book, possibly because it conflicted with the general view that the speed of light was a constant; and the concept of a 'black hole' was relegated to the status of a far-fetched speculation, where it remained until recent times.

The structure and history of the universe of stars

While attempts were being made to measure the distances of our nearest neighbours among the stars, and to understand the changes taking place among them, speculators were trying to make sense of the cosmos as a whole. One of

them provoked Isaac Newton to attempt an analysis of the
structure of the stellar universe that was to have consequences for cosmology in our own day.

The Newtonian universe and the darkness of the night sky

If one reads Newton's *Principia* (1687) hoping to discover
the author's conception of the universe of stars, one will be
disappointed. He has in fact next to nothing to say about
the stars, either as individuals or as a whole. To him they
were of limited interest: despite nearly two thousand years
of observation there was not the least evidence to contradict the ancient Greek belief that the stars were 'fixed',
motionless relative to one other.

One seldom sees what one expects not to see: Newton
himself followed current practice in using for 'star' the
Latin word *fixa* (that is, *fixa stella*), and this very term must
have helped close his mind to the possibility that the stars
might move. Although he was the first person in history to
grasp the enormity of the distances that separate us from
even the nearest stars, it never occurred to him that this
might undermine the supposed fixity of the stars – that the
stars might have seemed motionless, not because they were
truly at rest, but because they were so very far away that
their movements had so far escaped detection.

Nor had it then occurred to him that the fixity of the
stars posed a threat to his law of gravity: for he claimed that
gravity was a universal force, and forces generate motions –
yet every one of the stars was apparently at rest. It was a
young theologian, Richard Bentley (1662–1742), who in
1692 forced Newton to face up to this problem. Bentley had
been commissioned to preach a series of sermons – in
effect, lectures – on the compatibility of science and religion. He knew that the Lucasian Professor at Cambridge had
written a work with major implications for cosmology, but
the book was impenetrably mathematical, and so he
plucked up his courage and wrote direct to the author.

Bentley rejected the Cartesian conception of a god who
created the universe, set it in motion, and then left it to run
its course without taking any further interest in its well-
being. But he wanted to see what could be said in support of
such a view, and so he asked Newton what would happen if
matter were spread uniformly throughout infinite space,
and thereafter allowed to move freely under the action of
gravitational attraction. Newton, thinking that Bentley
intended by 'uniformly' a more or less regular distribution
of matter, replied that in any place where there was more

matter than usual the force of gravity would be greater than elsewhere, and so the surrounding matter might be pulled in, so adding to the existing concentration. This could lead to the formation of stars.

Bentley, however, had intended 'uniformly' in the absolute, mathematical sense; and Newton had to concede, when pressed, that symmetry would then ensure that the matter remained motionless, there being no reason why it should move one way rather than another. But such a universe, he admonished, would be artificial in the extreme – as implausible as having infinitely many needles all balanced on their points and standing on an infinite mirror. Bentley, thinking of the infinitely many stars all apparently at rest, very reasonably retorted: 'is it not as hard, that infinite such Masses in an infinite space should maintain an equilibrium . . .?'

The correspondence came to an end, but Newton could no longer close his mind to the challenge: if gravity is universal, how is it possible for the stars to be at rest? He was currently at work on material for a second edition of the *Principia*, and in a succession of drafts of a new theorem we can watch him contriving an answer to the challenge.

His solution was that Providence in the beginning devised an infinite system of motionless stars, a system that was (almost) symmetric and hence (except in the very long term) stable; and when, after a lapse of time, the lack of perfect symmetry led to movements that became sizeable and threatened to bring about the destruction of the original order through what we would now term 'gravitational collapse', Providence intervened and pushed the stars back to their original positions. In this way, Newton maintained his belief in God as the great clockmaker, whose universe was a machinery that endured from age to age. Indeed, Newton was grateful to God for enabling him and other students of the Book of Nature to appreciate how – far from turning his back on his creation as the Cartesians supposed – he providentially intervened from time to time, maintaining the machinery of the stellar system, just as he maintained the machinery of the planets (page 144). God, in Newton's view, had entered into a servicing contract with his creation.

In order to justify his assertion that the system of stars was (almost) symmetric, Newton had to test this model of the stellar universe against the evidence contained in star catalogues. For each listed star, the catalogues gave its observed position (that is, its direction from Earth) and its apparent brightness. The model, however, dealt in three-

dimensional positions, the equivalents of direction from
Earth and (relative) distance from Earth. The model could
not therefore be tested against observation without some
assumption connecting brightness and distance. Newton
accordingly assumed that the stars were physically
uniform, so that brightness and distance were directly
related; specifically, he assumed that stars of 1st, 2nd, 3rd,
. . . magnitudes lay at 1, 2, 3, . . . units of distance from the
solar system.

The observer engaged in compiling a star catalogue was
looking out from the solar system into the surrounding
space, inevitably giving the solar system a special significance in the resulting observational data on the universe of
stars. To make the test feasible, therefore, Newton had to
modify his symmetric model of the universe so as to give
the solar system a comparable significance. He substituted
one in which the Sun was imagined as surrounded by
spheres of radius 1, 2, 3, . . . units, on which were located
stars of the 1st, 2nd, 3rd, . . . magnitudes respectively, each
star on any given sphere being required to be at least 1 unit
distance from its nearest neighbours on the sphere.

The nearest (and brightest) stars were then 1 unit from
the Sun, but they had also to be at least 1 unit from each
other. Newton knew that on a sphere of radius 1, there
could be 12 (or was it 13?) points separated in this way, and
he therefore expected there to be 12 or 13 stars of the 1st
magnitude. The next sphere, of radius 2 units, had 4 times
the surface area and so room for 4 times 12 or 13 (say 50)
points; and Newton therefore expected there to be some 50
stars of the second magnitude. Similarly, on the spheres
with radii 3, 4, 5 and 6 units there was room for some 110,
200, 310 and 450 points respectively, and Newton expected
there to be these numbers of stars of the 3rd, 4th, 5th and
6th magnitudes.

In the drafts for the theorem, we find a Newton at first
so confident of his model that he sees no need actually to
check the evidence: the numbers, he thinks, can be filled in
later. But when eventually he did check the evidence, he
found there were 15 or 16 stars of the 1st magnitude, 63 of
the 2nd, 220 of the 3rd, 500 of the 4th – at which point it
dawned on him that these numbers were far too large.

His first reaction was to stop and cross out all mention
of 5th and 6th magnitudes. His excuse would be that the
star catalogues were incomplete when listing faint stars; yet
the problem was that they were, one might say, too complete! It then occurred to him that his assumption that an
n-th magnitude star lay at n units of distance was overhasty

(and indeed the modern definition would place a 6th magnitude star at 10 units of distance). He therefore adopted a more flexible approach; and after a little creative manipulation, he persuaded himself that he had rescued his theorem.

Newton's speculations about the structure of the universe of stars were known only to a few intimates, but his belief that God intervened to forestall the collapse of both the planetary and the stellar systems was hinted at in 'Queries' that he appended to the 1706 Latin edition of his *Opticks*. This aroused the ire of his great German contemporary, Gottfried Wilhelm Leibniz (1646–1716). Leibniz agreed that God was a clockmaker. But a perfect clockmaker, he argued in a famous correspondence (1714–16) with Newton's spokesman Samuel Clarke, would make a perfect clock, one that had no need for repair and servicing. Newton's divine interventions were to Leibniz the miracles of a god driven to desperate remedies, and so he condemned as utterly inadequate Newton's conception of God. But to the Newtonians the interventions by Providence were not emergency, *ad hoc* miracles motivated by panic, but part of the divine plan intended from the beginning.

The exchanges between Clarke and Leibniz, like most such controversies, continued until the death of one of the parties. Eventually, the unchanging clockwork universe would fall out of favour, to be replaced by one in which developmental changes, brought about by gravitational and other forces, would be seen as natural and expected. But that was for the future, and meantime Newton's symmetric universe was being challenged from a most unexpected quarter.

In drafting his theorem in the 1690s, Newton had persuaded himself that in the neighbourhood of the Sun, there was reasonable accord between model and observation. He had however ignored the fundamental observational fact concerning the stellar universe in the large; namely, the concentration of stars in the Milky Way. (Newton was not alone in this: contemporary astronomers and speculative cosmologists alike displayed a strange lack of interest in the Milky Way.)

This observation-based objection to Newton's symmetric universe of stars was eventually put to him face-to-face by – of all people – the young physician and antiquarian, William Stukeley (1687–1765). Stukeley pictured the Sun and the other stars that we see as individuals as forming a spherical cluster. Surrounding this cluster, and separated from it by empty space, was a flattened ring in which the stars of the Milky Way were collected together. Stukeley's

star system, then, seen from outside, would have looked
not unlike Saturn (the spherical cluster) and Saturn's ring
(the Milky Way). When Stukeley proposed this model of the
universe to Newton in conversation in or about 1720, the
great man responded by hinting cautiously at the merits of
his own model, of a universe symmetrically populated with
infinitely-many stars.

Stukeley, however, was about to set cosmology on the
path towards what has become known as 'Olbers's Paradox',
by focusing not on the effect of *gravity* (something impossible to investigate in Stukeley's universe, for his God was
forever extending the Milky Way by new creations of stars),
but on the *light* the stars collectively emitted. He put it to
Newton that if the system of the stars were symmetric and
infinite, 'The whole hemisphere [of the sky] would have
had the appearance of that luminous gloom of the milky
way.' Having no answer to this difficulty, Newton, it would
seem, did what all sensible people do in such a situation,
and made no comment.

Early in 1721, Stukeley breakfasted with Newton in the
company of Edmond Halley, the newly-appointed
Astronomer Royal, and they discussed astronomical topics.
What Stukeley contributed to the discussion we are not
told, but it would have been natural for him to mention his
theory of the universe. A few days later, Halley read to the
Royal Society the first of two short papers on cosmology,
remarking that 'Another Argument I have heard urged, that
if the number of Fixt Stars were more than finite, the whole
superficies of their apparent Sphere would be luminous' – a
form of words so close to Stukeley's that the 'urging' must
surely have come from him at their recent breakfast.

Halley then set out a (fallacious) mathematical argument
purporting to show that, in a symmetric universe, the
distant stars would – despite their numbers – send us only a
negligible quantity of light. But more important than the
details of his confused response is the fact that, with the
publication of Halley's two papers in *Philosophical
Transactions*, the discussion of the Newtonian model of a
symmetric stellar universe at last emerged – anonymously –
into the public domain.

A succinct and accurate analysis of the behaviour of light
in such a universe appeared in 1744, by the Swiss
astronomer J.-P. L. de Chéseaux (1718–51). In a symmetrical
universe of stars, Chéseaux pointed out, at twice the distance of the nearest stars from Earth there was room for
four times as many stars (since the surface of a sphere is
proportional to the square of the radius); but this increase in

number was exactly offset by the fact that each star would appear only one-quarter as bright (because light diminishes with the square of the distance). That is, there was room for four times as many stars, but each was reduced to one-quarter of the previous apparent brightness. In aggregate, therefore, the nearest stars would contribute to the night sky exactly the same amount of light as the stars at twice the distance.

A generalization of this argument showed that the same was true of stars at each of the successively greater distances. The reader was, therefore, to imagine the night sky gradually filling up with light as the light of stars at greater and greater distances was taken into account, until at length the entire sky was ablaze with the equivalent of sunlight.

A glance at the night sky showed that this did not happen in the real universe. Why this should be so was obvious to Chéseaux: he had assumed that *all* the light setting out from a star reached its destination. If however the transparency of space was less than perfect – and surely this must be the case in the real universe – then a certain fraction of starlight would be lost as it travelled a given distance. A similar fraction of what remained would be lost over the next such distance, and a fraction of what then remained would be lost over the next, and so on. In consequence, even if the loss of light over one such distance was small, when this loss was repeated over and over again it would be enough to cause nearly all the light from remote stars to be lost *en route*; and such stars therefore added little to the brightness of the night sky.

A similar view was taken by the retired German physician and amateur astronomer, H. M. W. Olbers (1758–1840), writing in 1823 and this time in a volume of the widely-read *Berliner astronomisches Jahrbuch*. Olbers showed that even if only 1 part in 800 of the light was lost in its journey from one star to the next, this loss would be sufficient to explain the appearance of the night sky.

By the mid-nineteenth century, enough was understood of the conservation of energy for physicists to realize that if light was absorbed by an interstellar medium, then this medium would itself heat up and begin to radiate light. In consequence, the Chéseaux–Olbers explanation was no longer adequate. But several other possible explanations were to hand – for example, the existence between one star system and the next of an etherless vacuum, across which no light could pass. And so it was that in the nineteenth century, as in the eighteenth, the darkness of the night sky was easily explained. Only in our own times has it come to play a significant role in cosmological thinking.

Speculative cosmology in the eighteenth century

To Stukeley and other speculators in Royal Society circles early in the eighteenth century, it was not easy to make sense of the universe of stars. Except in the Milky Way, the stars seemed to be scattered across the sky without rhyme or reason. But perhaps the disorder was apparent rather than real. After all, the bizarre behaviour of the planets had been shown by Copernicus to be nothing more than orderly motions viewed from an unhelpful location, a moving Earth. So William Whiston, Newton's successor at Cambridge, remarked that 'it is very rational to conclude, that some regular Order hath Place also amongst the Fixed Stars. There may be a certain orderly and harmonious Disposition of the Fixed Stars amongst themselves, when they are beheld from some other proper Place, altho' that Order appears not when they are seen from this Earth.'

But what form might this order take? Stukeley, in private, made more than one attempt to identify it. Another attempt was made by an equally unlikely individual: Thomas Wright (1711–86) of Durham in the north of England. Wright had been apprenticed to a clockmaker at the age of thirteen, but fled his master after a scandal. He taught himself navigation, and in turn taught the subject to seamen in the port of Sunderland. This launched him on a career as an itinerant teacher, and it was in this capacity that he prepared in 1734 a public lecture on cosmology.

It was also something of a sermon. Wright was convinced that the Sun and the other stars occupied a space that had the form of a spherical shell, with the Abode of God located in the midst of the shell. Outside the shell was the Outer Darkness, where the damned got at best a distant glimpse of the stars clustered around the Abode of God.

Had the stars been motionless, their system would have collapsed under their mutual gravitational pulls, and they would have fallen into the Abode of God. This being out of the question, the Sun and the other stars must avoid such a fate by forever travelling in orbit, circling this way and that around the Abode of God.

To illustrate his conception, Wright prepared a vast visual aid in the form of a cross-section of the universe; and to bring home to his audience their personal involvement in this universe, he allowed himself the artistic licence of portraying the region immediately around the solar system as it in fact appears to us on Earth. The handful of nearby

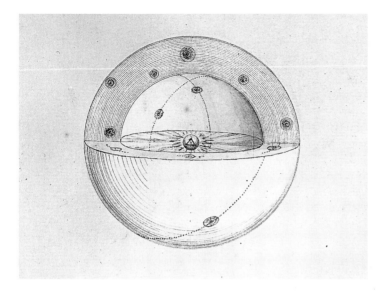

Wright's earliest conception of the universe. All creation was centred on the Abode of God (the Trinity, indicated by a triangle). The Sun, and the other stars, were distributed around this centre on every side, occupying a volume of space in the shape of a spherical shell. If any star had been motionless when first created, the gravitational pulls of the other stars would immediately have caused it to fall inwards into the Abode of God. To prevent this, at the Creation each star had been put into stable orbit, around the Abode of God. Wright's attempts to convey this conception are among the earliest drawings to portray stars in motion. Reproduced with kind permission of the University Library, University of Durham.

stars he therefore drew as bright and isolated, but the numerous stars at the limit of human vision he showed as merging to form 'a faint circle of light': the Milky Way.

After some years he realized the fallacy in this explanation of the Milky Way. The cross-section of the universe that he had drawn was merely one of many possible cross-sections, whereas the Milky Way follows a unique circle across the sky. Eventually he saw a way in which he could adapt his previous model to meet the difficulty, and Wright made the revised model the centrepiece of his handsome volume, *An Original Theory or New Hypothesis of the Universe*, published in 1750. The Sun was again one of innumerable stars that together formed a spherical system surrounding the Divine Centre (or rather, since there were now many such systems and centres, our local Divine Centre). The spherical shell of space occupied by the stars of our system had an immense radius – so immense that the part of the shell immediately around us, whose stars astronomers could actually see, looked flat.

The shell was also thin, with the result that when, from their location within the shell, Earth-dwellers looked either inwards or outwards, their gaze quickly emerged from the layer of stars into empty space. The result was that in these directions they saw only a few stars, which were near and therefore bright. By contrast, when they looked around within the shell itself, their line of sight encountered first the nearby stars, and then more and yet more stars at ever-increasing distances; and the light of these innumerable

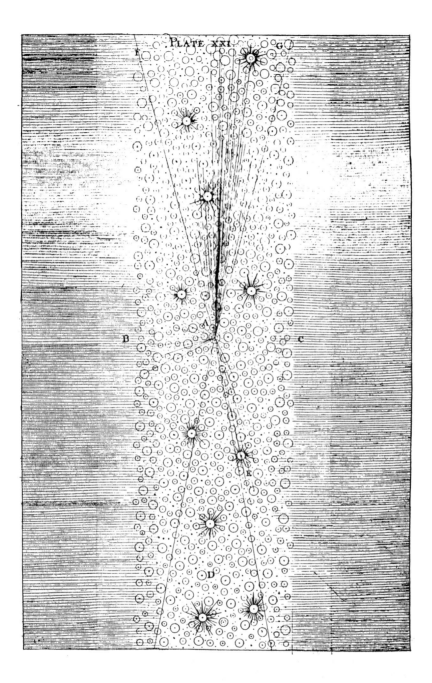

PLATE XXI

A sketch used by Wright to help readers understand the explanation of his preferred model of
the universe. In this sketch, the Sun and the other
stars are sandwiched between two parallel planes
(which his readers view edge-on, as vertical
boundaries to either side of the stars). An Earth-
based observer at *A* would see only a handful of
nearby stars when looking in directions such as *B*
or *C*. But he would see innumerable stars when
looking in directions within the layer, for
example towards *D*, *E*, . . ., and the light of these
stars would merge to create the appearance of a
milky way across the sky. From Thomas Wright,
An Original Theory of the Universe, 1750.

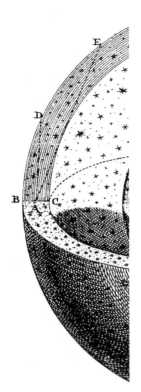

Wright's preferred model. Stars near enough to be seen by us form a small segment of a vast spherical system, whose radius is so great that the boundaries of the segment approximate to parallel planes. Again, an observer at *A* sees only a handful of nearby stars when looking towards *B* or *C*, but milkiness when looking towards *D, E, . . .* From Thomas Wright, *An Original Theory of the Universe*, 1750.

stars merged to give the appearance of the Milky Way. The plane of the Milky Way, then, was the tangent plane to the spherical shell at the point where the solar system was located.

Wright recognized that there was an alternative model that could explain the Milky Way, one in which our star system formed a flattened ring that surrounded our Divine Centre. The stars actually visible to us on Earth would then occupy a disc-shaped space located to one side of the ring. But this model offered no explanation as to what had motivated God to select the particular plane in which the ring lay. As the spherical symmetry of its rival left no such loose end, it was the spherical model that Wright preferred.

Torn out of context, Wright's explanation of the Milky Way as the optical effect of our immersion in a layer of stars sounds strikingly modern. But this insight was embedded in a theological view of the cosmos that was little short of bizarre, and Wright's book would have had even less

influence than it did, had it not been summarized in a
Hamburg periodical a few months after publication.

This summary chanced to come to the notice of the great
German philosopher Immanuel Kant (1724–1804). Now,
even with Wright's illustrations to hand, modern historians
have found it difficult to make sense of what Wright was
attempting to convey; without the illustrations the task
was well-nigh hopeless. Kant could not believe that Wright
was seriously proposing a universe with innumerable
Divine Centres, each surrounded by its star system; instead,
he thought Wright must intend a single Divine Centre
remote from our own star system – and therefore irrelevant
to discussion of the structure of this system. Kant accordingly believed that, to explain the appearance of the Milky
Way, Wright was offering two alternative explanations,
both entirely in the natural order: one involving stars occupying a spherical shell of space, the other stars occupying a
flat ring. To Wright, the centre of this ring was necessarily
void of stars, because that was where our Divine Centre
was located. Kant, misunderstanding Wright on this, saw
no reason why the stars should not continue across the
centre, so converting the ring into a continuous disc.

But was our star system spherical or disc-shaped? Kant
believed there were other such systems in the universe, and
that some of these had been observed (by the Frenchman P.
L. M. de Maupertuis) to have an elliptical outline. Discs do
have an elliptical outline when viewed slantwise on, but
spheres always appear circular. The systems observed by
Maupertuis, therefore, were disc-shaped rather than spherical, and the same must be true of our star system, or
Galaxy. By such reasoning was the first essentially-correct
model of the Galaxy arrived at!

Meanwhile, another cosmological speculator was at
work. Johann Heinrich Lambert (1728–77), an Alsatian who
spent his life on the fringes of the scientific community,
was a convinced believer in a stable, 'clockwork' universe.
When he developed his conception of how our universe is
structured, he had not as yet come across Halley's 1718
paper announcing the discovery of the proper motions of
certain stars. But, unlike Newton, he appreciated that the
immensity of the distances separating the nearest stars
from us had already undermined their supposed 'fixity'.

Lambert's universe had a hierarchical structure, with a
large but finite number of steps in the hierarchy: moons,
planets, stars, groups of stars, galaxies, . . . At any given
level, the unit consisted of a finite number of members
from the level below, each in stable orbit about a central

body. The Milky Way was composed of groups of stars, each group circulating about the (dark) central body of the Milky Way as do the planets about the Sun.

None of these speculations achieved wide circulation, and so their influence is not easy to assess. Yet the appearance in rapid succession of three works dedicated to speculative cosmology, shows that while professional astronomers were still preoccupied with the solar system, some outside their ranks felt a need to understand the universe 'in the large'. The man who was to reorientate astronomy so as to answer this need was then in the north of England, scratching a living as a refugee musician.

William Herschel and the construction of the heavens
William Herschel (Friedrich Wilhelm Herschel, 1738–1822) grew up in Hanover, and at the age of fourteen joined his father in the band of the Hanoverian Guards. Following the French victory in the Seven Years War, Herschel (who as a boy was not under oath, and so free to leave) fled to England in 1757, where he supported himself, first by copying music in London, and then as organist and teacher of music in the north of England.

In 1766 his fortune changed for the better when he was appointed organist to a fashionable chapel in the spa resort of Bath.

His new security gave him the chance to broaden his interests. In particular, he explored the classic two-volume textbook of *Opticks* by Robert Smith of Cambridge, a work that instructed the reader in the theory of optics and in the practice of constructing telescopes and microscopes. Smith had concluded his work with a chapter on 'Telescopical Discoveries in the Fixt Stars', and these few paragraphs served to focus Herschel's developing ambitions in astronomy.

In 1772, Herschel visited Hanover. His father had died, and the surviving family had reduced his talented sister Caroline (1750–1848) to little more than a household drudge. Promising to pay for domestic help, he persuaded them to allow Caroline to return with him to Bath. On the way they passed through London, 'of which', Caroline later recorded, 'I only remember the optician shops, for I do not think we stopt at any other'.

Caroline became mistress of his household (until his eventual marriage in 1788). She hoped for a career as a singer, but her brother was already obsessed with astronomy, and Caroline increasingly found herself drafted to be his assistant. Eventually she was to earn fame as an

astronomer in her own right, when her brother set her to
watch out for newly-arriving comets; between 1786 and
1797 she became the discoverer (or co-discoverer) of no
fewer than eight of these.

In 1773 Herschel bought a copy of *Astronomy Explained
upon Sir Isaac Newton's Principles*, a best-seller by the
leading popularizer of astronomy, James Ferguson
(1710–76). It added little to his fragmentary knowledge of
stellar astronomy, but it introduced him to some of the
wider issues in cosmology. Ferguson, like many of his contemporaries, held that all planetary systems associated with
stars are 'provided with accommodations for rational inhabitants'; and that even comets (once thought by William
Whiston to be 'so many hells for tormenting the damned
with perpetual vicissitudes of heat and cold') were probably
peopled with beings capable of appreciating God's handi-
work. Herschel was to extend this populating of the universe even to the Sun itself, which he came to believe was a
cold body like a planet, but surrounded by clouds that
protect the inhabitants from the exterior shell of fire;
sunspots were glimpses we have of the clouds seen through
gaps in the shell of fire.

Ferguson was not a Newton, dogmatically confident that
Providence would intervene whenever chaos threatened,
and restore the structure of the solar system or that of the
universe of stars: the universe, he wrote, 'will last as long as
was intended by its Author, who ought no more to be found
fault with for framing so perishable a work, than for making
man mortal'. Herschel learned the lesson well.

To view the objects described by Smith and Ferguson,
Herschel first experimented with refracting telescopes. But
the aperture of refractors was severely limited by the great
difficulty (and cost) of manufacturing lenses of appropriate
quality, and a lens of the size to match Herschel's ambitions was a technological impossibility. For his cosmological artillery Herschel therefore turned to reflectors, in
which the light falls on a mirror at the base of the tube and
is reflected back to a focus. Mirrors were altogether more
promising, and he had Smith's book to tell him how to do
the grinding and polishing.

By November 1773, he had placed orders for a number of
discs, one of which was for a reflector of 5½-foot focal
length. And it was when using this instrument, on 1 March
1774, that he decided to open an observing book. The first
page of this book must be the most portentous beginning to
any career in observational astronomy.

Six of the milky patches in the sky known as 'nebulae'

had been listed by Edmond Halley in *Philosophical Transactions* in 1716, and these were discussed by Smith in his *Opticks*. The term 'nebula' referred simply to the object's milky appearance, and did not prejudge its physical nature.

On this there was a long-standing dispute. It was obvious that a distant cluster of stars would appear nebulous when seen in a telescope of insufficient power to 'resolve' the cluster into its component stars. The question was, were there also true nebulae, formed of some sort of diffuse luminous fluid; or were all nebulae merely apparent, nothing more than star systems whose true nature was disguised from the Earth-based observer by their great distances? Halley took the former view: nebulae 'in reality are nothing else but the Light coming from an extraordinary great Space in the Ether; through which a lucid *Medium* is diffused, that shines with its own proper Lustre'.

Observation could contribute to the debate in two ways. First, a more powerful telescope might succeed in resolving into its component stars a star system that had appeared nebulous when viewed with lesser instruments. Second, if a nebula was seen to alter shape from one decade to another, or even one century to another, then the nebula could not be a star system. After all, a star system so extensive as to appear to the observer as spread across the sky, and yet so distant that the component stars could not be detected, must be vast indeed – too vast to alter shape so rapidly.

Herschel was familiar with the crude sketch of the Orion Nebula that Smith had reproduced in his *Opticks*, from a drawing made by Christiaan Huygens in 1656. Looking at the nebula with his home-made reflector, Herschel decided it must have changed, and he at once saw the implications: '. . . from this we may infer that there are undoubtedly changes among the fixt stars, and perhaps from a careful observation of this Spot something might be concluded concerning the Nature of it.'

But these were crowded years, and Herschel – organist, composer, conductor, and teacher of music – devoted most of the limited time he could spare for astronomy to improving his telescopes. Sometimes, when he was grinding a mirror and dared not take his hands from it and allow it to cool, Caroline had to put food into his mouth. But in 1779 he decided it was time he familiarized himself with the brighter stars, and so he systematically examined them one by one, using a portable reflector he had made himself of 7-foot focal length. He then embarked on a second such 'review', much more thorough and this time with an

additional goal: the identification of double stars that might
be of use in the detection of annual parallax (pages 177–8).
He was then, as we have seen, unaware of Michell's paper
that showed that most double stars were in fact binaries
and so useless for this purpose. He harvested 269 double
and multiple stars from this review, and a further 434 from
a third review, thereby multiplying many times the number
known to astronomy and introducing into the science a
new methodology.

Those who had received an orthodox education in the
science agreed that it was the job of an astronomer to study
the familiar celestial bodies – Sun, Moon, planets and their
satellites, comets, bright stars, each with its personal name
and individual characteristics. Herschel, knowing no better,
was beginning to play the natural historian, collecting
specimens in great numbers, counting and classifying them.
Soon he would be ordering nebulae according to the stage
they had reached in their life-cycle.

Meanwhile, in the late 1770s, word of this extraordinary
organist had begun to spread in astronomical circles. The
Astronomer Royal and other leading astronomers called on
him, recognized his great talent, and did what they could to
smooth his path. A Bath neighbour communicated papers of
his to the Royal Society in London. But some Fellows of the
Royal Society – mingling incredulity at his claims with
contempt for his ignorance of basic procedures and conven-
tions – declared him fit for the mad-house.

However, his discovery in 1781 of the planet Uranus (see
page 159), in the course of his second review, was a triumph
that none could gainsay, and soon every astronomer in
Europe had heard of Herschel. King George III, himself a
Hanoverian, granted him a life pension that would allow
him to give up music and devote himself to astronomy. The
pension was modest – £200 per annum – but it gave him
independence, his only duty being to reside near Windsor
Castle and be available to show the heavens to the royal
family when required. The King subsequently contributed
two grants of £2000 each, and an annual allowance of £200,
towards the cost of the monster 40-foot reflector with
mirrors 4-foot in diameter that Herschel completed in 1789.
Two years earlier Caroline had been granted her own
pension of £50 in recognition of her work as her brother's
assistant.

Finding a new planet had been far from Herschel's mind
on that fateful evening, but its discovery was no accident:
he was systematically searching the sky, and his dual skills
as telescope maker and observer enabled him to recognize

at a glance that the object was no ordinary star. Telescope
making would always be the foundation for his success as
an astronomer, and he now became a professional maker,
supplementing his pension by manufacturing reflectors for
sale. Although his price list offered a complete 40-foot for
8,000 guineas, the largest instrument he ever sold, and the
finest he ever made, was a reflector of 25-foot focal length
and 2-foot aperture made for the King of Spain; it was
erected in Madrid in 1804, only for the mounting to fall
victim to the Napoleonic troops four years later.

A successful reflector embodied three key components:
first, a well-shaped mirror – in fact, two mirrors, so that one
might be used while the other was being reground to
remove the tarnish caused by exposure to the night air;
second, a range of eyepieces for magnification; and third, a
stable yet adaptable mounting. Even when at Bath,
Herschel's ambitions to have large mirrors of great 'light-
gathering power', to permit the study of objects that were
very distant and therefore faint, had outrun the capacity of
local foundries to cast the blanks. Nothing daunted, in
August 1781 he had converted the basement of his own
home into a foundry, and had twice attempted to cast a 3-
foot disc, which would have made this organist the owner
of the largest telescopic mirror in the world. On the first
occasion the mirror cracked while cooling; on the second,
molten metal poured out onto the flagstones which,
expanding, began to fly about in all directions. At this even
Herschel admitted temporary defeat.

In the making of eyepieces, success seems to have come
easily to Herschel. Indeed his magnifications – of hundreds
and even thousands – were cited by him without special
comment, although to many these well-justified claims
were simply incredible.

It was in the mounting of large reflectors that Herschel
proved most innovative. A 20-foot reflector he had made
himself in 1776 had been slung from a pole, like the very
long refractors of past generations. But once he was free to
devote himself to astronomy, it took Herschel only a year
to build himself a new 20-foot. This time the mirrors were
18 inches in diameter rather than 12 inches; but more
importantly, the mounting was stable, and the observer
stood in safety on a secure platform.

Herschel was now equipped to tackle the riddle of the
nebulae. Despite his early look at the nebula in Orion, by
December 1781 he had seen only three more. It was then
that he was given a catalogue of some 68 nebulae and star
clusters. It had been assembled by the French comet-hunter

William Herschel's 'large' 20-foot reflector, completed in 1783. Its mirrors were bigger than those of the 'small' 20-foot, but the major advances were in the stable and secure mounting, and in the controls available to the observer. From an undated print published by William Herschel, private Collection.

Charles Messier (1730–1817), who had found these diffuse
objects an unwelcome distraction in his searches for
comets. (In fact Messier had already published an enlarged
catalogue with just over 100 nebulae, and this catalogue is
used today by astronomers when they refer to a prominent
nebula by an 'M', followed by its Messier number.)

Herschel made the momentous decision to use the new
20-foot to sweep the entire sky visible from England, in
order to collect as many specimens of nebulae as possible.
On nights when the 'seeing' was good, the telescope was
turned to the south, and the tube raised to some particular
angle. Herschel then let the sky drift past, so laying an
ambush for any nebula that came into the field of view.
When he saw one crossing the meridian, he would call out
its description for Caroline to copy down, along with the
angle and the time by the stars. For two long decades the
work continued. When the job was done, the team of
brother and sister had increased the number of known
nebulae from 100 or so, to 2,500.

But were all nebulae merely vast star clusters at great
distances; or were some truly nebulous, and formed of a
luminous fluid? Just before sweeping began, Herschel had
confirmed (as he thought) the variability of the Orion
Nebula, which he found 'surprizingly changed'; this, then,
must be a true nebula. But how in general was one to distinguish true nebulae from distant star clusters?

Herschel had noticed that some nebulous-seeming
objects had a uniformly milky appearance, while others
were mottled. The former, he decided, were true nebulae,
while the mottled nebulosity of the latter indicated to him
that they were 'resolvable' – in other words, they were clusters that with a sufficiently powerful telescope would be
resolved into their component stars.

In June 1784, a paper outlining his current work on
nebulae was read to the Royal Society. Within days,
Herschel came across two nebulae that contradicted the
very theory he had just published, for each seemed to
contain both forms of nebulosity. Indeed, in one of them he
believed he could also see stars mixed with the resolvable
nebulosity. He interpreted these stars as being in the region
of the nebula nearest to the observer, and the resolvable
nebulosity as composed of stars that were further away, a
little too distant to be individually visible. Must not then
the milky nebulosity simply consist of stars at a still
greater distance – rather than of the luminous fluid, or 'true
nebulosity', that Herschel had hitherto postulated?

Ignoring the changes he believed he had earlier observed

in the Orion Nebula, he now abandoned his belief in luminous fluid, and concluded instead that all nebulae were star clusters at great distances.

But clusters imply clustering, the assembling together of stars as a result of their gravitational pulls on each other. Accordingly, in an epoch-making paper 'On the Construction of the Heavens' published in *Philosophical Transactions* in 1785, Herschel examined the subject 'from a point of view at a considerable distance both of space and of time'.

He imagined a universe in which the stars were at first distributed with fair regularity; and he went on to outline how, in time, gravitational pulls would be likely to cause stars to assemble in places here and there, forming clusters of various types, examples of which he had already observed. He even envisaged the possibility that the continuing action of gravity might eventually lead to gravitational collapse, followed by some form of renewal. 'These clusters may be the *Laboratories* of the universe . . . wherein the most salutary remedies for the decay of the whole are prepared' – a far cry from the stable clockwork universe of the early years of the century.

In his 1785 paper Herschel also pioneered the use of statistics in astronomy, by showing how the natural historian of the heavens can derive insights simply by counting stars. The problem he set himself was to determine the shape of our star system, the Galaxy.

Like Wright, Kant and Lambert, he realized that the Milky Way was the optical effect of our immersion in a layer of stars. But what was the precise shape of this layer? Clearly, Herschel could not attempt to answer this question unless he was allowed to assume that his telescope could reach to the limits of the Galaxy in every direction. But how then to proceed? Herschel decided that the way forward was to assume next that throughout the Galaxy, the stars were distributed uniformly: that the galactic space was uniformly stocked with stars. Obviously the assumption was not true in the literal sense; Herschel hoped it would be true enough for his purpose.

Granted this, the number of stars in Herschel's field of view in a given direction was proportional to the volume of galactic space within that field of view – that is, to a cone-shaped volume of space, whose vertex was at the telescope, and whose axis was the line of sight from the observer to the border of the Galaxy. Herschel would then count the stars to get a number proportional to the volume of the cone, and a simple calculation would then give him the (rel-

ative) length of the axis. Time did not permit him to implement this programme in full, but to illustrate his method
he made the counts for a circle around the sky, and
sketched the resulting cross-section of the Galaxy.

In later life he was to abandon both of the assumptions
on which this famous figure was based: the cumbersome
40-foot reflector completed in 1789 brought many more
stars into view, and so his 20-foot had not after all penetrated to the borders in every direction; and increasing
familiarity with star clusters brought home to him how
very non-uniform is the distribution of the stars. But
astronomers, like Nature, abhor a vacuum. Herschel's
cross-section might have been disowned by its creator, but
for long there was nothing to take its place, and so we find
it being reproduced in books late into the nineteenth
century.

In November 1790 Herschel was sweeping as usual for
nebulae, when he came across 'a most singular phaenome-
non! A star of about the 8th magnitude, with a faint luminous atmosphere . . .'. It was in 1782 that he had
encountered his first 'planetary nebula' – an object that was
faint like a nebula but had the disc-shaped outline of a
planet – and he had found several more since. What he had
now encountered was in fact another of the class; but this
one was unusually large in appearance, and he could see its
structure and in particular its central star. The object must,
he decided, be a 'nebulous star': a star surrounded by a
cloud of (true) nebulosity, out of which the star was in the
process of condensing. Faced with this new evidence,
Herschel instantly reversed the position he had held since
1784, that all nebulae were nothing else but star clusters at
great distances. Nebulosity existed after all, and it represented a pre-stellar stage in celestial development.

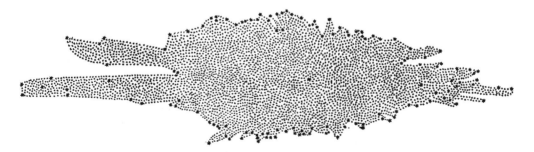

The cross-section of the Galaxy resulting from
Herschel's star counts. The Sun is the star near
the centre, and the (relative) distances to the
stars shown bordering the Galaxy are inferred
from the star counts: the greater the number of
stars in a given direction, the greater the distance. Note the bifurcation to the left. From
Philosophical Transactions, **75**, 1785.

These sketches by Herschel show objects from his catalogues of nebulae and star clusters, arranged in sequence to illustrate the stages by which stars condense out of nebulosity, and then gather together to form star clusters that become more and more concentrated as time goes on. From *Philosophical Transactions*, **104**, 1814.

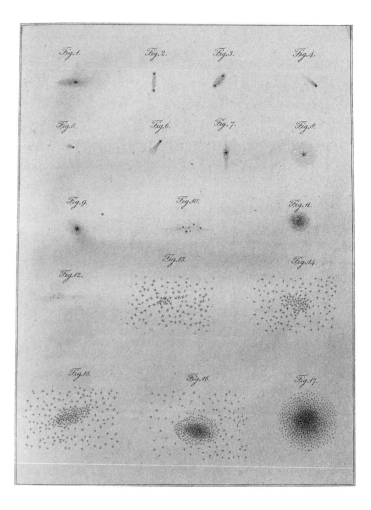

This change in belief was doubtless a contributory factor to the disappointing results Herschel obtained from the 40-foot reflector. The theory that all nebulae were distant star clusters could be tested against the success or failure of a powerful new instrument in 'resolving' additional nebulae into their component stars; but with Herschel's abandonment of this theory, such a *raison d'être* no longer applied. Also, the instrument was cumbersome in use, and the mirrors, which had had to be made of a modified alloy so that they would not distort under their own weight, tarnished easily.

Herschel's final theory of the evolution of the universe began with diffuse clouds of nebulosity, which gradually condensed here and there under gravity to form more concentrated nebulae, out of which in time individual stars began to form. These in turn would gather, at first into widely-scattered clusters, and then into more condensed ones. From the cataclysmic collapse of such clusters, and also from the light sent out into the universe from all luminous bodies, came the material to form new diffuse clouds of nebulosity, so that the cycle might repeat itself.

In 1811, Herschel illustrated how – despite their own brief life-spans – astronomers might reach an understanding of how nebulae develop. They should select a number of specimens that were currently at different stages of development. They should then order and classify them by age, each class representing one of the successive stages in the life story of a nebula: 'there is perhaps not so much difference between them [the nebulae in his consecutive classes], if I may use the comparison, as there would be in an annual description of the human figure, were it given from the birth of a child till he comes to be a man in his prime.'

The status of our Galaxy was also changed by the 1790 observation. In the mid-1780s, Herschel had believed it to be a star cluster of known and therefore limited extent, and all nebulae to be similar star clusters, though of varying shapes and sizes. The Orion Nebula, therefore, which he saw extended across the sky despite its being (supposedly) so distant that the individual stars escaped detection, had to be vast; indeed, it 'may well outvie our Milky Way in grandeur'. But in his post-1790 theorizing, the Orion Nebula reverted to being a nearby cloud of nebulosity located well inside our Galaxy, while the Galaxy became 'the most brilliant, and beyond all comparison the most extensive sidereal system'.

'A knowledge of the construction of the heavens', Herschel wrote in 1811, 'has always been the ultimate object of my observations . . .'. Herschel had been able to make this into a true science because he combined to a unique degree the three talents necessary: those of instrument builder, observer, and theorist. He built instruments that were ideal for his self-imposed task: they incorporated large mirrors, eye-pieces that magnified hundreds of times, stable mountings – all made with his own hands or under his direct supervision. With these instruments he played the natural historian of the heavens, collecting double stars by the hundred and nebulae by the thousand, in observational campaigns extending over many years. And, unusually for one so dedicated to assembling facts, he saw it as his clear duty to speculate too much rather than too little.

Among his contemporaries, Herschel's impact was mixed. The nebulae in particular were his private domain:

no one else had telescopes to equal his, so no one else had
access to the evidence. Few therefore knew what to make of
him and his speculations. But his papers in *Philosophical
Transactions* were readily available to the next generation
of astronomers, who would be more receptive to his ideas.
Prominent among them was his son, John.

John Herschel and the southern skies

William Herschel had been a German-born provincial musician when he cut his astronomical teeth. His only son, John
(1792–1871) – born when his father was already 53 – carried
the most famous name in astronomy, and from his under-
graduate days in Cambridge was a member of the scientific
establishment. After a flirtation with law, John Herschel
settled down to a teaching career in Cambridge. But in 1816
his father, whose strength was failing, prevailed upon him
to return home, so that William could hand on his skills as
a telescope maker and observer before it was too late. 'My
heart dies within me', wrote John as he left Cambridge. But,
once he had made the sacrifice, the son saw himself as
entrusted with a sacred mission, to complete his father's
work and bring it to perfection.

Double stars were the most obvious place to begin: his
father's telescopes had been designed as massive instruments of discovery, whereas the observation of double stars
was best done with instruments of precision. Fortunately,
among John Herschel's many scientific friends was James
South (1785–1867). South's wealth through marriage had
allowed him to give up surgery for astronomy, and he was a
skilled observer and the owner of two exceptionally fine
'equatorials' (see page 187). From 1821 to 1823, though with
interruptions, the two friends collaborated, often observing
the same object with different instruments and then comparing notes. Their efforts resulted in a catalogue of 380
doubles, fully detailed, and conveniently ordered for the use
of observers.

The principal legacy of Herschel's father, however, had
been his studies of 'the construction of the heavens', and
the catalogues of nebulae and clusters on which these
studies were founded. These catalogues listed the objects by
class rather than position, and were therefore highly inconvenient for other observers to use. In addition, there had
been ample scope for error when Caroline had copied down
the observations shouted out by her brother, as the various
nebulae came into view. Fortunately, John had himself
completely refurbished his father's 20-foot reflector under
the old man's supervision, so he possessed the ideal instru-

ment with which to re-examine his father's nebulae.
Caroline painstakingly rearranged her brother's catalogues
into a format that would facilitate John's re-examination, a
task for which she was later awarded the gold medal of the
[Royal] Astronomical Society. Writing in 1825 to thank her,
John told her: 'These curious objects . . . I shall now take
into my especial charge – nobody else can see them.' His
immediate efforts resulted in a catalogue of 2,306 nebulae
and clusters, published in 1833 in *Philosophical
Transactions*; it became the standard reference work, and
helped transform the study of nebulae from one of the
maverick William's exotic pursuits, into a component of
mainstream astronomy.

So far John Herschel's experience of nebulae had been
little different from that of his father: the same skies, and
much the same instrumentation. Not surprisingly, his theoretical stance was also the same. Only one significant new
item of evidence had come his way: he had several times
had a good view of the nebula M 51, which he saw as being
composed of stars in the form of a central cluster surrounded by a divided ring. The sky as seen by an observer
within the central cluster would, as he immediately realized, be strikingly similar to the sky we see from Earth: on
all sides a number of nearby (and therefore bright) stars and,
in and near the plane of the ring, a divided milky way of
innumerable faint stars. 'Perhaps', he remarked, 'this is our
Brother System'.

His father had seen only the skies visible from England,
and to complete his father's work John Herschel would
need to set up his telescopes south of the equator. And so,
declining official offers of financial support, he set sail in
November 1833 for the Cape of Good Hope. For deep-sky
exploration he had the 20-foot; for precision measurements,
he had an equatorial he had bought from South; and for a
preliminary reconnaissance, he had a 'comet-sweeper' his
father had made for Caroline.

The southern skies have much more to offer the observer
than their northern counterparts, for they contain the dense
Milky Way star clouds that lie in the direction of the centre
of the Galaxy, as well as such wonders as the two
Magellanic Clouds. However, because astronomy developed
in Europe, and travel south of the equator was difficult in
past centuries, no public observatory existed in the southern hemisphere until Fearon Fallows (1789–1831) arrived in
1821 at the Cape of Good Hope as His Majesty's
Astronomer. Before then, knowledge of the southern skies
had depended on sailors' lore, and on two expeditions, by

Edmond Halley in the seventeenth century and Nicolas-Louis de Lacaille (1713–62) in the eighteenth.

Halley had made a reputation for himself in Royal Society circles when no more than a youth. Early in 1676, and still not yet twenty years old, he was corresponding about suitable sites for an expedition south of the equator; but some of these sites were in foreign hands, and his choice eventually fell on the south Atlantic island of St Helena, used as a waystation by the (British) East India Company. The King was persuaded to request the Company to give free passage to Halley and his colleague, and Halley's father agreed to contribute to the costs of the expedition.

Halley arrived at the island in February 1677, and stayed for nearly a year. He was equipped with a range of instruments, prominent among which was a sextant of 5½-foot radius with telescopic sights, for measuring the angular distance between pairs of stars. The climate was less favourable than Halley had hoped, but he managed to compile a catalogue of 341 stars, listing their positions relative to two of Tycho Brahe's fundamental stars.

He also observed three 'nebulae', including the scattered cluster of stars now known as M 7, and the fine spherical ('globular') cluster Omega Centauri. The Magellanic Clouds, he wrote, 'reproduce exactly the whiteness of the Galaxy, and, examined through a telescope, they exhibit here and there small clouds and a few stars, from the concourse of which their white colour, like that of the Galaxy, is now believed to be produced'. On 7 November he saw Mercury move across ('transit') the face of the Sun.

It was to be the middle of the next century before the southern skies again came under scrutiny, during the visit to the Cape by Lacaille in 1751–53. In less than two years he determined the positions of nearly ten thousand stars. He achieved this extraordinary level of productivity by attaching a small telescope of wide aperture to a mural quadrant, and placing a rhomboidal diaphragm in the field of view of the instrument. The telescope would be fixed at a particular elevation in the meridian. As a star drifted through the field of view, Lacaille noted the (sidereal) times at which it entered and left the rhombus; the average of the times gave one of the star's coordinates and the difference in the times gave the other.

In 1757 Lacaille published the positions of nearly four hundred of the brightest of these stars, so establishing the framework for southern-hemisphere astronomy. His observations of ten thousand southern stars appeared post-

humously, in 1763, but he left the majority of his position
measurements in their raw ('unreduced') state, and a definitive catalogue of his southern stars was not published until
1847. He also assembled information on forty-two nebulous
objects, and assigned names to southern constellations,
most of which are in use today.

This, then, was the extent of knowledge of the southern
skies prior to John Herschel's arrival in January 1834. Over
the next four years, and with no Caroline to help him,
John explored the southern skies with a dedication surpassing even that of his father. The resulting volume, *Results
of Astronomical Observations Made During the Years
1834, 5, 6, 7, 9 at the Cape of Good Hope*, appeared, after
considerable delays, in 1847. Arguably the greatest single
publication in the whole history of observational astronomy, it bore a proud subtitle: 'Being a completion of a telescopic survey of the whole surface of the visible heavens,
commenced in 1825', for John Herschel was and would
remain the only astronomer in history systematically to
examine the entire sky with a major telescope. The book
listed over 1,700 nebulae and clusters and over 2,100
double stars, as well as thousands of star counts, extensive
sequences of the comparative brightness of stars, and much
else besides.

In March 1838, his duty to his father's memory nobly
discharged, Herschel took ship for England. His future
work in astronomy would be done sitting at a desk: in
1864 he published a consolidated catalogue of over five
thousand nebulae and clusters, the ancestor of the *New
General Catalogue* or NGC that astronomers use today. In
any case, within a year or two of his return from the
Cape the Herschel monopoly of great reflectors would
come to an end, and with it the period in which, as
Wilhelm Struve put it, the study of the nebulous heavens
had seemed 'almost the exclusive domain of the
Herschels'.

The Leviathan of Parsonstown
In 1839 William Parsons (1800–67), future third Earl of
Rosse, built a large reflector of 3-foot aperture in the
grounds of his castle at Parsonstown (now Birr) in central
Ireland. Its first mirror was assembled in sixteen segments,
but the following year Rosse succeeding in casting a solid
mirror with the help of local labourers. The telescope was
mounted in the manner of Herschel's 20-foot, but its
mirrors had twice the diameter and four times the surface
area. Scarcely was this instrument completed when Rosse

began work on a monster with mirrors an incredible 6-foot in diameter. With its south-facing tube crudely slung between two huge walls of masonry, the 'Leviathan of Parsonstown' was designed for a final assault on the classic problem of nebular astronomy: Are all nebulae merely distant star clusters?

The Leviathan first saw light early in 1845. John Herschel declared its completion to be 'an achievement of such magnitude . . . that I want words to express my admiration of it'. Within a few weeks Rosse was able to announce the notable discovery that the nebula M 51 is spiral in shape. A comparison of the superb sketch Rosse made of it with John Herschel's modest drawing provides convincing proof of the Leviathan's power.

Yet in many ways it never fulfilled its potential. A terrible potato famine struck Ireland in the late 1840s when the

The 'Leviathan of Parsonstown'. The first mirror was cast in 1842, work on the mounting began late the same year, and the telescope first saw light in February 1845. By April Rosse had discovered the spiral structure of M 51. Reproduced with kind permission of the Museum of the History of Science, Oxford.

instrument was in its prime; Rosse, a leading public figure and landowner, could scarcely devote himself to the stars while his tenants lay dying. In addition, for Rosse the construction of the great reflector – with all the technical challenges involved – was an end in itself. Rosse was no Herschel. He observed from time to time, his visitors were welcome to use the instrument, and he employed worthy if uninspired assistants; but this was no substitute for Herschelian dedication.

The decisive test came when the instrument was turned to the Orion Nebula. If every nebula was a star cluster, surely this nebula above all – so extensive as seen by us, and therefore relatively near – would reveal its starry nature when interrogated by the Leviathan.

In March 1846, Rosse announced that numerous stars were indeed to be seen in the nebula. In fact the stars that Rosse described are genuine enough; but they are embedded in what is generally a gaseous nebula. This was not appreciated at the time, with the result that the supposed resolution of this and other prominent nebulae persuaded all but a few sceptics that the nebular hypothesis – that true nebulosity exists and condenses into stars – had been discredited.

But sceptics there were. Like William Herschel back in the 1770s, Mikhail V. Lyapunov of the university at Kazan, far to the east of Moscow, had convinced himself – and the influential Wilhelm Struve – that major changes were occurring in the Orion Nebula. This being so, the 'alleged miracles of resolution', as Struve termed them, 'are nothing but illusions'.

The reality of such changes was indeed debatable, but the complete disappearance of a nebula was not. In 1852 John R. Hind (1823–95), astronomer at George Bishop's private observatory in Regent's Park, London, reported the discovery of a small nebula in Taurus. The nebula was observed several times in the following years; but in October 1861 the German-born astronomer Heinrich Louis d'Arrest (1822–75) of the University of Copenhagen, who was making a careful study of the positions and appearances of nebulae, could find no trace of it whatever. News spread of the vanishing of what would become known as 'Hind's wonderful nebula'.

This was an example at last of change of unimpeachable authenticity. Other such claims began to be taken more seriously, and a new hesitation crept into accounts equating nebulae with star clusters. But the decisive proof, in 1864, that true nebulae exist – that the light of certain nebulae originates in gas rather than in stars – was to come, not from traditional methods of observation, but through the use of a technique of laboratory physics that was opening a new chapter in the history of astronomy.

8

The message of starlight: the rise of astrophysics

David Dewhirst and
Michael Hoskin

From Antiquity to the Renaissance, the principal task of
the astronomer had been to devise geometrical models
that would reproduce the movements of the planets
against the unchanging background of the 'fixed' stars. In
the seventeenth century, Kepler taught astronomers to
investigate the forces causing these movements; and in the
aftermath of the publication of Newton's *Principia* in
1687, the agenda of 'celestial mechanics' was to demonstrate that these and all the other movements within the
solar system had a single physical cause – the pulls resulting from the inverse-square law of gravitational attraction.
By contrast, even as late as the mid-nineteenth century,
the astronomy of the stars still consisted mainly in the
patient compilation of ever more accurate and comprehen-
sive catalogues of positions and magnitudes, a humdrum
task that had been begun by Hipparchus and Ptolemy in
Antiquity.

All this was to change in the second half of the century,
when prisms were fitted to telescopes and the resulting
rainbow-like spectra analysed for the information they contained. The astronomer of the Middle Ages had been a
geometer, the astronomer of the eighteenth century a celestial mechanician; now, in the late nineteenth century, the
astronomer had also to master the skills of the laboratory
physicists specializing in the measurement and analysis of
light. The theories and techniques used by physicists were
adapted to the study of starlight, and this enabled the 'astrophysicist' to penetrate the secrets of the physical processes
at work among the stars, and to determine their motions in
three dimensional space. The stars, and the mysterious
nebulae, now became the focus of interest, and even the
fringe discipline of cosmology was assimilated into mainstream astronomy.

Publishing astronomy, organizing astronomers, 1665–1950

A comparable revolution was taking place in the organization of astronomy. The invention of printing in the mid-fif-
teenth century had made possible the rapid publication of
broadsheets and pamphlets expounding the latest astronomical wonder; but major advances in knowledge continued to
be published in books, and books usually took years to
write and to see through the press.

The pace of scientific advance accelerated dramatically
with the publication from 1665 of the *Philosophical
Transactions* of the Royal Society of London. This grew out
of the international correspondence conducted by the
Secretary of the Society, Henry Oldenberg, but was soon
taken over by the Society as its official monthly publication.

The discovery that lies at the heart of spectroscopy provides a good illustration of the speed with which scientific
news could now be disseminated. Early in 1672 Newton
was persuaded to share with the readers of *Philosophical
Transactions* his discovery (see page 224) that sunlight is
not simple but a combination of many colours. In no time
at all, a surprised and outraged Newton found himself
under widespread attack, from critics at home and overseas.

Ephemerides (tables of the positions of heavenly bodies
calculated for some years ahead) had long been an essential
tool for astronomers and astrologers alike. Publication of
the first year-book dedicated to astronomy, and definitely
eschewing astrology, began in Paris in 1679 with the title
La Connoissance des Temps (with the early French spelling
of the present-day *Connaissance*). Its primary purpose was
to publish tables of forthcoming solar, lunar and planetary
positions, and other tabular material of interest to observers
and navigators, but it soon began to include topical articles
on related subjects. The similar *Nautical Almanac and
Astronomical Ephemeris*, from London, started in 1767.
Before long, volumes of more permanent interest started to
appear elsewhere, as the public observatories established in
the late seventeenth and eighteenth centuries came to see it
as their duty to disseminate, in published form, the
observations that were their *raison d'être*. By the nineteenth century it was normal for an observatory to present
copies of its publications to other observatories of the
appropriate standing, and to receive their publications in
exchange.

A monthly journal, *Monatliche Correspondenz*, designed

for the rapid dissemination of scientific news in general and astronomy in particular, was founded in 1800 by Franz Xaver von Zach (1754–1832), astronomer to the Duke of Saxe-Coburg. The first rapid-publication journal dedicated exclusively to astronomy was *Astronomische Nachrichten*, begun in 1823 by the German astronomer and surveyor, Heinrich Christian Schumacher (1780–1850). Schumacher has been called 'The Postmaster General of Astronomy', who 'wrote to everybody and sent copies of everybody's letters to everybody else'. To ease the burden of his correspondence, he had it set in type by a local printer and circulated it to an international mailing list every few weeks. The journal, with the same title, has continued publication to this day, though telegrams and, more recently, electronic mail have taken over its most urgent duties.

The Royal Society of London had begun in 1660 as an amateur society catering for scientific (and quasi-scientific) interests of all kinds, and remained predominantly amateur until the mid-nineteenth century. The social status of a member, or 'Fellow', and his ability to pay the contributions, were as important as his knowledge of science. By the early years of the nineteenth century, the number of committed astronomers in England – academics such as the professors at Oxford and Cambridge, professional observers like the Astronomer Royal, dedicated amateurs like the Herschels – had grown to the point where there was a need for the type of forum that only a specifically astronomical society could provide. In the teeth of opposition from the then President of the Royal Society, who was fearful of the competition, the Astronomical Society of London was founded in 1820, receiving its royal charter and so becoming the Royal Astronomical Society eleven years later. The RAS held regular meetings at which the latest discoveries were presented and discussed, its premises housed a research library, and outstanding achievements were recognized by the award of the Society's medals. The Society began publishing its *Memoirs* within months of its foundation, to provide a vehicle for the dissemination of tables of observational results and material of similar calibre. There was also a need for a forum for the rapid publication of astronomical news, lesser research notes and the like, and for this the Society instituted its *Monthly Notices* in 1827.

Similar societies and journals, both national and local, began to spring up throughout Europe and in the New World. In America the *Astronomical Journal* was founded in 1849 by Benjamin Apthorp Gould (1824–96), who had taken his doctorate in astronomy the previous year under

C. F. Gauss at Göttingen and who had then returned home with the express intention of promoting astronomy in the USA. As the *Astronomical Journal* was not the organ of an established society, it was vulnerable to outside events, and publication was interrupted by the American Civil War; but Gould himself eventually revived it, after a gap of a quarter of a century.

On the continent of Europe, the Astronomische Gesellschaft was founded at Heidelberg in 1863. Although primarily German, its membership was explicitly open to other nationals, and the society took a lead in projects that were international in character. From 1881 it was involved in the publication of *Astronomische Nachrichten*, and in 1898 it helped in the foundation of the annual *Astronomischer Jahrsbericht*, which aimed to abstract every publication in astronomy. This invaluable aid to research (known as *Astronomy and Astrophysics Abstracts* since 1969, when the German publishers bowed to the dominance of the English language among astronomers) now publishes over 6,000 pages each year.

The corps of twenty-four 'celestial police' organized at Lilienthal in 1800 to search for a planet in the Mars–Jupiter gap (see page 157) was perhaps the first true example of international cooperation in astronomy. The second half of the nineteenth century was to see several more schemes to achieve results that were beyond the resources of any one institution. Between 1859 and 1862, F. W. A. Argelander (1799–1875) of Bonn published a three-volume catalogue of nearly one-third of a million stars, known as the *Bonner Durchmusterung*, and a forty-plate atlas. It was the work of a tiny group of Bonn observers, and although the *Durchmusterung* ('Survey') was to prove invaluable to future observers, the listed positions were necessarily imprecise. In 1867, therefore, Argelander proposed to the Astronomische Gesellschaft that a project be organized to measure – this time with great accuracy – the positions of the *Durchmusterung* stars down to the ninth magnitude. The task was eventually shared among sixteen observatories, all operating to a common plan.

Work however proceeded slowly. Not for the last time did some contributing institutions fail to honour their undertakings; and the project itself was to some extent overtaken by a new development, the application to astronomy of photography (see page 233). In 1885 this led the director of the Paris Observatory, Admiral E. B. Mouchez (1821–92), to suggest the possibility of a great photographic star chart of the whole sky. First responses from foreign col-

leagues were encouraging, and so the Paris Academy of
Sciences issued invitations to a meeting in Paris in April
1887. Fifty-six scientists from nineteen nations attended,
and a permanent commission was set up to promote a
Carte du Ciel – a photographic map of the sky (of stars to
the fourteenth magnitude) – and, from the measurement of
the photographs, a precision catalogue of stars to the
eleventh magnitude. In order to secure the necessary uniformity, many decisions were required, some of which
called for intensive research; and until the First World War
the Permanent International Committee of the *Carte du
Ciel* occupied a unique position in the international astronomical community. The work itself, however, ran into difficulties, as some observatories dropped out and others
produced sub-standard results, and publication of the catalogue was not completed until 1964.

In 1904 a major advance in international co-operation
came about with the formation, at the instigation of the
American solar physicist, George Ellery Hale, and with the
active support of the US Academy of Sciences, of the
International Union for Co-operation in Solar Research.
Experience of the benefits resulting from the work of this
organization and of the *Carte du Ciel* Committee in their
respective fields made astronomers receptive to the proposals for organized international cooperation, advanced by
scientists of many disciplines in the aftermath of the First
World War. In 1919 the International Astronomical Union
was formed, and came to play a central role in the organization of astronomy. It is organized into numerous 'commissions', each dedicated to a field of astronomy and
comprising the IAU members working in that field. These
commissions encourage collaboration in research, and adju-
dicate on standards and terminology. For example, the
constellations, which formerly merged vaguely into each
other, were given precisely defined boundaries in 1930.
Every three years there is a General Assembly of the IAU at
which astronomers of almost every nation meet to discuss
the latest research in their field, and as a complement to
these large gatherings, the IAU has developed the sponsor-
ship of frequent specialist symposia and colloquia.

The IAU can speak for astronomy on the international
scene, defending the interests of the science against (for
example) excessive pollution of the night skies by city
lights, or the cluttering of the outer atmosphere of the Earth
by unnecessary debris that interferes with research. After a
beginning marred by the exclusions of astronomers from
the defeated nations, and despite numerous problems in

securing the necessary finance from the adhering countries, the IAU is the focus of the organization of astronomy worldwide.

The sun and the origins of astrophysics, 1800–90

Until the publication in 1672 of Isaac Newton's 'New Theory about Light and Colours', it was thought that white light was simple and basic, and that colours were 'modifications' of white light – that something happened to white light to turn it into red light, or into blue light.

There were various theories as to how this happened. Curious to test them, Newton bought a prism in Cambridge, retired to his room in Trinity College, and set up an experiment. He closed the shutters, except for a circular hole; the sunlight that entered the hole passed through the prism, and the resulting spectrum fell on a screen that Newton had placed at a considerable distance.

There was nothing novel about the rainbow-like colours of a spectrum of sunlight; they were, as Newton put it, 'celebrated'. It was the shape of Newton's spectrum that drew his attention, for it was incompatible with the usual theory of modifications. After a series of further experiments, Newton concluded that white light, far from being simple, was in fact compounded from the colours; and he went on to demonstrate that his spectrum could be reassembled into white light by the use of a lens.

In 1802 the English chemist William Hyde Wollaston (1766–1828) repeated some of the Newton's work but in a more refined way: for Newton's round hole he substituted a narrow slit only one-twentieth of an inch wide. The spectrum of the light, now freed from the overlapping colours, showed him some seven lines that were more or less dark and which he regarded as natural boundaries between the colours.

That this simple explanation was inadequate became evident when the Bavarian instrument-maker, Joseph Fraunhofer (1787–1826), in trying to improve the design of lenses, developed the first simple 'spectroscope'. Sunlight was allowed to enter through an extremely narrow slit; it then passed through a prism, after which the spectrum was examined by Fraunhofer with the aid of a telescope focused on the slit. He was astonished to see, not Wollaston's seven lines, but many hundreds. He counted some 600 of them and drew a map, giving to the darkest and most prominent of them letters of the alphabet starting with A, B, C, . . . from the red end – letters that are still convenient to use

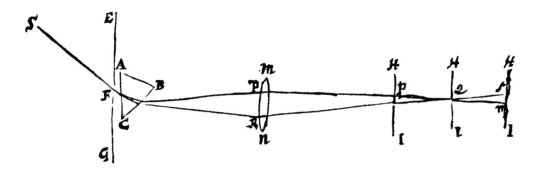

today. The spectrum of the Sun came to be known as 'the Fraunhofer spectrum'.

Fraunhofer was a practical optician, not a 'scientist'. He noted with interest that the D line in the yellow seemed to coincide exactly with a bright line that appeared in the light of many flames. That there was such a ubiquitous yellow light of very pure colour had been observed in the middle of the eighteenth century by a remarkable young Scotsman, Thomas Melvill, who died in 1753 at the early age of twenty-seven. Although Melvill's discovery was published, it appeared in an obscure collection of essays and went almost unnoticed. The discoveries of Fraunhofer, by contrast, aroused widespread interest.

In the many investigations that resulted, crucial experiments were carried out by, among others, the Heidelberg chemist Robert Bunsen (1811–99) and his colleague, the physicist Gustav Kirchhoff (1824–87). The first of their discoveries, announced in 1859, was that particular sets of lines are associated with individual chemical elements. William Miller (1817–80) of King's College London had already noted that the light emitted by electric arcs struck between two metal rods differed from metal to metal. By making many measurements in the laboratory, Bunsen and Kirchhoff identified lines with metals, and dramatically confirmed their results by identifying two new elements – caesium (a Latin word that means 'bluish grey') and rubidium (red) – with lines in those parts of the spectrum. A perplexing problem for the many investigators had been that even when quite pure specimens of metals were introduced into the flames, Fraunhofer's D line appeared in the spectrum. It only gradually became apparent – and the discovery was itself an important one – that the spectrum test was of great sensitivity, and that even the most minute traces of sodium, present for example in common salt, would produce that line. The great purity of the metals and

Newton's diagram showing how he combined the colours to make white light once more. Light from the Sun (left) enters the hole *F* and is separated into colours by the prism. The lens brings the colours together again. They fall on the screen *HI*, and if this is at *Q* they produce 'whiteness'. From *Philosophical Transactions*, **80**, 19 Feb. 1672.

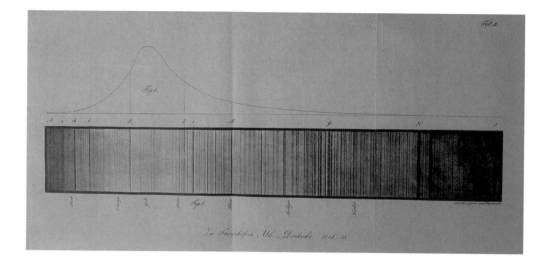

Part of a diagram of the solar spectrum (Fig. 5), published by Fraunhofer in 1817. He carefully mapped the positions of over 300 lines, and put in the rest by eye. Above it appears a curve indicating the brightness of the spectrum as judged by the eye (Fig. 6), peaking in the yellow-green. Courtesy of Institute of Astronomy, Cambridge University.

compounds coming from Bunsen's chemical laboratory was an essential feature of this work.

A further experiment at Heidelberg, carried out in 1859 by Kirchhoff in the physics laboratory, led to an understanding of how the lines were produced. He looked at a spectrum of the Sun through a yellow sodium flame, expecting the bright light of the flame to mask the dark line in the Sun: instead, it became even darker. A similar experiment, substituting an intense white incandescent lamp for the Sun as a light source, also showed a dark line. He inferred, and his interpretation was generally accepted, that Fraunhofer's dark D line arose because sodium vapour existed in a glowing atmosphere surrounding the Sun and absorbed light of that particular wavelength. Such lines became known as 'absorption lines'.

Here again Kirchhoff had been partly anticipated ten years earlier by very similar observations carried out by the French physicist J. B. L. Foucault (1819–68). Foucault, however, did not attempt a physical explanation of them at the time.

It followed that spectrum analysis could be applied to the Sun: the lines that were bright in the flames and arcs in the laboratory (that is, 'emission lines') were the same lines that appeared dark in the spectrum of the Sun – and Fraunhofer himself had been able to see a few of the most prominent lines, including the D line, in the spectrum of the very brightest stars. Earlier in the nineteenth century the French philosopher Auguste Comte in his *Cours de philosophie positive* (1835) had cited the chemical composition of celestial bodies as an example of things that were

inherently unknowable. Now it *was* knowable: several
metals had been identified in the Sun, and within a few
years the list was to be greatly extended.

Julius Plücker (1801–68) studied also the spectrum of rar-
ified natural gases contained in narrow glass tubes subjected to an electrical discharge. These glowing tubes had
been invented by his colleague, and skilled glass-worker,
Johann Geissler (1815–79) of Bonn University – the familar
red neon advertising sign is but a modern development.
Plücker was the first to propose that Fraunhofer's C and F
lines were due to hydrogen. A difficulty at the time was the
lack of any accurate and agreed scale of the wavelengths of
spectrum lines that would enable workers in different laboratories to correlate their observations. This was largely
resolved by the work of Anders J. Ångström (1814–74) of
Uppsala, whose monumental *Recherches sur le spectre
normal du Soleil* of 1868 provided much improved wavelengths for some one thousand Fraunhofer lines. Although
these were still further refined in later years, the Ångstrom
Unit (1 Å = 10^{-10} metre) became universally accepted for the
designation of the wavelengths of lines. For example, his
value for the red C line of hydrogen was 6561.8 Å, differing
from the modern determination by only one part in seven
thousand.

This assimilation, of the spectroscopy of the Sun to spectroscopy in the laboratory, between about 1850 and 1870,
was one of the great achievements of astrophysics. By 1891,
thirty-six elements, mostly metals, had been identified in
the Sun.

Heinrich Schwabe and the discovery of the sunspot cycle, 1843

We have seen in Chapter 5 that spots on the Sun (observed,
rather dangerously, on the heavily reddened disc at sunset)
were among the first objects studied with the newly
invented telescope, and surprisingly early in the seventeenth century a number of basic facts about sunspots
became established. They were not objects in interplanetary
space but attached to the Sun in some way – being either
dark clouds in the Sun's atmosphere, or rafts of cinders
floating on its incandescent surface, or perhaps the dark
surface of the Sun seen through gaps in a luminous atmosphere that was the source of its light. The Sun turned about
an axis fixed in direction in space, with a rotation period
variously estimated to be within a day or two of twenty-six
days, and the spots rarely appeared more than 30° north or
south of the Sun's equator.

But sunspots were sporadic phenomena from which not much more could be learned; indeed already by 1618 they seemed to have become rather infrequent. Thereafter, except for occasional renewed interest during the eighteenth century when particularly large or numerous spots appeared, little systematic study of them was made.

A change was brought about by Heinrich Schwabe (1789–1875) of Dessau, a small town to the southwest of Berlin. He became interested in astronomy at Berlin University, while studying pharmacy as a preparation for joining the family business. On returning to Dessau he bought a small telescope from Munich in 1826, and with advice from his friend K. L. Harding of the Göttingen Observatory began regular observation of the Sun whenever weather and time permitted. His interest was to look for a possible undiscovered planet between Mercury and the Sun. Eventually such a body must surely pass over the disc of the Sun as a small black dot, but only constant observation would trap it, and only regular recording would distinguish it from a small sunspot. He was able to dispose of his business interests in 1829 and devote the rest of his life to astronomy and other scientific studies.

With commendable patience Schwabe observed and recorded. After seventeen years, he had still discovered no new planet, but he was able to announce that sunspots seemed to come and go in a fairly systematic way, the disc being more covered with spots at about ten-year intervals, with corresponding minima in between. His modest note in *Astronomische Nachrichten* in 1843 attracted relatively little attention, but he continued collecting observations, and in 1851 the famous German traveller and natural historian, Alexander von Humboldt (1769–1859), drew attention to their significance in his *Kosmos*. In 1857 the amateur of Dessau was awarded the Gold Medal of the Royal Astronomical Society, then as now an accolade in the astronomical world, and in later years he presented to the Society his meticulous records covering forty-two years of observing. In the long term the average length of the sunspot cycle is in fact a little over eleven years, but the interval is not exact and was nearer to ten years when Schwabe made his early observations.

Sunspots and the Earth's magnetic field
To understand the importance Humboldt attached to Schwabe's discovery we must retrace our steps to the time when Schwabe was starting his observations, and to

natural events that at first seemed to have nothing to do with astronomy at all. The north-seeking property of the needle of the magnetic compass had been used by seamen for centuries. But the needle points to a magnetic pole that is not the same as the north geographic pole, and further-more the magnetic pole moves slowly over the Earth's surface. Nor are these the only complications: as early as 1722 it was known that there is a smaller but daily varia-tion.

In 1828 Humboldt set in train a great international study of terrestrial magnetism, based at the Observatory in Göttingen University under the direction of C. F. Gauss, and within a few years a world-wide net of magnetic obser-vatories was in operation. By 1851 John Lamont (1805–79), the Scots director of the Munich Observatory, discovered in the magnetic observations between 1835 and 1850 distinct evidence for a variation of the magnetic constants with a period of about 10⅓ years. Quite independently, and using different data, Sir Edward Sabine (1788–1883), a British army officer and future President of the Royal Society, found that the occasional sudden and more violent vibra-tions of the needle (which Humboldt had called magnetic storms) were more violent and more frequent at intervals of about ten years.

It was Sabine who announced in 1852 that between the sunspot cycle and these magnetic activities there was a remarkable coincidence, which he had found by patient deduction from many hundreds of thousands of observa-tions collected for more than twenty years. His discovery elevated the study of the Sun from curiosity to practical utility, for if the Sun could influence a magnetic needle, what else might it not influence? The weather? The ability to predict famines from crop failure in India?

The existence of solar–terrestrial relationships was dramatically confirmed by two independent observations of the Sun made by the English amateur astronomers R. C. Carrington (1826–75) and Richard Hodgson (1804–72) one morning in September 1859. An outburst of two bright patches of white light near a large spot group, lasting for five minutes, was so remarkable that both noted its appear-ance carefully. Simultaneously there was a violent mag-netic storm, telegraphic communication was interrupted, and that night one of the most magnificent auroral displays ever observed danced across the skies.

With the spectroscopic discoveries of the same year, solar physics had arrived. The physical explanation of how visible causes on the Sun resulted in effects on the Earth –

that solar disturbances emitted X-rays, extreme ultraviolet
radiation and interplanetary streams of charged fundamental particles – was to be unravelled only a century later.

Expeditions to observe eclipses of the Sun
The explanation that Kirchhoff had given of the dark
absorption lines in the solar spectrum required the existence of some atmosphere, presumably faint and tenuous,
of hot gases surrounding the Sun. The knowledge of physics
at the time, however, made it very difficult to understand
the nature of the Sun's surface. In the laboratory, the
continuous spectrum of white light was produced only by
white-hot solid bodies or by liquids like molten metals, and
the white light of the Sun presumably came from such a
surface. Yet many of the things seen on the Sun's disc in
white light – sunspots, the rather brighter areas surrounding
them named faculae ('little torches'), and the fine details
seen within the sunspots themselves – seemed to be of an
atmospheric nature.

The resolution of these difficulties was to come from
observations made during total eclipses of the Sun. By a
remarkable coincidence in nature, the Moon and the Sun –
the one very much smaller than the Earth, and the other
very much larger – are at such distances from the Earth that
their apparent sizes in the sky are very nearly the same.
Because the orbits of the Moon (about the Earth) and of the
Earth (about the Sun) are both elliptical, these apparent
sizes change with time, and under favourable circumstances the Moon's disc is sufficiently larger than that of
the Sun to eclipse the Sun for as much as seven minutes or
so. As a total eclipse approaches the sky is darkened, and, at
the onset of totality, the Moon slowly obscures the outer
layers of the Sun seen at the edge of its disc, only to reveal
them again progressively at the end of totality. This most
beautiful and awe-inspiring event is rare at any one place on
the Earth's surface, but every few years the narrow zone tra-
versed by the Moon's shadow crosses some reasonably
accessible part of the Earth.

By accidents of history, the latter half of the nineteenth
century saw a number of favourable eclipses, and travel was
then becoming easier and cheaper. Accordingly, it became
usual for astronomers to set up temporary observatories,
with increasingly elaborate instruments, along the line of
the eclipse track, to study in scientific detail what had previously been regarded as little more than a spectacle.

The eclipse of 8 July 1842 brought astronomers from
several European countries to sites in central and southern

Europe. Most had never seen a total eclipse before. They were all astonished at the extent and brightness of the white 'corona' surrounding the eclipsed Sun, and noted with even more surprise three small protuberances (later to be termed 'prominences') looking like mountains projecting from the edge of the Sun or the Moon, and of a striking colour generally described as red tinged with lilac.

Later favourable eclipses were to follow in Sweden and Norway in 1851, North America and Spain in 1860 and India and Malaya in 1868, to mention only the earlier and more significant ones.

Each eclipse brought important discoveries. It gradually emerged that all the significant phenomena arose in the various layers of the Sun's atmosphere (and none, as was once suspected, in a supposed atmosphere of the Moon). The red flames were seen to be sporadic extensions of a lower layer which surrounded the whole Sun, named the 'chromosphere' in 1868 by the English amateur (as he then was), Norman Lockyer (1836–1920); and the spectroscope showed that its lilac-red colour arose from several bright emission lines, especially those of hydrogen in the red (Fraunhofer's C line) and the blue. At the 1870 eclipse the American C. A. Young (1834–1908) of Dartmouth College made the crucial spectroscopic discovery of an even lower and narrower 'reversing layer': in the last few seconds as the Sun's edge disappeared, the dark lines and the continuous spectrum also disappeared, and for the next two or three seconds hundreds of bright lines flashed out, apparently in the places in the spectrum where the dark Fraunhofer lines had been.

Such observations gradually built up a quite complicated picture of the structure of the Sun's outer layers. But puzzles still remained. A curious one was that with good telescopes under good observing conditions, the edge of the Sun at high magnification is always seen to be perfectly sharp – one might have expected it to fade out rather fuzzily into space. The true explanation of the nature of the 'photosphere' – the apparently thin boundary from which the white light of the Sun was emitted – would depend on a quite detailed understanding of the atomic processes that occur in the outer layers of stars generally, and this was not arrived at until well into the twentieth century. Briefly, the transition from the gaseous layer that is 'foggy' to the layer above that is clear is so abrupt that the edge of the Sun looks solid – rather as does the edge of a distant sunlit thunder cloud.

Two near-contemporaries in the early years of solar

physics deserve particular mention. J. Norman Lockyer was born in 1836 in Rugby, England, and became a clerk in the War Office in 1857. He began to devote his spare time to spectroscopy and eventually his rise to scientific prominence despite his amateur status led to government appointment in 1885 as the director of the Solar Physics Observatory in South Kensington, London, a position he held until 1911. He was indefatigable in soliciting support for this innovative institution and made important contributions to stellar as well as solar physics. In 1869 he founded the still flourishing weekly journal *Nature*, and edited it for half a century, retiring only the year before his death in 1920. He was knighted in 1897.

The rather older P. J. C. (Jules) Janssen (1824–1907) was born in Paris and later studied science at that university. Kirchhoff's discoveries directed his interests to spectroscopy and astrophysics. His notable work led the French government to establish in 1876 a new observatory for physical astronomy at a favourable site in Meudon, in the southwestern suburbs of the capital, and this he directed until his death. His life was a very remarkable one. Despite lameness following a childhood accident, he travelled extensively on scientific expeditions (he escaped from the siege of Paris in a balloon to travel to the 1870 eclipse of the Sun) and also established high altitude spectroscopic observatories on Mt Blanc.

At the 1868 eclipse in India, Janssen had been so impressed by the brilliance of the bright lines in the spectrum of the prominences that he was struck by the idea that, by use of a spectroscope of high dispersing power (which would spread out the continuous spectrum of the photosphere but not the monochromatic bright lines), it should be possible to see them without an eclipse at all. The same idea had also occurred, independently, to Lockyer. The two papers by Lockyer and Janssen describing the discovery were read at the same meeting of the Paris Academy of Sciences, which commissioned from the French mint a fine commemorative medal bearing their two portraits. Tests showed their idea to be well founded, and from then on it was possible to study the chromosphere and prominences from day to day. Soon such work was under way at a number of solar physics observatories.

The resulting ability to make accurate measurements at leisure, rather than in the hectic minutes of an eclipse, led to the discovery that the brilliant orange line which had been assumed to be the D line of sodium had in fact a rather shorter wavelength and could not be identified in any laboratory spectrum: it was attributed to an unknown element, which was given the name 'helium', from the Greek word for the Sun. The element was eventually isolated in the laboratory in 1895 by the Scots chemist William Ramsay (1852–1916), as a gas extracted from radioactive minerals where it is formed as a decay product.

Similar unidentified bright lines in the spectrum of the corona were to prove less tractable: they were attributed to an unidentified 'coronium', but did not yield to explanation until 1941. It was then that the Swedish astronomer Bengt Edlén (1906–   ) showed by theory that they are in fact due to iron atoms, but in extremes of high temperature and low density that are not reproducible in the laboratory. Indeed the nature of the corona remained an enigma for a similar length of time: its light is now known to be sunlight scattered partly by a true atmosphere of free electrons round the Sun, and partly by meteoritic dust particles in the space between the Sun and the Earth.

Photography enters astronomy

Some of the improved understanding about the nature of the Sun came about by the gradual addition of photography to the tools of research available to astronomers. Indeed the very brightness of the Sun made it an attractive subject for the very slow and inefficient early photographic processes. The first daguerreotype of the solar disc was made by Foucault and his fellow French physicist A.-H.-L. Fizeau (1819–96) in Paris in 1845. The faster collodion process, invented by the English sculptor F. S. Archer (1813–57) in 1850, was used by his fellow-countryman, the photographer and amateur astronomer Warren De la Rue (1815–89), to photograph the sunspots on the solar disc with a specially constructed camera at the Kew Observatory, near London. This he did in 1858, only a few years after Humboldt's publicity of Schwabe's discovery of the sunspot cycle. The 'autographic' recording of the daily state of the solar disc started there almost immediately, and was continued until 1872, when it was further extended by large programmes of almost continuous recording on a world-wise basis, not only of the spots on the photosphere but also of the outer chromosphere over the disc and at the edge of it.

The greatly improved dry gelatine plate became available in the late 1870s. It was sensitive enough to record the flash spectrum of the Sun, so enabling the lines to be measured later in the laboratory. More importantly, these plates were at last sensitive enough for them to be applied to telescopes

A 1907 photographic
montage showing a
spectroheliogram of the
solar disc combined with
one of the prominences at
the edge of the disc.
Courtesy of Institute of
Astronomy, Cambridge
University.

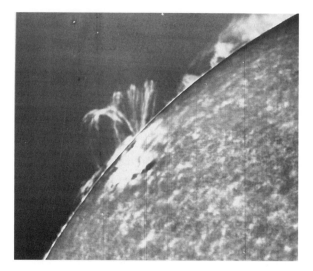

at night, to photograph the very much fainter stars and
nebulae.

The source of the energy of the Sun

Despite the improvement in knowledge of the workings of
the outer layers of the Sun, a more fundamental question
remained only partly answered: what was the source of the
energy that the Sun poured out so prodigiously in its heat
and light? By the middle of the nineteenth century, partly
as a result of the studies of physicists and partly from the
interest shown by engineers in the efficiency of steam
engines, there was a fairly good understanding of the flow of
heat and how to measure it. The German physician, Julius
R. Mayer (1814–78) of Heilbronn, asked himself such questions about the Sun, and he calculated that even supposing
the Sun to be made of coal and supplied with unlimited
oxygen to burn it, the solar furnace could last for only a few
thousands of years. He suggested another source of energy:
the continuing infall onto the Sun of solid particles from
interplanetary space, which would give up their energy on
colliding with it.

The Irish-born physicist William Thomson (1824–1907),
later Lord Kelvin, satisfied himself that this also was unten-
able, for even if sufficient material were available it would
so alter the mass of the Sun that its greater gravitational
pull would shorten the length of the year by weeks within a
few thousand years, which clearly was not happening. An
uneasy solution, of a more tenable nature, was worked out
by the German physicist Hermann von Helmholtz

(1821–94) in 1854 and was also studied by Thomson: the
mass of the Sun was not changing, but the star was slowly
contracting; the resulting change in the size of the Sun
would be too slow to be measurable, and the output of
energy could be maintained at least for several millions of
years. But even this timescale was inadequate to meet the
requirements of the geological record, and the problem
remained unsolved until the discovery of sources of nuclear
energy in the twentieth century.

Developments in telescope making

The principal business of public – that is to say, state and
university – observatories in the early nineteenth century
continued to be the determination of the exact positions of
the stars in the sky, to provide a framework against which
the apparent motions of the Sun, Moon and planets could
be measured. The instruments used (see page 129) were
'transit circles', refracting telescopes of modest size
mounted on massive masonry piers and fitted with accurately graduated circles to make the necessary angular
measurements. But the interests of amateur astronomers
like William and John Herschel, Lord Rosse, and William
Lassell (1799–1880) called for larger telescopes than the professional telescope makers could supply, and, as we have
seen, they developed the reflecting telescope, using mirrors
of speculum metal (an alloy of copper and tin) shaped in
their own workshops. These were, however, temperamental
and difficult instruments: unlike glass lenses, the mirrors
tarnished easily, and to maintain them the astronomer
needed to be skilled in the grinding and polishing of optical
surfaces.

Fortunately a solution offered itself. Under pressure from
the opticians and seeing some profitable new business, glass
manufacturers (particularly in Switzerland and later in
France) began to experiment with the fabrication of larger
discs of good quality optical glass for telescopes, and the
opticians in turn developed improved ways of figuring the
perfectly spherical surfaces of the lenses. As we have seen
(on page 186), before Fraunhofer's death in 1826 his workshops had completed for Dorpat Observatory a very fine
refractor of 24 centimetres (9½-inches) aperture. From this
time onwards, many of the growing number of newly constructed observatories in Europe, and soon in America,
began to include in their shopping list of desirable instruments as large a refracting telescope as their endowment or
the generosity of their benefactors could afford. Sizes crept

up a few inches at a time, from the 15-inch refractor by
Merz & Mahler (Fraunhofer's successors) for the Pulkovo
Observatory near St Petersburg in Russia and a twin for the
recently founded Harvard College Observatory in 1847, to a
culmination in the 40-inch diameter lens for the telescope
of the Yerkes Observatory of the University of Chicago in
1897. Many of these great telescopes still exist, now no
longer suited to most of the needs of modern astronomy but
preserved as superb examples of the instrument-maker's
craft.

These great refractors were not to be without their problems when photography began to enter into stellar and
planetary astronomy and into solar physics, with the rapid
development of the dry plate process from the 1870s. The
big doublet lenses (see page 128), although called achromatic ('without colour'), did not bring the light of all wavelengths to the same focus. They were designed for visual
use and gave sharp images in the yellow-green part of the
spectrum. But each star was seen surrounded by a coloured
halo of out-of-focus red and blue light, the blue image to
which the photographic emulsion was sensitive was less
sharp and not at the same distance from the lens, and there
was no position in which all the light from a star could be
imaged to pass through the narrow slit of a spectrograph.

A discovery in chemistry presented a way out of the
problems. In about 1853 the German chemist Justus von
Liebig (1803–73) developed a process in which a very thin
layer of metallic silver could be deposited on a clean glass
surface from an aqueous solution containing dissolved
silver nitrate. This was used by K. A. Steinheil in Munich,
and by Jean Foucault in Paris, to manufacture the first
reflecting telescopes in which the mirrors were made, not
of the customary speculum metal, but of glass with silver
deposited on the front, optical surface.

The invention was at first viewed with suspicion: the
silver film was fragile and also tarnished quickly, but its
reflectivity was greater and the chemical process did not
affect the optical figure of the mirror. The glass was easier
to shape than speculum metal, it did not need to be of high
quality since the light did not pass through it, and there
was only one optical surface to be figured, as against at least
four in an achromatic lens.

The last large telescope with a speculum metal mirror
was a 48-inch made by the Dublin firm of Thomas Grubb
in 1862 for the Melbourne Observatory. It proved difficult
to keep in good order, and thereafter the silver-on-glass
reflector became the preferred instrument for astronomical

The 30-inch refractor installed at Pulkovo Observatory in 1885. The mounting was by Repsold & Sons of Hamburg, with optics by Alvan Clark & Sons of Cambridge, Massachusetts. The telescope was briefly the largest refractor in the world. Courtesy of Institute of Astronomy, Cambridge University.

photography and spectroscopy. The 36-inch Crossley reflector, donated by an English amateur to the Lick Observatory in California in 1895, first demonstrated the power of reflectors for the photography of nebulae, and the 60-inch reflector (commissioned in 1909) and the 100-inch (1917) of the Mount Wilson Observatory near Los Angeles were to establish the superiority of this form of telescope for the astrophysics of the twentieth century.

An important figure in these developments was George Ellery Hale, whose roles in the International Union for Solar Research and in the International Astronomical Union have already been discussed. The son of a wealthy Chicago business man, he was born in 1868 and studied at the Massachusetts Institute of Technology. As a young man he further improved the 'artificial solar eclipse' method of Janssen and Lockyer by the invention of the spectrohelioscope, and while an associate professor of astrophysics at Chicago he founded the *Astrophysical Journal*, by common consent still the leading journal in the subject. A persuasive man with influential connections, he was successful in obtaining major support from magnates like C. T. Yerkes

The eye end of the Pulkovo 30-inch refractor. The complex array of controls for setting, focusing, adjusting, and recording are brought within reach of the observer. With such instruments, the management of the telescope reached a new level of sophistication. Courtesy of Institute of Astronomy, Cambridge University.

and J. D. Hooker, and from the Carnegie and Rockefeller
Foundations, to finance what were in turn the four largest
telescopes in the world – the 40-inch refractor, the 60- and
100-inch reflectors, and finally the 200-inch reflector on
Palomar Mountain, which was completed after the death of
Hale in 1938. Arguably he did more for the promotion of
astronomy than any other person in history.

Solar system studies

As well as being suited for the exact measurement of
double stars (page 188), the large refractors of the later nineteenth century were well adapted to the study of the physical nature of objects in the solar system. It chanced that
this period witnessed the appearance of a larger than
average number of spectacularly bright comets.

Comets: heads and tails

The nature of comets had always been enigmatical; now
there were both comets and the telescopes to observe them.
The 'great' comets, strikingly visible even to the casual
observer and dominating the night sky for days or even
weeks together, are nearly always those with orbital periods
about the Sun of thousands of years; there are therefore no
records of their previous appearances, and their re-appear-
ances are unpredictable. There had been such comets in
1811 and 1843, but these were at least equalled by the
comet first seen in the telescope by the Florence
astronomer Giovan Battista Donati (1826–73) early in June
1858. As it approached the Sun in the following weeks it
brightened to naked-eye visibility in September; by early
October the head was brighter than the nearby star
Arcturus, and the tail curved across some 30° of the sky. It
was well placed for the observatories of the northern hemisphere and was minutely observed (see figures below).

As in other bright comets, the comet's tail and head both
had complicated structures which changed from day to day
and even from hour to hour. There was a straight tail,
streaming away from the direction of the Sun, and a series
of more or less curved tails. H. W. M. Olbers had suggested
in 1812 that these shapes in the head and the tail were a
consequence of different forms of material being ejected
from the head of the comet as it approached the Sun and
being swept away from the Sun along different paths that
depended on the speed of ejection, the path of the comet in
space, and the forces acting on the particles.

F. A. Bredikhin (1831–1904), director of the Moscow and

A drawing of the head of Donati's Comet of 1858, made by G. P. Bond of Harvard on 29 September. The comet was particularly well placed in the sky for the observatories of Europe and America. It still had to be observed visually, but telescopes of the calibre of the 15-inch refractors of Pulkovo and Harvard were available. Harvard produced in 1862 a volume of 362 pages devoted to the comet, with many accurate engravings. These skilful drawings are still of value in understanding the interaction between the material of the comet, the solar radiation, and the interplanetary material. Courtesy of Institute of Astronomy, Cambridge University.

A lithograph of Donati's Comet as seen over the bridges of Paris. Intended as a work of art rather than of science, this frontispiece to A. Guillemin's 1875 book, *Les Comètes*, remains one of the finest representations of a 'great' comet as seen by the naked eye. It is scientifically accurate: the bright star near the head of the comet is Arcturus, and the several components of the tail are correctly drawn. Courtesy of Institute of Astronomy, Cambridge University.

later the Pulkovo observatories, collected drawings of the brighter comets and developed a classification of the shapes of their tails that is still in use today, and his essentially correct theory was hampered only by lack of knowledge of the forces involved. In fact the detailed working out of these productive ideas was to take a long time. The main repulsive forces acting contrary to the solar gravitational attraction are the pressure of radiation on small solid particles, not measured in the laboratory until 1901, and the effect of the 'solar wind' of ionized particles emitted by the solar corona, the existence of which came to be recognized by solar physicists in the 1950s.

The heads of these bright comets were bright enough to enable the light to be examined with the spectroscope. This was first done by Donati on 5 August 1864, when he examined a comet that was then of the second magnitude. It had been thought that comets shone by reflected sunlight and would therefore have the same Fraunhofer spectrum as the Sun: Donati made the important discovery that there were three bright bands separated by wider dark spaces, suggesting that some of the light came from glowing gas of unknown nature in the comet itself. The bright comets of following years and the increasing use of the spectroscope both at the telescope and in the laboratory added considerably to knowledge about the nature of comets: by 1880 it was generally accepted that the light of the head of a comet came from reflected sunlight in part, but was dominated by broad emission bands (as distinct from narrow bright lines) which could be simulated in the laboratory by electrical discharges in tubes containing mixtures of carbon dioxide and simple hydrocarbons like methane and ethylene. The organic chemistry of the time was inadequate to understand exactly how the light arose, and only in the twentieth century was it understood that the bands arose from simple combinations of a few atoms, variously of carbon, oxygen, nitrogen and hydrogen, that are not stable under terrestrial conditions. It was sufficient to cause alarm, however, in the public press when it was learned that the Earth would pass through the tail of Halley's Comet in 1910, and that the tail presumably contained cyanogen, the poisonous compound of carbon and nitrogen.

Shooting stars

Through most of the history of Western science, the Greek word *meteor* meant, generally, anything that happened in the atmosphere – clouds, lightning, whirlwinds – and hence the name of the science of meteorology. Shooting stars, specifically called meteors since about 1840, and the still brighter fireballs, were also regarded as phenomena of the atmosphere and thought to be of terrestrial origin. That the bits of matter that caused them came with high velocity from outer space was so novel a concept that it impinged only slowly even on the nineteenth-century mind. Two telling events were the falls of meteorites at Agen in France in 1790, and near Weston, Connecticut, after a brilliant fireball over New England in the early hours of a morning in 1807; both were well-attested. Ernst Chladni (1756–1827), better known for his work on acoustics, thought that the likely source of the material was debris left in interplanetary space after the planets were formed. The essential correctness of this idea was to be accepted only very slowly.

Interest in meteors was stimulated by the remarkable meteor storm observed in the early hours of 13 November 1833 over the eastern states of America, when the meteors were so numerous as to defy counting. Denison Olmsted (1791–1859) of Yale collected and analysed the many accounts and saw the importance of an observation made by several observers, that the apparent paths of the meteors across the sky could be traced back to a point close to the stars in the head of the constellation Leo. He correctly deduced that in space the meteors were travelling in parallel paths from a source far above the atmosphere. It was recollected that the naturalist Alexander von Humboldt and others had seen a very similar display from South America at the same time in the month in November 1799.

Only after Olmsted's study did other astronomers begin to take meteors seriously. The first person to recognize that many bright meteors seen in early August 1834 also had a radiant point, this one near the star Algol in Perseus, seems to have been John Locke (1792–1856) of Ohio, and in 1836 Adolphe Quetelet, the Brussels mathematician, hypothesized that this shower was annual in the days about 9 August. Later studies traced records of these August meteors in occasional years back to 1781. After a considerable interval yet another annual shower, in April, from a radiant point in Lyra, was recognized.

Hubert A. Newton (1830–96) of Yale collected historical data about the November meteors and correctly predicted the probable recurrence of the major storm of 1833 in November 1866; in consequence this was widely observed, this time in western Europe.

Several astronomers recognized that the position of the radiant point, combined with the date in the year of the maximum activity, provided precise information about the

orbit in the solar system that the particles were following:
John Couch Adams in Cambridge followed up the work of
H. A. Newton and calculated a unique orbit for the
November meteors, and others did the same for the August
Perseids and the April Lyrids. These confirmed a prescient
recognition by Giovanni Virginio Schiaparelli (1835–1910)
of Milan that the paths in space of meteoric particles were
much like those of comets, and within a short time the
identity was recognised of the orbits of the Perseids and the
bright comet of 1862, of the Leonids with Tempel's Comet
of 1866, and of the Lyrids with the comet of 1861.

The exact physical relationship between the numerous
minuscule members of the solar system – the comets and
asteroids, the meteors, and the larger meteorites which
crash through the atmosphere to land on the Earth's surface
– long defied interpretation. It became recognized gradually
that meteorite falls did not, as might have been expected,
occur during great swarms of shooting stars and that meteorites were perhaps more closely related to the asteroids.
But the studies of the latter half of the nineteenth century
were important foundations for later ideas about the history
of the solar system itself.

Moons of Neptune, Uranus, and Mars
The increasingly powerful telescopes, sited, in the later
years of the century, in better observing locations, added
gradually to the number of known members of the solar
system. The first satellite of the newly discovered Neptune
was found by the Liverpool amateur William Lassell only
seventeen days after the first observation of the planet; once
Triton's orbit about Neptune had been determined, this
gave a value for the mass, and hence the density, of the new
planet. Shortly afterwards, in 1851, Lassell added two satellites, Ariel and Umbriel, to the planet Uranus; these were
interior to the brighter Oberon and Titania that had been
discovered, as was the planet itself, by William Herschel.

In 1877 Mars was closer to the Earth than usual. Until
then it had had no known satellites; but Asaph Hall
(1829–1907) searched with the 26-inch refractor of the
Naval Observatory in Washington, DC, and discovered two
faint satellites (Deimos and Phobos) close to the planet, the
inner indeed so close that its orbital period is shorter than
the rotation period of Mars itself. This was a unique circumstance, and the newly discovered satellites not only provided an accurate value for the mass of the planet, but were
recognized as being more alike in size to the smaller of the
asteroids than were the other, larger, planetary satellites.

Martian 'canals'

Mars itself was also attracting interest. As early as 1783 William Herschel had commented that it was the most Earth-like of the planets, and by the mid-nineteenth century it was generally accepted that Mars – smaller than Earth but with a very similar rotation period and an axis of rotation that was similarly inclined to the plane of the solar system – had white polar caps that changed with the Martian seasons, and more or less permanent surface markings, some of which could be identified on the first telescopic drawings of the disc made in the seventeenth century.

Attention to these markings began to be paid by both professional and amateur astronomers from about 1870 onwards. Both good telescopes and a steady atmosphere were necessary; and also patience, for in some years Mars is not well placed for observation, and the closest approaches to Earth occur only at fifteen-year intervals. Even then the disc is less than one-seventieth of the apparent diameter of the full moon.

Although the newly-invented dry photographic plate became available at about the same time, it proved of limited use in planetary photography. Because of turbulence in the air even at good mountain sites, minute details at the limit of the telescope optics were blurred during even a short exposure. The human eye and artistic skill remained the best way of depicting planetary surface markings, and this continued so until the era of interplanetary probes and space telescopes a century later.

We have already met G. V. Schiaparelli in connection with his work on comets and meteors. He had graduated from Turin, and after further training at the Berlin and Pulkovo observatories had returned to his native Italy to join the staff of the Brera Observatory, Milan, in 1860. After 1877, when he was provided with a good Merz refractor, he turned to planetary observations, especially of Mars. He embarked on a mapping of the surface at the close approach in 1877, and he noted that the dark markings (supposedly oceans) were the sources of many fine, barely perceptible, dark lines, running across the brighter, orange (continental?) areas. He named them *canali* ('channels'), claiming for them no more than a natural, topographical character.

He unwittingly launched planetary astronomy on a most curious controversy, that was to be resolved only a century later. From the start, some astronomers, with larger telescopes than Schiaparelli's, could not see the channels, while others could. The word *canali* was erroneously translated

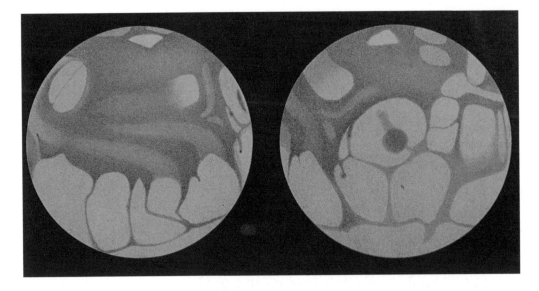

Maps of two aspects of the surface of Mars, drawn by G. V. Schiaparelli in September 1877, when the planet was close to the Earth. From the dark 'oceans', natural *canali* ('channels') supposedly ran across orange areas. Courtesy of Institute of Astronomy, Cambridge University.

into English as 'canals', implying an artificial construction by intelligent beings, with the corollary that there was intelligent life on Mars.

The controversy attracted the attention of Percival Lowell (1855–1916) of Boston, a member of a powerful and wealthy New England family. After graduating from Harvard in 1876, Lowell devoted himself to business affairs and travel, but in the 1890s his interests turned increasingly to astronomy. In 1894, with another favourable opposition of Mars approaching, Lowell abandoned most of his business operations and with ample private means at his disposal selected a high-altitude observing site near Flagstaff, Arizona, and installed telescopes and a residence there, later employing privately engaged staff. He soon convinced himself that there were canals made by intelligent beings, and his books *Mars and its Canals* (1906) and *Mars as the Abode of Life* (1908) attracted wide attention from the public, though at best only cautious acceptance from most of the astronomers accustomed to using large telescopes.

It took seventy years, until the series of Mariner spacecraft close approaches to Mars between 1965 and 1969, to show that Lowell was wrong: there are no artificial canals. But in a broader sense his work had great influence: his observatory made other discoveries that demonstrated the importance of good equipment placed at carefully selected observing sites; it kept alive an interest in planetary studies (leading, for example, to the discovery there of the planet Pluto); and as we shall see later, it made important contributions to observational cosmology in the twentieth century.

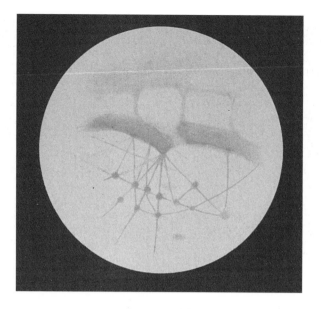

Mars as drawn by Percival Lowell in 1894/95, when Mars was in opposition to the Sun and favourably placed for observation. Lowell believed that intelligent beings on Mars, faced with a water shortage, would have irrigated the planet. He accordingly drew artificial canals, rather than the natural *canali* of Schiaparelli. Courtesy of Institute of Astronomy, Cambridge University.

The outer solar system

After the mass of the planet Neptune had been determined in 1847 using the orbit of its satellite Triton, it was clear that the zone of the asteroids divided the solar system into two: the four inner Earth-like 'terrestrial' planets, of similar size and mean densities 4 or 5 (where the density of water is 1), and the four outer giant planets of much greater size but much smaller mean density, that of the body of Saturn being even less than that of water. That much of the exterior parts of these outer planets must be gaseous seemed likely, and was confirmed by telescopic observations of Jupiter, whose changing dark and light belts parallel to its equator, and varying light and dark spots, were evidently cloud structures. Painstaking observations by Julius Schmidt (1825–84) at the Athens Observatory and others from about 1880 onwards revealed that although the rotation period of the planet was very close to the $9^h 56^m$ first determined by J. D. Cassini in the seventeenth century, various types of markings gave rotations periods differing by a few minutes, with a generally slower rotation period near the poles of the planet.

James E. Keeler (1857–1900), who was to become director of the Lick Observatory in 1898, used the new 36-inch refractor in 1889 to make many drawings of the planet, but remarked with frustration on the 'wealth of detail which could not be accurately represented'.

Although showing much less detail on its disc, Saturn

A drawing of Saturn, made in 1872 at Harvard College Observatory. It is one of a series of the best such planetary drawings, published by Harvard in 1877. Courtesy of Institute of Astronomy, Cambridge University.

was unique in having a system of rings in its equatorial plane; that Uranus and Neptune had similar but very much more tenuous rings remained unknown until the Voyager spacecraft missions of the late 1980s. The physical nature of Saturn's rings had puzzled astronomers ever since their recognition in the mid-seventeenth century. As a visual spectacle arguably the most remarkable object in the sky, Saturn was a natural object of study for each of the new telescopes that came into operation, as observers used their increased optical power in the search for elusive detail. In 1850, William Cranch Bond (1789–1859) of Harvard, with the new 15-inch refractor, discovered the faint inner 'crepe' ring, seen independently a few days later by the Rev. W. R. Dawes in England. George Phillips Bond (1825–65), the son of William, suspected that the features of the system of rings changed slowly, and thought the rings must be fluid, not solid.

An important advance was made in 1857 by the young Scottish mathematician, James Clerk Maxwell (1831–79). The problem of Saturn's rings had been set as a prize essay at Cambridge, and Maxwell showed convincingly in his essay that neither solid nor fluid rings could exist, and that the rings were made up of millions of small particles, each pursuing its individual orbit in the plane of the planet's equator. His deductions from the mathematical theory accounted for most of the known observational features and was immediately accepted – though the crucial observation that the outer parts of the ring were moving more slowly than the inner, as the theory required, was made only in 1895, by Keeler when working at the Allegheny Observatory, in Pittsburgh.

Several astronomers speculated in the early twentieth century that there might be other major planets beyond Neptune; there were still very small discrepancies in the motions of Uranus and Neptune that might be caused by an unknown 'Planet X', and other indications from families of comet orbits that such a body or bodies might exist. Percival Lowell embarked upon a perturbation analysis similar to the one that had led Le Verrier and Adams to Neptune, and published this and his predictions of the place of the supposed planet in 1915. His observatory conducted an extensive photographic search for the planet, which was rewarded only in 1930 when the young staff astronomer Clyde Tombaugh (1906–97) discovered a planet that was given the name Pluto. The agreement between the predicted and the observed orbit was surprisingly closer than might have been expected from what Lowell had admitted

to be the exiguous data he had used, but the discovered
planet was much too small to give rise to the perturbations
attributed to it, and the apparent agreement came to be
regarded, after much discussion, as fortuitous. Nevertheless
the dedicated search did lead to the discovery of a trans-
Neptunian planet by Lowell's observatory, and Tombaugh's
exhaustive scrutiny of much of the sky to faint limits
makes it unlikely that any major planet does exist beyond
Neptune.

Theories of cosmogony

The term 'cosmogony', strictly speaking, refers to a theory
of the origin of the entire cosmos, but for the last two centuries it has more commonly been applied to theories of
how the solar system was formed. As we saw earlier (on
page 156), in 1796 Laplace proposed what became known as
the 'nebular hypothesis', in which he explained the pattern
of movement from west to east among the planets and
then-known satellites of the solar system, as resulting not
from chance or the *fiat* of the Creator, but from the rotation
of the giant nebula out of which they had formed.

In all, Laplace published four further editions of his
Exposition du système du monde, the last in 1824; even in
1796 he was aware of things that did not quite fit – for
example, that the plane of the orbits of the satellites of
Uranus was nearly perpendicular to that of the ecliptic. The
new knowledge acquired during the nineteenth century
that we have just described, combined with the theories of
the nature and evolution of the stars to which we shortly
turn, led astronomers to question the adequacy of the
nebular hypothesis. The major problem, for which Laplace
had no explanation, was that although the greater part of
the mass of the solar system resided in the Sun, most of the
angular momentum was carried by the outer planets – put
simply, the Sun should be rotating more quickly than it
does.

Already in 1796 Laplace had reminded his readers of an
earlier theory, due to the respected naturalist G. L. Leclerc,
Comte de Buffon (1707–88), that a large comet had crashed
into the Sun and that the ejected fragments from the impact
had condensed into planets; at a time when the mass of
comets and the nature of the Sun were unknown, the
theory was at least plausible.

A new theory recalling that of Buffon was developed by
the American geologist Thomas C. Chamberlin (1843–1928)
and his younger colleague, the astronomer Forest Ray
Moulton (1872–1952), in the very first years of the twenti-

eth century. Both were then professors at the recently established University of Chicago.

As finally presented, the Chamberlin–Moulton theory
proposed two stages. Chamberlin had noted that under
normal conditions solar prominences were readily projected
to great distances above the photosphere. The close
approach of another star might thus be expected to expel
large amounts of the outer layers of the Sun, in two diamet-
rically opposed tidal streams (reminiscent of the spiral arms
of the nebulae). This material would then cool and slowly
aggregate into 'planetesimals', from which in turn the
larger planets might have aggregated.

Although offering a possible solution to the angular
momentum problem, but raising almost as many questions
as they answered, modified tidal interaction theories of this
type were espoused by other theorists and remained tenable
alternatives to the Laplace theory at least until the late
1930s. It may be remarked that 'nebular' theories would
make the existence of planetary systems around stars
common, whereas close approaches of stars would be rare.
The consequences for the possibly ubiquity of life on Earth-
like planets in the universe, or the uniqueness of life on
Earth, was always implicit but rarely referred to in the published papers, and this perhaps influenced the preference of
the theorists one way or the other.

A shift of opinion back to Laplace-type theories became
possible from about 1960 when new ideas about magnetic
fields in the early solar system were introduced by theorists
such as H. Alfvén (1908–95). If the condensing Sun had a
general magnetic field extending through the electrically
conducting outer parts of the hot solar nebula, the field
would link the Sun and the nebula together, slowing down
the Sun and transferring angular momentum to the outer
nebula.

Through the following years of the twentieth century,
almost every new major discovery about the physics and
chemistry of the solar system has produced new theories of
its origin. The present consensus blends features of both the
nebular and planetesimal theories: infrared observations
show the presence of a cool condensing nebula about newly
forming stars, while interplanetary spacecraft have revealed
the heavily cratered surfaces of many of the planets and
their satellites that are believed to remain as the relics of
the impact of planetesimals. Even the origin of the
Earth–Moon system itself is uncertain. The pair of bodies is
more like a twin planet than the large planet and small
satellite associations elsewhere in the solar system. In 1879

G. H. Darwin (1845–1912) developed a detailed mathematical theory of the fission of a rapidly spinning and viscous single Earth, but alternative explanations, including major collisional processes, have since been postulated.

The stars and the nebulae

The improved knowledge of the spectra of molecules, and refined optical methods of measuring the polarization of light reflected from planetary surfaces, led to further astrophysical studies of the planets in the early years of the twentieth century. These led to the discovery of methane in the atmospheres of the giant planets, and also of ammonia in those of Jupiter and Saturn. But increasingly the interest of the astrophysical observatories was turning to the more exciting problems of the stars and the nebulae. With a few exceptions, planetary observations were dropped. Partly because it was difficult to see what more could be done, and partly as a consequence of the Mars controversy, the study of the fascinating and ever-changing markings on the surface of planets like Mars and Jupiter was left, by tacit consent, to amateur astronomers.

The impact of photography in the late nineteenth century

We have already seen how rapidly the use of the spectroscope and of photography improved understanding of the Sun. It was to do the same for the stars and the nebulae, but later. When the limited light of even a bright star was passed through the narrow slit of a spectrograph and then spread out by the prism into the spectrum, it was barely visible to the eye, and much too faint to be recorded on the early photographic plates. In 1872 the New York amateur astronomer Henry Draper (1837–82) managed to photograph a spectrum of the bright star Vega, and in 1879 the London amateur William Huggins did the same for a few more stars, extending the recorded spectrum beyond the violet region into the ultraviolet, to which the photographic emulsion was sensitive and the human eye not. After about 1880 the newly invented dry plate was improved with every passing year, as the commercial manufacturers vied with each other to supply the rapidly widening trade in portrait and landscape photography with better products, and this increased the possibilities for the photography of stellar spectra.

Undoubtedly a key event was the photography of the great comet of 1882, which in September was well placed for observation in the southern hemisphere. The Scots astronomer David Gill (1843–1914), at the Cape Observatory, being anxious to secure a picture of the tail of the comet stretching across the sky, sought the help of a local portrait photographer. A studio camera was strapped to the counter-weight of one of the telescopes, and the telescope was then used to follow the comet during the time exposure. Gill was astonished by the result: the photograph not only had a superb image of the comet but also showed, on the single plate of a very large area of sky, the images of thousands of stars whose measurement and recording by the older methods of visual observing would have taken many weeks of night-time work.

Almost from this moment photography began to alter the very character of astronomy, and to change the nature of observatories too. On the one hand the large telescopes, both refractors and reflectors, were fitted with plateholders and spectrographs to photograph small areas of sky in great detail, and examine the spectra of individual objects. On the other, new types of instruments, developed from the studio camera, were introduced. Although by the standards of the big telescopes these had relatively small lenses – 5 or 10 inches in diameter – they had wide fields of view. Finally, by putting a prism of small angle in front of the camera lens, each star image could be spread out into a little spectrum, so that a single photograph could give useful spectra of some hundreds of the brighter stars on the one plate. Whatever the instrument, the astronomer was freed from the limitations of the human eye, and from trying to make difficult and delicate measurements in the cold and the dark of a telescope dome. When the photograph had been developed, the images could be measured more accurately, by day and in comfort.

It is coincidental that these improvements in photographic technology, commercially driven, came just at the time when a number of observatories were built specifically for the study of astrophysics rather than the old astronomy of position, while some of the older observatories created astrophysical departments. By 1900 astronomers were provided with a new armoury of techniques that were to provide answers, over the next forty years or so, to problems that had tantalized them for centuries.

Angelo Secchi and the classification of stellar spectra
The first photographic maps of the sky were at first only more complete and more accurate extensions of those that had been compiled laboriously by visual methods in earlier

years. They were still no more than maps of the stars on the
apparent surface of the sky, as only for a few stars had the
distances – the third dimension – been determined by
laborious and imprecise visual measurements.

But one difference between the visual and photographic
maps soon became apparent. Although the positions of the
stars agreed precisely, the apparent brightnesses of the stars
differed: two stars that seemed equally bright to the eye in
the telescope might differ quite strikingly on the photograph. It was known that the photographic plate was less
sensitive to yellow and orange light than the human eye:
evidently the stars differed in colour.

This was not a new discovery. As long ago as 1798
William Herschel, out of curiosity, had examined through a
prism the light of six of the brightest stars. 'The light of
Sirius', he noted, 'consists of red, orange, yellow, green,
blue, purple, and violet . . . Arcturus contains more red and
orange and less yellow in proportion than Sirius . . .' But
these were facts for which he could offer no interpretation.

However, nineteenth-century physicists were well aware
that as a piece of metal is heated in the laboratory it
becomes first red hot and then white hot. This implied that,
by examining the spectra of stars, one could learn something about their temperature as well as their chemical
composition. Many astronomers contributed to this study,
the most notable being William Huggins, Father Angelo
Secchi (1818–78) of the Collegio Romano, H. C. Vogel
(1841–1907) of the Potsdam Astrophysical Observatory, and
E. C. Pickering (1846–1919) of the Harvard College
Observatory.

In 1863–67, before the application of photography, Secchi
had patiently examined the spectra of some 400 stars, and
at the meeting of the British Association for the
Advancement of Science in 1868 he proposed that most
stars could be grouped into four types of spectra. Vogel at
Potsdam arrived independently at a very similar grouping.
We no longer use Secchi's classification of Roman numerals, but his groupings are essentially correct and can be recognized in the later schemes. Secchi's types are:

I.   The 'Sirian' (like Sirius), whitish or bluish, with a few
     dark bars due to hydrogen, and only faint metal lines.
II. The solar type – stars like Capella and Arcturus,
     sharing a predominance of light in the middle, yellow,
     part of the spectrum and with innumerable fine dark
     lines.
III. Red (and often variable) stars of long and irregular

period like Betelgeuse (in Orion) and Mira (in Cetus),
with broad bands of light each brighter at the red end
and fading out at the blue end, and crossed by regularly
spaced shadings giving a fluted appearance.

IV. Stars rather similar to III, but of an even redder colour
and with the bands differently arranged. These stars are
the smallest group, and none of them is bright in the
sky; seen visually in the telescope they 'gleam like
rubies among the other stars'.

These early studies of stellar temperatures and spectra led
to the first attempts at a scheme of stellar evolution. In his
book about measuring the brightness of stars published in
1865, the Leipzig astronomer Friedrich Zöllner (1834–82)
suggested that stars were first formed hot and, as they naturally cooled, passed through the solar type to the red stars.

William Huggins and the discovery of 'true nebulosity', 1864

William Huggins (1824–1910) as a young man was forced by
circumstances to devote himself to his family's modest
mercer's business in London. But in 1854 he succeeded in
divesting himself of it, so that he could indulge his passion
for astronomy.

On learning of Gustav Kirchhoff's 1859 discovery that
the chemical composition of the Sun was revealed by its
spectrum, he realized that the method could be applied to
the stars and nebulae, and he formed a collaboration for this
purpose with W. A. Miller, a professor of chemistry. In 1875
he married, and thereafter his young wife Margaret was his
devoted partner in his researches. Huggins quickly became
a world leader in the 'new astronomy', and was a pioneer in
the field, working at his home and private observatory at
Tulse Hill, south of London.

Huggins and Miller proceeded on rather different lines
from Secchi and Vogel, examining and interpreting the
spectrum of individual objects selected for their particular
interest rather than surveying many objects. Huggins had
already resolved, by one crucial observation, one of the
great dilemmas of observational astronomy. We have seen
that the great telescopes of William Herschel, Lord Rosse,
and William Lassell had been built largely to determine the
nature of the nebulae: were they distant, unresolved clouds
of stars, or were they clouds of 'true nebulosity' – glowing
gas from which stars might condense? On the night of 29
August 1864 Huggins turned his telescope to a planetary
nebula in Draco which he thought bright enough to yield a

spectrum. It did, but showed only a few bright lines in the green part, one certainly due to hydrogen, the others unidentified. By analogy with the spectra of gases in tubes in the laboratory the conclusion was immediate: there were two kinds of nebulae, gaseous (of which the planetary nebula was an example) and stellar, and the spectroscope could distinguish between them.

By 1868 Huggins had examined the spectra of some seventy nebulae of various shapes and sizes. About one-third of these were gaseous; two-thirds had faint continuous spectra which were probably the light of unresolved stars. Huggins's discovery was a diagnostic test of great value, and ranks as one of the great discoveries of the early period of astrophysics.

Of course there were complications. From the outset there were a few objects, some of them quite bright, that did not fit into the scheme. They were certainly stars as seen in the telescope, and with continuous spectra that sometimes had dark lines; but the spectra also included bright, sometimes spectacularly bright, emission lines. Many of these objects were known variable stars; some were subject to spectacular outbursts like that of Eta Argus (now Eta Carinae) observed by John Herschel at the Cape in the 1830s. They were to provide problems for the future. Were they perhaps stars that had most of their surface covered by the bright faculae seen on the Sun, or objects in transition from a nebular to a stellar state, or vice versa?

The first radial velocities of stars

Spectroscopy also gave information of a rather unexpected nature: about the movement of stars in space. In 1842 Christian Doppler (1803–53), of the University of Prague, proposed that the colour of starlight would be affected by the velocity of a star; the wavelength of the light of a receding star would be lengthened and hence the light made redder. The Doppler Effect exists, but Doppler (who was not an astronomer) wrongly interpreted it by grossly over-estimating the velocity of stars in space. A more correct interpretation was arrived at by Fizeau and later and independently by the physicist Ernst Mach (1838–1916) in 1860. The typical velocity of a star relative to us (say 20 km/sec) is so much smaller than the velocity of light (300,000 km/sec) that the effect on the colour of the star is imperceptible; but the small displacement of sharp defined features like the dark absorption lines can be measured, and this yields a velocity of the star towards or away from the Earth.

This measurement was first attempted by both Secchi and Huggins but it proved too difficult for their spectroscopes. The first measurements that agree with modern determinations were made on a small number of bright stars by Vogel and Julius Scheiner at Potsdam, and by James E. Keeler at Lick Observatory in California, in the years around 1890. They were accurate to within a few kilometres per second. Such measurements always remained of great difficulty, but their value was recognized by astronomers like E. B. Frost of the Yerkes Observatory (who went to Potsdam to study Vogel's methods) and by W. W. Campbell at Lick, who persevered with the work. The difficulty lay in the smallness of the displacement of the lines. An error of only a few thousandths of a millimetre in measuring resulted in an error of several kilometres per second. During the exposure of an hour or two the moving spectrograph sagged under its own weight, and contracted as the temperature fell in the dome. It was necessary to design the frame of the spectrograph carefully and enclose it in a jacket with the temperature kept within a fraction of a degree of constancy. The patient work of a relatively small number of astronomers slowly accumulated lists of line-of-sight ('radial') velocities, and by 1920 more or less reliable radial velocities were available for some 2,000 stars measured at some fifteen observatories around the world, and the number was growing by two or three hundred a year. The data were soon to become an important resource for the study of the structure of the Galaxy (see page 281).

Spectroscopic binary stars

The Doppler Effect however explained a discovery made at Harvard in 1887 by E. C. Pickering, the director, and followed up by Antonia Maury (1866–1952). It concerned changes in the spectrum of the bright star Mizar in the tail of the Great Bear: sometimes the familiar sharp absorption lines were split into close twins. Mizar, it transpired, was not one star, but two. Studies of binary stars (pairs of stars in orbit about their common centres of gravity) had continued through the nineteenth century, but here was a new class: two stars so close together that visually they invariably appeared as a single star, but which, as one star approached the Earth and the other receded in the course of the orbit, the spectroscope revealed to be double. Other such close 'spectroscopic binaries', like Capella, were soon discovered and were later to prove of great importance. In special cases, where the Earth lies in the plane of the orbit of the close double star, one star can pass between the Earth

and the other star. These are 'eclipsing binaries', like the
star Algol (page 172), and a detailed analysis of the variations in the brightness and spectrum of what always seems
visually to be a single star can yield information about the
sizes, temperatures, separations, and masses of the individual stars, that can be obtained in no other way.

The measurement of the distance of individual stars of
particular interest was becoming increasingly desirable.
After the successes in the 1830s (see pages 186–9) and 1840s
for a very few of the nearest stars, the visual methods began
to prove almost impossibly difficult, and for the same star
different observers would get discouragingly different
results. The combination of photography and the long focus
refractor again came to the rescue, by transferring the
process of measuring from the telescope to the laboratory.
But until well into the twentieth century the determination
of stellar distances remained on the verge of the possible,
and even for stars only 75 light years away there were errors
of 20 percent or more. Knowledge of distances was limited
to a tiny local pool of stars within the as yet unfathomed
ocean of the Galaxy.

One remaining quality of the nature of the light of a star
called for more exact measurement, namely the brightness
of the star as seen in the sky. Since Classical Antiquity
stars had been assigned a magnitude, the brightest in the
sky being of the first magnitude and the faintest visible on a
clear night of the sixth; the five intervals between were
estimated. When the first telescopes revealed many still
fainter stars, the scale was extended by what was little
more than guesswork; a star classed by one astronomer as
eighth magnitude might be described by another observer as
eleventh.

This clearly would no longer do. Progress was made in
two steps. Physics laboratories were experimenting in the
measurement of the brightnesses of lights in the laboratory,
and gradually such methods were adapted to the needs of
the astronomer, for example by projecting into the field of
view of the telescope an image of an artificial star, and then
adjusting the brightness of that image until it matched that
of the star being measured. The second step was taken by
the English astronomer Norman Pogson (1829–91) in 1856.
He recognized (as had Edmond Halley as long ago as 1721)
that stars of the first magnitude were roughly one hundred
times brighter than those of the sixth. By proposing that 5.0
magnitude steps should be taken as corresponding exactly
to a ratio of 100 in brightness, he defined the scale precisely.

Pogson made his proposal almost as an aside in a paper on variable stars, and at first it attracted little notice. Only twenty years later did a number of observatories, notably Potsdam, Oxford, and Harvard, embark on major programmes of stellar photometry, and their independent acceptance of the Pogson scale led to its eventual adoption worldwide. With the further acceptance of certain well observed stars as standards, an accurate and internationally accepted system of stellar magnitudes had been built up by the turn of the century.

This system of stellar magnitudes, in use today, remains a curious one and unlike almost any other used in the physical sciences. Its inverse logarithmic scale, in which the brightest stars have magnitude around −1, and the faintest detectable galaxies about +30, is in direct descent from the star catalogue compiled by Ptolemy in the second century AD.

Annie Jump Cannon and the Harvard catalogue of stellar spectra

These magnitudes were 'visual magnitudes', observed with the eye. We have already noted that the magnitudes measured on photographs were not the same, and a similar system of 'photographic magnitudes' was set up. Physics was far enough advanced for it to be understood that hotter stars emitted more of the blue light that was more easily recorded by the photographic plate. The difference between the two magnitudes, the 'colour index', measured the colour of the star's light, and this was directly related to temperature; a large colour index of say 1.5 magnitudes implied a red star of low surface temperature. Many refinements were introduced in later years, and in particular from the 1920s onwards sensitive photo-electric cells replaced estimates made by the human eye. But the principle was the same.

These developments made it possible to refine the rather simple classification of types of stellar spectra that had been introduced by Secchi and Vogel. The work was started by Henry Draper, who died prematurely in 1882. As a memorial to him, his family provided the Harvard College Observatory with funds for new instruments and additional staff, to continue Draper's work. Secchi's four types were extended to sixteen, lettered A, B, C, . . . With a better understanding of what was going on, these letters, which denoted the occurrence of various lines in the spectra, were gradually rearranged into what was seen to be more nearly a sequence of decreasing surface temperatures of the stars, and simplified a little. The eventual sequence of classes was

A typical photograph used in the preparation of the monumental Henry Draper Catalogue. The photographs were mostly taken with cameras of 8 inches aperture, through prisms of either 5° or 13° angle. This one was taken at the southern outstation in Arequipa, Peru, and is of a region of sky around the variable star Eta Carinae. Some of the spectral types have been labelled for this to appear as the frontispiece to the last volume of the main Catalogue, published in 1924. Courtesy of Institute of Astronomy, Cambridge University.

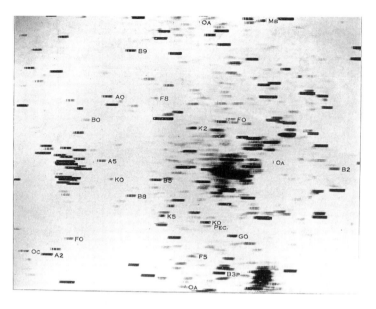

of types O B A F G K M R N S, and the description of the spectra was sufficiently minute to permit the division of some of the classes numerically, so that the Sun (for example) became a star of spectral type G2.

E. C. Pickering, the Harvard director, thereupon embarked upon a spectroscopic survey of the whole sky (Harvard set up outstations in the southern hemisphere). The work involved the use of wide-field cameras with an objective prism as already described. Pickering was fortunate that his skilled assistants included Annie Jump Cannon (1863–1941), for her prodigious industry was crucial in the publication, between 1918 and 1924 (with some later extensions), of the Henry Draper Catalogue. The catalogue, which contains the spectral type and magnitudes of some 225,000 stars, is still of value, 'the greatest single work in the field of stellar spectroscopy'. It is interesting to note that Zöllner's early speculation that the sequence followed also the life history of the stars was still thought useful, if not accepted as certain. The O B A . . . stars were referred to as 'early type' and G K M . . . as 'late type', and this usage curiously continued among astronomers long after matters were known to be much more complicated.

Plotting magnitude against spectral type: the H–R diagram

The Harvard spectral classifiers had noted subtle differences between otherwise identical spectra; in particular, in the

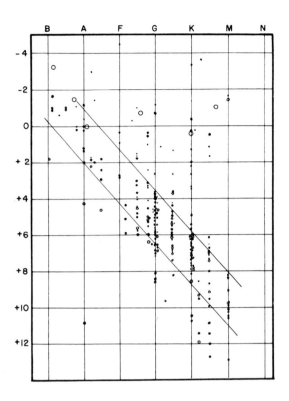

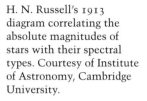

H. N. Russell's 1913 diagram correlating the absolute magnitudes of stars with their spectral types. Courtesy of Institute of Astronomy, Cambridge University.

later spectral types some stars had very narrow lines. In 1905 Ejnar Hertzsprung (1873–1967) noted that these stars tended to have very small proper motions, and so were probably distant and highly luminous.

By about 1910 two separate pieces of information were beginning to be available for a significant number of individual stars. The first was the spectral type, or the colour index, or the surface temperature of the star, all closely related to one another. The second, determined with greater difficulty, was the distance of a star, which enabled one to turn the (apparent) magnitude into a calculated 'absolute magnitude' or 'luminosity'.

The question to be asked was: can a star of a particular absolute magnitude be of any spectral type or surface temperature, or are only certain combinations of these quantities permitted in the universe? In other words, if one plots on a graph for each star the value of the absolute magnitude against (say) the spectral type, are the points scattered all over the diagram, or do they occur only in certain regions of it? Such a plot was made in 1913 by Henry Norris Russell (1877–1957) of Princeton University, using all the stars for which he felt there were reasonably reliable distances. Russell had been born in Oyster Bay, in the state

of New York, and he had gained his first degrees at
Princeton and then had gone to Cambridge, England where
he worked as a post-doctoral assistant measuring photographic stellar parallaxes. He returned to the Princeton
Observatory in 1905, and was its director from 1912 to
1947. Despite indifferent health Russell was an indefatigable researcher in several major fields in astrophysics,
and had an influence on the subject comparable only with
that of G. E. Hale.

In his diagram, Russell had been partly anticipated a year
or two before by H. O. Rosenberg (1879–1940) and
Hertzsprung, who had plotted similar diagrams for the stars
in each of the Pleiades and Hyades star clusters; the stars in
any one cluster are so nearly at the same distance that
apparent magnitudes are closely related to absolute magnitudes.

Such diagrams became of crucial importance in understanding the ways stars evolved: they became known as
Hertzsprung–Russell diagrams, or 'H–R diagrams' in the
jargon of working astronomers. Even on his first diagram,
Russell felt that he could say that Nature does restrict the
kinds of stars that can exist. There were two main bands
populated by stars. One became known as the Main
Sequence; it sloped down from highly luminous hot stars at
the top left to intrinsically faint cool stars at the lower
right. Across the top was the Giant Branch: the most luminous stars could be of any spectral type.

As the accuracy of measurements increased, these features of the H–R diagram were confirmed and more features
became recognizable. An important discovery, made in
1914 at Mount Wilson Observatory by W. S. Adams and
Arnold Kohlschütter, was that there were further subtle
differences between the spectra of the Main Sequence and
Giant Branch stars of the same spectral type, revealed by
the relative strengths of particular pairs of lines. Such
differences had again already been noted by the Harvard
classifiers, without being understood. Now, by a careful
examination of the spectrum of a star, it was possible to
identify to which part of the standard H–R diagram the star
belonged, read off its absolute magnitude, and so, the apparent magnitude being known from observation, determine
its distance. With careful refinement of the Harvard scheme
of spectral classification in the 1930s by W. W. Morgan and
his colleagues at the Yerkes Observatory, this inverse
method of using the H–R diagram to determine distances
from the observed spectral types and apparent magnitudes
of stars ('spectroscopic parallaxes') became a powerful tool

for determining the distances of stars far beyond the reach of the trigonometric method.

Atomic spectra and the chemical composition of stars
By about 1920 an enormous body of knowledge about spectra had accumulated. Meanwhile, in the physics of the Sun, new effects had been found. For example, the splitting of some spectral lines due to iron, when they originated in the laboratory in the presence of a strong magnetic field, was discovered in 1896 by the Dutch physicist Pieter Zeeman (1865–1943); this was quickly used to identify and measure the strength of magnetic fields in sunspots.

Now the spectra of hot blue stars showed only a small number of very black lines due to hydrogen and helium; in the cooler stars these were much weaker, but the lines due to metals like iron and chromium were innumerable. Did this mean there was no chromium in hot stars? And why, in the outer layers of the Sun, presumably made throughout of the same gases mixed together, did the cooler sunspots have a spectrum so different from that of the photosphere?

The explanation of this large body of empirical knowledge was at hand, but it came from parallel discoveries made in physics, about radiation and about the structure of the atom. In 1901 the Berlin physicist Max Planck (1858–1947) set out his quantum theory, by which light and other radiation like X-rays is emitted or absorbed in discrete packets of energy. In Manchester, in 1911, the New Zealander Ernest Rutherford (1871–1937) proposed a model of the atom, in which a positively charged nucleus was surrounded by a cloud of negative electrons. Two years later the Danish theoretical physicist Niels Bohr (1885–1962) combined features of both theories to develop a new model of the atom that was capable not only of explaining the sharp lines in spectra, but also of calculating their wavelengths and predicting the wavelength of new lines to be looked for. Although no more than a model, which served to calculate and predict, the 'Bohr atom' was enormously useful and continued to be used in the interpretations of astronomical spectra even when it had been realized, with the introduction of quantum mechanics and the wave theory of the electron, that the real atom was much more complicated.

In the Bohr model of the atom, the electrically charged nucleus was surrounded by a number of negatively charged electrons that exactly balanced the charge of the nucleus, so making the atom neutral. The electrons, arranged in shells, orbited the nucleus, and the number and arrangement of the electron orbits determined the pattern of lines in the spectrum of that element. When a gas was heated, however, an atom could lose one or more of its outer electrons to become 'ionized' and positively charged, and the changed electron orbits gave a completely different pattern of lines.

This kind of theory, combined with knowledge of temperatures and densities in the outer layers of the Sun and stars, was used by the Indian physicist Meghnad Saha (1894–1956) around 1920 to give a very satisfactory explanation both of the great differences between the spectra of different types of stars, and also of the subtler differences between the spectra of giant and dwarf stars of the same spectral type. In turn it underpinned the use of such methods as spectroscopic parallaxes, which previously had been based on empirical experience.

Cecilia Payne and the abundance of hydrogen in stars
Cecilia Payne (later Payne-Gaposchkin, 1900–79) was born in England and after graduating from Cambridge University went to Harvard College Observatory for doctoral research, and remained there for the rest of her career. Her 1925 thesis was later described by Otto Struve as 'the most brilliant Ph.D. thesis ever written in astronomy'. It established clearly the relationship between the temperature and the spectral class of a star, and went on to consider the relative abundance of the elements, suggesting that there was evidence that hydrogen was enormously more abundant in most stars than had been supposed. At first this idea was not accepted, but by 1929 H. N. Russell, for example, had been persuaded, and writing on the solar atmosphere concluded that it contained '60 parts of hydrogen (by volume), 2 of helium, 2 of oxygen, 1 of metallic vapors, and 0.8 of free electrons practically all of which come from the ionization of metals'. Later investigations would change these figures, but not the main conclusion, that much the most abundant element in the stars is hydrogen, followed by helium.

Further explanations followed from the use of physical theory. So, for example, in 1927 Ira S. Bowen of the California Institute of Technology showed that the lines supposedly due to the undiscovered element nebulium were in fact caused by oxygen in an unusual state of ionization, and (as already mentioned) in 1941 Bengt Edlén similarly explained the 'coronium' lines as caused by iron.

Eddington and the mass–luminosity relation
There were other fruitful interactions between observation and theory at about the same time. The sizes of stars

cannot be measured directly at the telescope by simply measuring their discs, for they are much too far away. They could be inferred in two ways, either by studying eclipsing binaries, or by calculating what the surface area (and hence diameter) of a star must be to emit the observed amount of energy (absolute magnitude) at the observed surface temperature (colour index). To determine the density of the star, and the attraction of gravity at its surface, required a knowledge of its mass – how much matter the star contained.

There was one, and only one, way in which the masses of stars could be determined, by the application of Newton's law of gravitational attraction to the measured orbits of binary stars. This was why the study of these stars was very important, the information from the close, spectroscopic binaries being particularly valuable. Such information was available for a relatively small number of stars. Nevertheless, it became apparent that stars could have a surprisingly large range of mass. In 1924 A. S. Eddington (1882–1944) of Cambridge collected the available information and found stars of as little as one-fifth the mass of the Sun, and as large as twenty-five times. More importantly, he showed convincingly that these values were quite closely related to the absolute magnitudes of the stars: there was a mass–luminosity relation in the sense that the most massive stars were the most luminous – stars of 25 solar masses were emitting about 4,000 times the energy of the Sun.

The source of stellar energy

Where did all this energy come from? At last the answer to this question, which had so perplexed the physicists of the nineteenth century, seemed to be in sight. It was becoming generally agreed that the clue lay in Albert Einstein's theory of General Relativity (1915), which held that mass could be transformed into energy. In a semi-popular exposition, in his book *Stars and Atoms* (1927), Eddington speculated that four atoms of hydrogen (each of mass 1 unit) might be combined to form one atom of helium (with the slightly smaller mass of 3.97 units). 'To my mind the *existence* of helium is the best evidence we could desire of the possibility of the *formation* of helium . . . I am aware that many critics consider the conditions in the stars not sufficiently extreme to bring about the transmutation – the stars are not hot enough. The critics lay themselves open to an obvious retort; we tell them to go and find *a hotter place.*'

Eddington was right: the nuclear reaction that he sug-

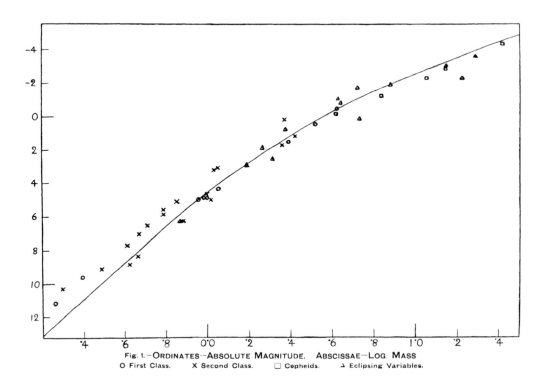

Fig. 1.—Ordinates—Absolute Magnitude.   Abscissae—Log. Mass
O First Class.   X Second Class.   □ Cepheids.   ⅃ Eclipsing Variables.

A. S. Eddington's 1924 mass–luminosity diagram for stars. The graph plots absolute magnitude (vertical scale) against the logarithm of the mass (horizontal scale), with each symbol representing an individual star. The Sun (not shown) would have absolute magnitude of about 5 and a mass of exactly 1 (logarithm=0). The curve is drawn from theory. Courtesy of Institute of Astronomy, Cambridge University.

gested was taking place in the hot central cores was eventually proved to be the source of energy of the Main Sequence stars. But at the time he could do no more than speculate, and he thought there were other possibilities. In the same passage he continued, '. . . the evidence, however, is not very coherent, and I do not think we are in a position to come to a definite decision'. It required more information from nuclear physics to understand exactly how the reactions could operate. It evidently had to be some step process, since the probability of four hydrogen atoms instantaneously colliding to form helium seemed very small. Several physicists and astronomers studied the problem through the 1930s.

R. d'E. Atkinson of England and F. Houtermans of the Netherlands, working together in Göttingen, had provided in 1929 an initial theory of nuclear interactions between the lighter elements in stellar interiors. At Rutgers University in New Jersey in 1931, Atkinson learned of the recently accepted great abundance of hydrogen in stars, and considered the mechanisms by which fusion of hydrogen into helium might proceed. Later, in Germany, C. F. von Weizsäcker worked out how further interactions could produce elements heavier than helium.

Finally, in 1939, the German-born H. A. Bethe (1906–   ), then at Cornell University, proposed a mechanism that was generally accepted as feasible and was consonant both with the known nuclear physics and with the accepted conditions of density and temperature at the centres of stars, stemming from Eddington's work in his classic book of 1926, *The Internal Constitution of Stars*.

But that was 1939, and ironically many of those who had contributed to the solution of the problem were soon to find themselves engaged in using their knowledge to more practical, and grimmer, military ends.

Unusual stars

In a volume of space around the Sun – say, a sphere of radius 50 light years – stars of much the most common kind are less luminous than the Sun itself. They are the dwarf stars at the lower end of the Main Sequence. On an H–R diagram, therefore, the giant stars, whether blue giants or red giants, were much over-represented, because although relatively rare they can be seen to very much greater distances. But whether rare or common, nearly all stars fell into the two main branches of the diagram.

A very few did not. Indeed there was one already in Russell's first plot of absolute magnitude against spectral type: an intrinsically faint star of spectral type A in the lower left of the diagram (see page 261). It is the fainter companion of a nearby double star in the constellation Eridanus. A few similar stars were known. The bright star Sirius, about 9 light years away from the Sun, had a fairly rapid motion across the line of sight, and in 1844 F. W. Bessell announced that the motion was irregular and that the star moves in a wavy line. The simplest explanation was that Sirius was a double star with a very faint or even dark companion. Some twenty years later the American telescope maker, Alvan Clark Jr, testing a new refracting telescope, first saw the faint companion 'Sirius B', which was then favourably placed for detection in its fifty-year orbit about the brighter star.

These stars became known as 'white dwarfs'. For those that are members of double star systems, the mass of the white dwarf companion could be determined, as we have seen. They turned out to be paradoxical stars, with masses not too different from that of the Sun, but sizes (derived from their surface temperatures and measured brightness) not much greater than that of the Earth. The implied density was enormous: as Eddington said, a ton of the material could be put in a match box. The situation around

The path of Sirius in the sky, 1850–1920. The smaller diagram shows the orbit of Sirius B relative to Sirius. Courtesy of Institute of Astronomy, Cambridge University.

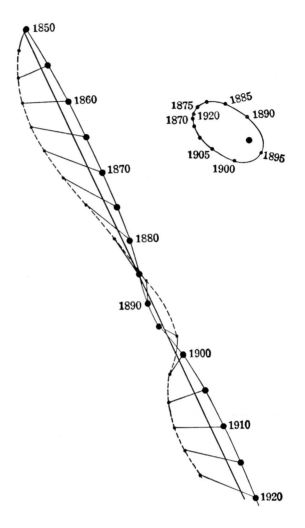

1914 was an embarrassment; astronomers felt that something must be wrong, but did not know what. The explanation had again to wait for a better understanding both of atomic physics and of the internal structure of stars; eventually, in 1926, R. H. Fowler of Cambridge used the recently developed theories of quantum physics to explain the existence of so-called degenerate matter of such high density.

In 1931 the young Indian mathematician, Subrahmanyan Chandrasekhar (1910–95), a student of Fowler's, calculated an upper limit (later named the Chandrasekhar limit) for the masses of such stars (see page 320). White dwarf stars are examples of objects for which observation preceded understanding. The first were discovered in the nineteenth century; further observations early in the twentieth century revealed their incredibly high density, but it required new

theories of the physics of the states of matter to explain
them and to confirm that the observations had been right,
and rightly interpreted.

But the alternative route to discovery can also occur. In
1932 James Chadwick identified in the laboratory a new
fundamental particle of atomic structure, the neutron; and
the Swiss-born astronomer Fritz Zwicky (1898–1974) and
his German-born associate Walter Baade (1893–1960), both
then working in Pasadena, California, suggested that its
known properties would permit the theoretical existence of
stars, made essentially of packed neutrons, that would have
a density many millions of times that of even the white
dwarfs. This speculation, published in 1934, attracted little
attention at the time, and it was not until 1967 (see page
321) that such objects ('pulsars') were found to exist, discovered not as a result of deliberate search but serendipitously,
in the course of unrelated researches in radio astronomy.

There were other classes of stars whose nature had also
been enigmatic, that yielded to the improved understanding
of the internal constitution of stars. In particular, the
classes of variable stars with remarkably regular light
curves and constant periods, of which the naked-eye star
Delta Cephei (see page 278) is an example, had long presented a problem. The American Harlow Shapley (see page
277) had shown in 1914 that any explanation of them as
spectroscopic binaries had to be abandoned, and suggested
that these were single stars that were pulsating. Theoretical
astronomers like Eddington developed this idea, and the
theory of these stars was worked out in increasing detail
during the 1920s and 1930s. Observations of individual
stars agreed well enough with the theory: as the star oscil-
lated in size in a regular cycle, the observed radial velocity
of the star (the approach of its surface towards the observer
as the star swelled), its surface temperature, spectral type,
and observed magnitude correspondingly changed through
the cyclical period of the pulsation. This understanding
gave astronomers increasing confidence that these stars, of
high luminosity, were indeed of great utility in measuring
distances in the universe at large.

The more catastrophic variable stars like novae continued to puzzle. Routine photography of the sky greatly
increased the rate of their discovery and of empirical
knowledge of their behaviour; and from 1917 onwards (see
page 289) faint stars with the same characteristics as novae
in the Milky Way were discovered in the larger, and presumably nearer, of the spiral nebulae. It was only much
later, in 1934, that Baade and Zwicky established that there

were two very distinct classes of novae: the ordinary ones, which occur in galaxies like our own at a rate of ten or twenty times a year, and the much rarer and even more remarkable supernovae, whose light can approach that of all the other stars in the galaxy combined. The probable identification of the Crab Nebula, in the constellation Taurus, as a gaseous remnant of a supernova in our own Galaxy recorded in Chinese annals in AD 1054 had been suggested by Edwin Hubble in 1928 and was supported by further studies in 1942. But the cause of these stellar explosions remained a mystery, even when the general sources of stellar energy were known, and remained so until quite recently.

The evolution of stars

Much the greater majority of stars, however, were unchanging, and could be classified (as we have seen) into a limited number of spectral types. As soon as the process of classification began, in the 1860s, it was recognized that the classification must contain information about how the stars were formed and how they evolved from their initial state. With little physical understanding beyond the feeling that the source of stellar energy lay in the Kelvin–Helmholtz contraction (see page 234), and virtually no knowledge of the sizes or masses of the stars or of why stars had such widely differing spectra, early attempts to understand the life histories of stars could be little more than speculative.

There were two main schools of thought. One, following the suggestion of Zöllner in 1865 (see page 255), argued that the stars were formed as 'Sirian' – white, hot stars similar to Sirius. There was some support for this idea, for it was noticed that such stars sometimes had bright emission lines as well as dark lines, suggesting that the star still contained some evidence of its nebular origin. The star, then cooling and contracting, passed through the stages of stars like the Sun to the red stages and a presumed extinction. This was essentially the view taken by H. C. Vogel of the Potsdam Astrophysical Observatory.

Rather later, in the late 1880s, Norman Lockyer developed a somewhat more elaborate theory which was based on two distinct ideas. The first was his 'Meteoritic Hypothesis', eventually expounded in his book with that title published in 1890; the idea was that 'all self-luminous bodies in the celestial space are composed either of swarms of meteorites, or of masses of meteoritic vapour produced by heat' – in other words, the primordial matter from which

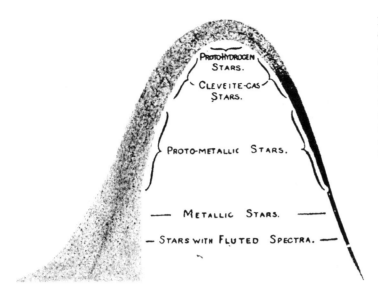

PROTO-HYDROGEN STARS.

CLEVEITE-GAS STARS.

PROTO-METALLIC STARS.

— METALLIC STARS. —

— STARS WITH FLUTED SPECTRA. —

Norman Lockyer's theory of stellar evolution. This is only an illustrative diagram, but it can be regarded as a graph with temperature increasing upwards and the age of a star increasing from left to right. Courtesy of Institute of Astronomy, Cambridge University.

the stars formed was composed of chunks of rock, not gas. The second was an attempt to explain why the same chemical element, in laboratory spectra, gave different sets of lines in the electric arc (hot) and in the spark (still hotter). This postulated that, at high temperature, elements were dissociated into simpler building blocks he called 'protoelements'. Both ideas, though wrong in detail, foreshadowed later accepted ones: there are solid particles in interstellar space; and Lockyer's protoelements are akin to the ionized atoms of modern atomic theory. Lockyer's theory of stellar evolution (see figure) started with large, cool stars condensed out of the interstellar material. These increased in temperature as the young stars contracted, until the rise in surface temperature was balanced by the loss resulting from radiation; thereafter a final stage of cooling set in.

Astronomers of the next generation, including both Hertzsprung and Russell, were quick to see that the distinction between giant and dwarf stars, apparent in the early H–R diagrams (see page 262), provided important new clues to stellar evolution. In a semi-popular article in 1914 Russell accepted, with everybody else, that the source of stellar energy was contraction, so that evolution proceeded in the direction of increasing density and decreasing size. Stars formed as red giants of spectral type M, moved along the Giant Branch in the direction of increasing temperature in the reverse order of the spectral types to classes A or B, and then moved down the Main Sequence, now in the order of the spectral types, to reach a cool, dwarf, state at spectral type M again. Although based upon a different and better established set of data, Russell's theory of stellar evolution had features not unlike that of Lockyer – for example, the affirmation that a star is hottest during the middle of its life history. It was simple, it accorded well with the observations, and it became generally accepted.

But the theory ran into trouble within ten years, when there were new ideas that we have already mentioned (see page 264) about stellar atmospheres, stellar structure, and the relation between mass and luminosity. In 1926 Russell produced a different theory, but this was more speculative and attracted little support. It did however contain a crucially important new idea: the life history of a star was primarily determined by its initial mass, which remained essentially unchanged, and stars were not constrained to move along the Giant Branch and Main Sequence but could move across them: these were rather resting places, where the star remained in a stable configuration for significant parts of its life. The absence of stars from some areas of the H–R diagram meant rather that stars moved through this combination of luminosity and spectral type quite rapidly, so that at any one time most stars were in the Giant Branch or Main Sequence areas of the diagram.

A fuller understanding of the life history of stars of different initial masses was to emerge only gradually from Russell's intuition of 1926, and Bethe's proposal of an acceptable source of nuclear energy in 1939. Even today it is not solved in detail, and there are still uncertainties in, for example, the ages of what are believed to be the oldest stars, in the spherical aggregations of stars known as globular clusters. But by about 1960 the main outlines of what turned out to be a very complicated situation had become understood, and have since been altered in detail rather than in fundamentals. The life history of the majority of stars may be summarized – in the framework of the H–R diagram – in four stages:

(1) The star condenses from the interstellar medium and settles fairly quickly to a stable configuration on the Main Sequence – the more massive the star, the more luminous it is and the earlier (e.g. Type F rather than Type M) its spectral type.

(2) The star spends much of its life in this stable configuration, deriving energy from the conversion of hydrogen to helium in the core; the more massive the star, the more rapidly it evolves.

(3) When the hydrogen in the core is exhausted the star

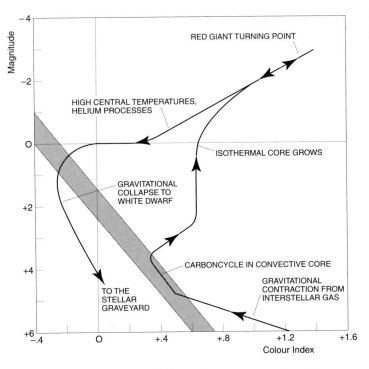

Courtesy of Institute of
Astronomy, Cambridge
University.

'migrates' rather quickly to the Giant Branch of the
H–R diagram, where energy is supplied by the further
conversion of the helium core into heavier elements.

(4) When finally these energy sources are exhausted the
track of the star moves across the H–R diagram to the
left, and the star collapses to the white dwarf state.

For the stars of the smallest masses, perhaps one-tenth of a
solar mass, the evolutionary sequence proceeds very slowly,
so that even the oldest stars are still in Stage 1; on the other
hand, the most massive stars evolve with great rapidity and
have total lifetimes of millions rather than billions of years,
and the final stages of their collapse are marked by extreme
phenomena like supernovae and the formation of exotic
objects like neutron stars. The diagram, by Bart J. Bok,
shows the evolution of a star with mass a little greater than
that of the Sun, as it was understood in 1958.

The structure of the Galaxy

In his attempt in 1785 to establish the structure of the
Galaxy (see page 208), William Herschel had, like earlier
speculators, taken the view that the Milky Way is the
optical effect of our immersion in a layer of stars. But his
bid to determine the outline of the layer had run into difficulties.

Nor had John Herschel fared much better. Greatly impressed by the Milky Way visible from the Cape of Good Hope, he had envisaged the Galaxy as formed of a thinly-populated central cluster surrounded by stars concentrated in meandering arms whose overall shape baffled him.

In 1845 Lord Rosse's great reflector revealed the spiral shape of some of the nebulae (see page 216). Might the Galaxy itself be a spiral? Somewhat surprisingly, John Herschel never suggested this possibility, but in 1852 Stephen Alexander (1806–83), professor at what is now Princeton University, published a discussion entitled 'The Milky Way – a Spiral'. In the spiral M 99, the second to be recognized by Rosse, four branches curved out from the central cluster. Alexander argued that if the Sun and the brighter stars formed the central cluster in a similar spiral, and if the paths of the four branches were suitably chosen, then inhabitants of the solar system would see much the same night sky as we in fact observe from Earth.

There was too much contriving in Alexander's model for the taste of his contemporaries, and it was largely forgotten. But with spiral nebulae being discovered in ever-increasing numbers, the possibility that the Galaxy has a spiral structure was inevitably mentioned from time to time, notably by the Dutch journalist and amateur astronomer Cornelis Easton (1864–1929), who around the turn of the century published a series of influential drawings of the Galaxy imagined as a spiral structure seen face-on.

Another who sought to reproduce in detail the complexity of the observed Milky Way was the respected English amateur and popularizer, Richard Proctor (1837–88). His 1869 model, it was remarked, 'resembled a bent and broken ring, with long, riband-like ends, looped back on either side of an opening'. But most astronomers found this excessively complex and therefore unsatisfying.

Even less attractive – but for opposite reasons – was the model that had been proposed in 1847 by Wilhelm Struve, on the basis of statistical analyses of the star counts of William Herschel and of later star catalogues. He envisaged a universe in which there was a central plane (that of the Milky Way) of unlimited extent, in which the stars were everywhere distributed with a uniformly high degree of concentration. To either side of this central plane, the density of the stars fell off systematically with increasing distance from the plane. To explain the difference between his unbounded model and the limited number of stars in the observable universe, Struve invoked obscuring matter that rendered very distant stars invisible.

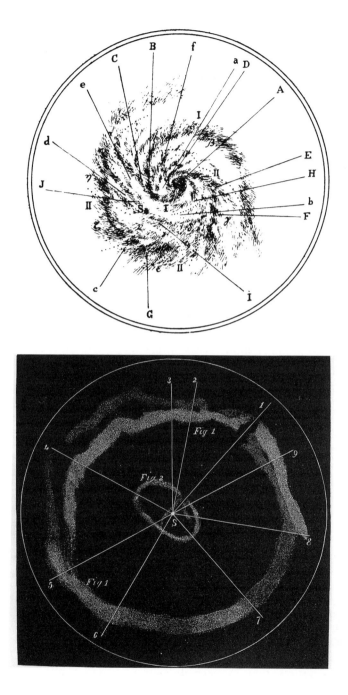

Part of Cornelis Easton's 1900 drawing, showing the spiral shape he ascribed to our Galaxy. He used the lettering to illustrate how various observed features of the Milky Way would arise, if the Galaxy had the shape as drawn. Private Collection.

Richard Proctor's sketch of the principal features of the Milky Way (Fig. 1), and his final attempt (Fig. 2), in the 1880s, to suggest a form of the Galaxy that could give rise to these features. For example, a bifurcation that is observed in the Milky Way (at 9 in Fig. 1) could be the result of the Galaxy's turning in on itself (near the Sun S, along the line S9, in Fig. 2). Courtesy of Institute of Astronomy, Cambridge University.

But whereas Struve was thought to have gone altogether too far in forcing the real star system onto the Procrustean bed of his model, many astronomers agreed that it was right to proceed statistically, working outwards from the solar system and examining first the nearer stars and then those

further away. They hoped this would allow them to estimate the extent of the system of those stars that were near
enough to be individually visible on telescopic plates, and
they prayed that the riddle of the Milky Way would
somehow solve itself. And so it was that by the turn of the
century, when C. A. Young (1834–1908), then a professor at
Princeton, came to write his textbook *Manual of
Astronomy*, a consensus had emerged among the more conservative minded. Most of the visible stars lay within a
round, flat disc of space, whose diameter was about eight or
ten times its thickness, and whose radius measured at least
ten or twenty thousand light years. At right angles to the
plane of the disc the stars were thinly scattered, eventually
giving way to the region of the nebulae. 'As to the Milky
Way itself, it is not certain whether the stars which
compose it form a sort of thin, flat, continuous sheet, or
whether they are ranged in a kind of *ring*, or in *spires*, with
a comparatively empty space in the middle where the Sun
is placed.'

Kapteyn's 'Plan of Selected Areas', 1906

Continuous sheet? A ring? Spires? The reason for the uncertainty was obvious: our location inside the Galaxy puts us
at a grave disadvantage when we try to picture the Galaxy
as a whole. Some astronomers – notably Hugo von Seeliger
(1849–1924), director of the Munich Observatory, and J. C.
Kapteyn (1851–1922), professor at Groningen in The
Netherlands – accepted that the secure way forward was
through study of the individually visible stars. The trouble
was that the necessary data concerning these stars – their
positions in the heavens, apparent magnitudes, and proper
motions – were accumulating much too slowly.

The basic problem, as we have seen (on page 258), was
that only the very nearest stars were close enough for their
distances to be found by straightforward triangulation, and
the method of spectroscopic parallaxes lay in the future.
However, it must in general be true that the nearer a star,
the larger its proper motion; and when sufficient proper
motion data were available, this truth could be exploited
statistically to provide a measuring rod capable of reaching
much further out into space.

How to acquire the necessary data? There was clearly
insufficient manpower in the existing astronomical community to permit the investigation of the entire sky in the
necessary detail, but the task could be reduced to manageable proportions by selecting representative samples of sky,
and sharing these out among a large number of observato-

ries. In 1906 Kapteyn published a *Plan of Selected Areas*, and he managed to persuade astronomers world-wide to make it the basis of an international campaign of data collection.

But there was a possible snag. The method involved the analysis of proper motions and apparent magnitudes, and it assumed that the apparent magnitudes could be relied upon – that they had not been distorted and the starlight dimmed by obscuring matter lurking in the interstellar spaces. The problem was recognized by those involved, but at the time the available evidence on obscuration seemed reassuring. However, it would eventually be found that dust was indeed present in the galactic plane, in such amounts that it not only affected the apparent magnitudes of the nearer stars, but concealed much of the Galaxy from inspection. Consequently, Kapteyn and his collaborators were unwittingly restricting themselves to a localized region within the Galaxy, while the great bulk of the galactic system lay hidden and beyond their ken. As a result, their investigations appeared to confirm the common view, that the Sun was located near the centre of the Galaxy, whose diameter was to be measured in terms of a few thousands of light years.

While this painstaking work was in progress, a daring young talent appeared on the American scene. Harlow Shapley (1885–1972) was born in Nashville, Missouri, and worked for a time as a reporter before resuming his education. In 1911 he received a fellowship to work with H. N. Russell at Princeton on eclipsing binary stars, and in 1914 he was recruited by G. E. Hale to the Mount Wilson staff. His career there, as observer and brilliant theoretician, was to be cut short by his appointment in 1921 to a position at Harvard, where the directorship was vacant. Russell had declined the post, Shapley greatly coveted it, and after a few months at Harvard was so appointed. He remained there until his retirement in 1952, creating a stimulating atmosphere and later playing a significant role in the international scientific community.

When he was appointed to Mount Wilson, Shapley had recently discussed his future plans with Solon I. Bailey of the Harvard College Observatory. Also on the Harvard staff was Henrietta Leavitt (1868–1921), whose duties in recent years had included the painstaking examination of variable stars on the photographs of the Small Magellanic Cloud taken at Harvard's southern station in Peru. To identify a variable star by examination of the differences in a pair of photographs was one thing; to determine its period – the number of days between one maximum brightness and the next – was altogether more difficult. However, though the task was onerous, the effort was especially worthwhile because all objects in the Cloud lay at about the same distance from Earth. A difference in apparent magnitude therefore corresponded to a difference in absolute magnitude (luminosity).

In 1908 Miss Leavitt had published a memoir listing 1,777 variables in the Cloud, and for sixteen of these she had been able to determine the periods. She remarked that the longer the periods, the brighter the stars. Four years later the sixteen had increased to twenty-five, and Miss Leavitt had found a mathematical relationship that linked the period with the apparent magnitude – and hence with the luminosity. This 'period–luminosity' relationship was to have momentous consequences for the history of astronomy.

In the variable stars in question (which Shapley had suggested were pulsating stars, see page 269), the brightness would rise rapidly to a maximum and then fade away slowly over a number of days, in the manner of the star Delta Cephei whose variability had been discovered by John Goodricke as long ago as 1784. Such 'Cepheid variables' are highly luminous and so very conspicuous, and their characteristic light curve makes them relatively easy to identify. They stand out from a cluster of stars much as a lighthouse beacon stands out from the ordinary house and street lights of a port. The 60-inch reflector at Mount Wilson was the most powerful telescope in the world for the study of distant and therefore faint objects, and Bailey – who had himself managed to identify a number of Cepheids in 'globular' star clusters – suggested that Shapley pursue this work with the 60-inch.

Globular clusters are striking objects, spherical assemblages of hundreds of thousands of stars. Edmond Halley had seen the brightest of them all, Omega Centauri, in 1677 when visiting St Helena (see page 214), and many more had been discovered by William and John Herschel in their sweeps for nebulae. Curiously, as John Herschel had noted, the globular clusters were by no means scattered uniformly across the sky. They were mostly in one half of the sky, and no fewer than one-third of them were concentrated in the Sagittarius region, which also contained some of the richest star clouds of the Milky Way. In 1909 the Swedish astronomer Karl Bohlin had suggested that the globular clusters formed a system that surrounds the centre of the Galaxy, which therefore lay far away in the direction of

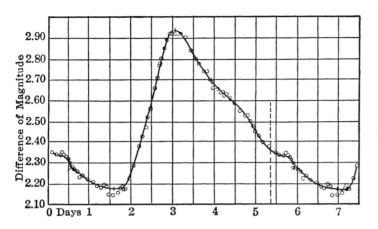

The light curve of Delta Cephei, prototype of Cepheid variables. A Cepheid goes through a cycle in which it increases rapidly in brightness, and then slowly declines. Courtesy of Institute of Astronomy, Cambridge University.

Sagittarius, at an enormous distance from the Sun. But cosmological speculations have a poor success rate and few took Bohlin seriously.

To measure the distances of the inconceivably remote globular clusters seemed a hopeless quest. But the resourceful Shapley saw a way of tackling this problem. Suppose Miss Leavitt's 'period–luminosity' relation for the Cepheids in the Cloud also held true for the Cepheids in the globular clusters. This being so, a Cepheid of a given period in one globular cluster was of the same luminosity as a similar Cepheid in a second cluster. Therefore, differences in the stars' apparent magnitudes must reflect differences in their distances – and therefore in the distances of the two clusters to which the stars respectively belonged.

But that would give only relative distances. To convert these to actual distances, it would be necessary to calibrate Miss Leavitt's relation – to establish the actual distance of one Cepheid somewhere. The only Cepheids near enough to offer any hope of this were among the isolated stars that are the Sun's neighbours in the Galaxy; and these galactic Cepheids would come into play only on the basis of a further assumption, that such Cepheids are likewise physically similar to those in the Cloud and in the globular clusters. Assumptions of uniformity of this type are inherent in any attempts in astronomy to measure great distances; the hope is that the assumption is true enough to give answers to otherwise unanswerable questions, and astronomers whose research depends upon distance measurements have their fingers permanently crossed. But such assumptions had never been made with the daring and virtuosity that Shapley was to display.

Unfortunately, even the nearest of the galactic Cepheids

were too distant for the usual methods of measuring stellar distances. Shapley used ingenious – but questionable – statistical procedures to overcome the difficulty, and so achieved the necessary calibration. This done, he set about exploiting this veritable yardstick for the universe. He was now able to obtain a distance for any globular cluster in which he could manage to determine the period and apparent brightness of a Cepheid variable.

This was easier said than done; even on a plate taken with the 60-inch, the individual stars in a globular cluster were very faint, if indeed they could be seen at all. To identify a star as a Cepheid variable, and to determine the number of days that it took to complete a cycle from one maximum brightness to the next, would require many such plates. With the dedication of a young man driven by ambition, Shapley worked and worked, until he had identified the periods of Cepheids in a dozen of the nearest globular clusters.

He then encountered the next problem: in the more distant clusters, Cepheids were too faint to be any longer visible. However, Shapley had noticed that the brightest stars in any one cluster were similar in luminosity to the brightest stars in other clusters. This he made the basis of another uniformity assumption, one that took him from clusters where he could see Cepheids in addition to the brightest stars, out to clusters where he could see only the brightest stars.

Finally, at distances where even the brightest stars could no longer be distinguished, he made yet another assumption, that the clusters themselves were physically uniform. Accordingly, comparison of the apparent diameter of a remote cluster with the apparent diameter of a cluster of known distance yielded the distance of the remote cluster.

By this succession of daring steps, Shapley in 1917 arrived at a distance for the most remote of the globular clusters, of some 200,000 light years. This was many times the accepted diameter of the entire Galaxy. To be so distant and still appear of significant size, the clusters themselves must be vast, perhaps (Shapley at first thought) comparable to the Galaxy itself.

But why were the clusters concentrated in the direction of Sagittarius, and why were they distributed symmetrically to either side of the galactic plane? Could it be that, as Bohlin had suspected a decade earlier, the Galaxy is enormously bigger than previously thought, and that its centre lies far away in the direction of Sagittarius, in the midst of the system of globular clusters? If so, the diameter of the

Galaxy must be a staggering 300,000 light years, and its volume a thousand times greater than other astronomers believed.

Whereas practitioners of the 'selected areas' approach were studying in painstaking fashion the stars near enough to be individually visible, Shapley had leapfrogged further and further into space, from globular cluster to globular cluster. His resulting model for the Galaxy, announced in 1918, was too daring for his contemporaries. To most of them, the distribution of the globular clusters to one side of the sky – the fact that had led him to his Big Galaxy – was a single anomaly in an otherwise coherent picture.

Shapley was to see himself as a latter-day Copernicus, who had dethroned man and banished him to the margins of the Galaxy. But such banishments had been common-place down the centuries, and had presented no problems. If anything, his estimate for the galactic diameter tended to make man special (and thereby added to the problems of acceptance); for by multiplying the volume of our Galaxy a thousandfold, Shapley made it the dominant structure in the known universe.

Orbital motions about the galactic centre

In 1925 the Swedish astronomer Bertil Lindblad (1895–1965) considered the consequences for the observed motion of the stars that would follow from Shapley's Galaxy in which the Sun was far removed from the galactic centre. In the solar system, the inner planets not only have less far to travel in their orbits than do the outer planets, but they move through space with greater velocities. The very same laws of dynamics apply to the stars. Suppose then that Shapley was right, and the solar system was located not at the centre of the Galaxy but at a great distance from it. The stability of the galactic system would require the Sun and the nearby stars all to be moving in orbit about the centre of the Galaxy. This being so, then because of the laws of dynamics, the stars in orbits inside that of the Sun would be travelling faster and so forging ahead of us, while the stars moving in orbits outside the Sun's would be falling behind. As viewed from the solar system, therefore, the inner stars would seem to be moving in one direction (as they forged ahead of us) and the outer stars in the opposite direction (as they fell behind); and the greater a star's distance from us in the direction towards (or away from) the galactic centre, the greater would be its apparent speed.

This theoretical pattern can easily be converted into an

equivalent pair of patterns, one of motions at right-angles
to the line of sight (the 'proper motions' that accumulate
century by century), and the other of motions in the line of
sight (radial velocities, as revealed by spectroscopic Doppler
shift). The demonstration that such patterns actually exist
in the observed stellar motions called for painstaking work,
carried out principally by Jan Oort (1900–92) of Leiden
Observatory.

Oort's conclusions, published in 1927, offered valuable
support to Shapley. Both believed that the Sun was located
far from the galactic centre. But because Oort was studying
stars that were nearer than Shapley's globular clusters, his
data were less affected by obscuration, and so the size he
derived for the Galaxy was markedly smaller than
Shapley's.

*A problem recognized: dark matter in inter-stellar
space*

The explanation for the discrepancy was not long in
coming. The application of photography to the mapping of
the sky had rapidly increased the number of irregular,
gaseous nebulae like the Orion Nebula that were found in
the Milky Way. They were regarded as individual objects in
space, and were given names and catalogue numbers, and
plotted on maps of the sky. The spectrograph showed them
to be made of glowing gas.

A new type of nebula was recognized in 1913. The star
Merope in the Pleiades had been found by Wilhelm Tempel
in 1859 to be surrounded by a faint nebulosity. Later photographs showed all the stars in the cluster to be involved in a
rather unusual nebula of faint, wispy streaks. V. M. Slipher
at the Lowell Observatory photographed the spectrum of
this nebulosity in 1912, and was surprised to find it a
continuous spectrum crossed by dark absorption lines,
which mimicked exactly the spectrum of the stars. The
nebula was evidently made of dust particles which reflected
the starlight; and more of these 'reflection nebulae' in the
vicinity of bright stars were soon discovered.

At about the same time, E. E. Barnard (1857–1923) of
Lick Observatory was using wide-field portrait lenses to
take a superb series of photographs of areas in the Milky
Way. They showed a complicated structure of star clouds,
and rifts and holes where there were few or no stars. At first
Barnard believed that this was the real distribution of the
stars, but as he continued his work he almost reluctantly
accepted that these were true clouds, not of bright gas but
of dark obscuring material, and in his two-volume *Atlas* of

photographs, published in 1927, he included a catalogue of the more prominent of them.

There was a curious reluctance on the part of astronomers to accept that the bright and dark nebulae might be indicative of a more general substrate of gas and dust in interstellar space; the general view was that these were isolated objects, and that the vast interstellar spaces were largely empty and quite transparent. The consequence of any significant amount of, in particular, absorbing dust, which would dim the light of distant stars and further complicate the determination of their distances and intrinsic brightnesses, would be very serious. Indeed, it almost seems that this was too great a problem to be contemplated. Eddington once remarked that astronomers were like the guest who refused to sleep in a reputedly haunted room, and who explained, 'I do not believe in ghosts but I am afraid of them'.

The evidence for the presence of a layer of obscuring material through the central plane of the Galaxy now seems to us overwhelming. In 1869, by painstakingly plotting onto sky charts the 4,000 'irresolvable' nebulae that had been catalogued by John Herschel, Richard Proctor had shown that the region of the sky close to the Milky Way contained very few of them. Most of these irresolvable nebulae later proved to be spirals. Heber D. Curtis (1872–1942) of Lick Observatory argued in 1920, in the 'Great Debate' with Harlow Shapley (see page 290), that the spiral nebulae that were seen edge-on had at least a peripheral band of dark matter, and that such a band in our own Galaxy would explain why the spiral nebulae appeared to avoid the Milky Way (see page 288).

The observations that finally convinced astronomers were the result of patient work by another staff member of Lick Observatory, Robert J. Trumpler (1886–1956), published as recently as 1930. Trumpler had for many years concentrated on the study of the hundreds of open star clusters, which are rather narrowly confined to the plane of the Milky Way, and of which the Hyades and the Pleiades are the nearest examples. He first grouped them into types characterized by similar structures and the similar shapes of their H–R diagrams. Within one type, the relative distance of the various clusters could then be measured by two distinct methods. On the one hand the Main Sequence stars of, say, spectral type F had the same absolute magnitude in all the clusters. The greater the difference in the observed apparent magnitudes of such stars in any two clusters, therefore, the greater was the difference in distance between the two clusters (the 'fainter means more distant' method). On the other hand, for clusters of the same intrinsic size, the angular diameters of the clusters on the sky was also a measure of their distance (the 'smaller means more distant' method).

The two methods did not give compatible results: the distances given by the apparent magnitude method were greater than those given by the angular diameters, because the starlight had been dimmed in its passage to the Earth. This showed the presence of a general absorption at all wavelengths.

Trumpler also investigated the phenomenon of selective absorption. If the dust particles are about the same size as the wavelength of light, the blue light is scattered out of the line of sight more than the red light, so the star appears to be redder than its spectral type would suggest. (The setting Sun appears red for just this reason.) This increases the measured 'colour index' of the star by an amount that is called the 'colour excess'. It was difficult to measure the quantities exactly on photographs, and only when photoelectric methods of measuring stellar magnitudes became widely used after about 1950 did measurements of the colour excess become a powerful method for correcting for the effects of the interstellar dust. Trumpler, summarizing his discovery in his 1930 paper, wrote: 'We are thus led to the conclusion that some general and selective absorption is taking place in our Milky Way system, but that this absorption is confined to a relatively thin layer extending more or less uniformly along the plane of symmetry of the system.'

No later discoveries were to change this view. Trumpler derived a general absorption averaged over different directions in the Galaxy of about 1 magnitude of dimming for every 5,000 light years, not much smaller than the presently accepted value. The concentration of the material to the plane of the Galaxy is very strong, so that the actual nucleus of the Galaxy some 30,000 light years away is invisible to the eye and the telescope, which explains the early failures to discern the spiral pattern of the Milky Way.

The Galaxy takes shape
In the years following, the essentials of the modern picture of the structure of the Galaxy finally took shape. It was summarized in 1938 by J. S. Plaskett (1865–1941), who had recently retired as Director of the Dominion Astrophysical Observatory in British Columbia. There was, he said, 'a great central disc of stars, irregularly distributed in groups or clusters, probably with a general underlying field of stars

and with possible spiral structure'. The diameter was some
100,000 light years, and the thickness increased to around
16,000 light years at the centre. 'A stratum of diffuse
absorbing matter strongly concentrated to the galactic
plane' had a thickness of some 1,000 light years, while
around the centre was a spherical assemblage of globular
clusters and stars of special types. After all the turmoil of
recent decades, Plaskett could at last take comfort in the
fact that 'the concept developed has a certain unity,
completeness and probability'.

The spiral nebulae

Lord Rosse's reflector had revealed the spiral shape of M 51
within weeks of seeing 'first light' in 1845, and in the
second half of the century, the number of known spirals
grew by leaps and bounds. In 1898 James E. Keeler began
systematic photography of nebulae with the Crossley reflector at the Lick Observatory. The distinction gradually

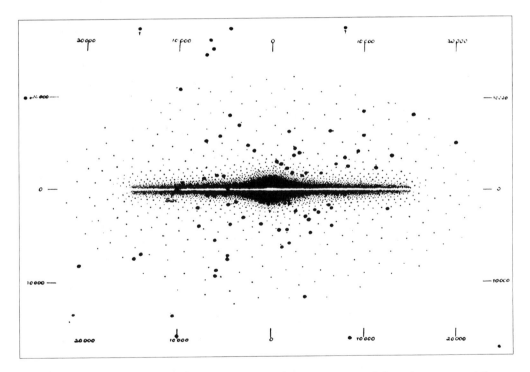

J. S. Plaskett's 1938 diagram to illustrate his
picture of the structure of the Galaxy, here seen
edge-on. The Galaxy has a diameter measuring
some 100,000 light years, and the central bulge is
16,000 light years in thickness. In the galactic
plane is a stratum of obscuring matter, while
around the Galaxy is a halo of globular star clusters and special stars. Courtesy of Institute of
Astronomy, Cambridge University.

became apparent between the irregularly shaped nebulae in the Milky Way and those with a more regular form. Moreover, in the parts of the sky where the spiral nebulae were common there were also many similarly sized nebulae which, although not showing a spiral structure, also had regular shapes, of circular or smoothly elliptical outline and, like the spirals, getting brighter towards their centres. It became common to call all of this class of regularly shaped nebulae 'spiral nebulae', even if a spiral structure was not recognizable. Keeler estimated that some 120,000 spiral nebulae were accessible to the Crossley, perhaps half of which would show an actual spiral form.

The 1900 consensus

What were these mysterious objects? Many, including Keeler, thought they were planetary systems in the making. If, he wrote, 'the spiral is the form normally assumed by a contracting nebulous mass, the idea at once suggests itself that the solar system has been evolved from a spiral nebula'. William Huggins, writing in 1889, had taken a similar view. Of a photograph of the Andromeda Nebula, he said that it 'shows a planetary system at a somewhat advanced stage of evolution; already several planets have been thrown off, and the central gaseous mass has condensed to a moderate size as compared with the dimensions it must have possessed before any planets had been formed'.

On the other hand, could the spiral nebulae be galaxies, vast star systems whose true nature was disguised from the Earth-based observer by their great distances? It was difficult to believe that such 'island universes' existed in their hundreds of thousands. And indeed there was compelling evidence to the contrary, some of it familiar in type, some new.

The long debate over the nature of the nebulae had witnessed many claims that change had occurred in a particular nebula, too rapidly for the nebula to be a distant and vast star system. One might have supposed that, once photographs superseded sketches made by observers of limited artistic ability working under difficult conditions, the reality or otherwise of such changes would be a question easily settled. Such was very far from being the case. As early as 1899, the leading English practitioner of nebular photography, the amateur Isaac Roberts (1829–1904), was claiming to have photographic evidence of the rotations of the Andromeda Nebula and of the Whirlpool Nebula, M 51 – rotations that we know to have been illusory.

There were also two important additional pieces of evidence against the identification of spirals with galaxies. One was the discovery by Richard Proctor (see page 283) that the region of the sky close to the Milky Way contained very few of them. If the spirals were galaxies comparable to our own, why did they distribute themselves in space so as to avoid our galactic plane?

Then, by a remarkable chance, in 1885 a star flared up in the Andromeda Nebula. It increased in brightness until it was equal to one-tenth of the entire nebula. If the nebula was indeed a galaxy containing millions of stars, then in a matter of days this one single star had increased its luminosity until it equalled the combined brightness of hundreds of thousands of ordinary stars. In fact this was exactly what had happened, for S Andromedae (as it is known) was a supernova; but no process known to nineteenth-century physics could have produced such dramatic celestial fireworks. It seemed far more likely that a single star had encountered a nebulous cloud of modest size, and flared up as it passed through the cloud.

The majority opinion among astronomers in 1890 was summarized by the highly respected historian of astronomy, Agnes C. Clerke, in her book whose title, *The System of the Stars*, significantly used 'system' in the singular:

> The question whether nebulae are external galaxies hardly any longer needs discussion. It has been answered by the progress of discovery. No competent thinker, with the whole of the available evidence before him, can now, it is safe to say, maintain any single nebulae to be a star system of coordinate rank with the Milky Way. A practical certainty has been attained that the entire contents, stellar and nebular, of the sphere belong to one mighty aggregation.

All scientists consider it their duty to work with theories as simple as the evidence will allow, and astronomers are no exception. The term 'nebula' had originated as a description of objects in the sky. Some (we know) are gaseous and so 'truly' nebulous, while others are stars systems disguised by distance; but over the centuries, astronomers had struggled to force them all into either one category or the other. Accordingly, when, in 1864, Huggins's spectroscope had demonstrated beyond doubt the existence of gaseous nebulae (see page 255), this had sown doubt in the minds of those who had hitherto believed in a multiplicity of island universes. Now these doubts had ripened into a widespread conviction that astronomy knew but one island universe.

1900–20: the consensus called into question

The pendulum had swung too far. Before long, new evidence suggested that the death of the island universe theory of spirals had been greatly exaggerated. Although the spirals were faint and what little light they sent was dispersed to the point of near-invisibility when passed through a spectrograph, by 1912 improvements in instrumentation and in photography allowed V. M. Slipher at the Lowell Observatory to secure spectrograms of a number of nebulae that were clear enough to reveal the more prominent spectral lines. The work was painstaking indeed, with twenty to forty hours of observing time required for each plate; but the results were sensational, and when he announced them to a meeting of the American Astronomical Society in 1914, he was rewarded with a standing ovation.

Not surprisingly, the Andromeda Nebula was the first that Slipher tackled. The continuous spectrum crossed by a few dark lines was characteristic of starlight, and by January 1913 he had secured four plates on which the shift in the spectral lines caused by the Doppler Effect (see page 256) could be measured. Stars were known to move with line-of-sight ('radial') velocities of perhaps 20 km/sec; Slipher found that the Andromeda Nebula was approaching at no less than 300 km/sec, by far the greatest velocity known for any object in the universe. Word of his discovery was greeted by some astronomers with incredulity, but soon his results were confirmed by other observers. By the time of the 1914 meeting, Slipher had radial velocities for fifteen spirals, and by 1917 the total had grown to twenty-five. No fewer than four of these were in excess of 1,000 km/sec.

Slipher's twenty-five spirals were not uniformly distributed across the sky. Most were located to one side of the Galaxy and were receding; but some, including the Andromeda Nebula, were on the opposite side and most of these were moving towards us. To Slipher the explanation was clear: the Galaxy was itself 'a great spiral nebula which we see from within', and it was drifting among the other spirals with a velocity of some 700 km/sec.

Meanwhile, at Lick Observatory, Heber D. Curtis was continuing the programme of nebular photography begun by Keeler, and (as we have seen, page 283) he was encountering nebulae that were clearly edge-on to us. These had at least a peripheral band of obscuring matter, and Curtis realized that similar obscuring matter in our own Galaxy would prevent our seeing spiral nebulae close to the galactic plane. He concluded that spirals did not in fact avoid the

Milky Way as Proctor and others had thought; spirals were present in those directions as elsewhere, but concealed from our view. This confirmed him in his belief that the spirals were 'inconceivably distant, galaxies of stars or separate stellar universes so remote that an entire galaxy becomes but an unresolved haze of light'.

His conviction received a further boost in 1917, when he discovered three novae on his photographs of spirals. At about the same time, G. W. Ritchey (1864–1945) of Mount Wilson found a nova on a photograph of a spiral taken only a few days earlier, and this nova was still visible and so available for study by astronomers worldwide. Encouraged by these successes, others joined in the hunt, and yet more novae were discovered. Without exception they were altogether fainter than S Andromedae, which suggested to Curtis that novae might fall into two distinct classes (as later proved to be the case, see page 270). Suddenly the inference drawn from the exceptional brilliance of S Andromedae began to seem less secure.

But not all the new evidence favoured the island universe theory of spirals. At Mount Wilson Observatory a Dutch astronomer, Adriaan van Maanen (1884–1946), had won himself a reputation as a meticulous measurer of photographic plates. In 1916, to study possible changes in the spiral nebula M 101, van Maanen used a machine called a blink microscope that enabled him to switch between two photographs of the nebula taken at different times. This made it relatively easy to detect alterations that had taken place in the interval between the two photographs. He concluded that the nebula was rotating. This being so, it could hardly be a distant galaxy of vast diameter; for as the spiral rotated, the outlying parts of the supposed galaxy would have had to travel through space with incredibly large velocities.

It is difficult to imagine any investigation in astronomy that could stay closer to the plain facts. The photographs represented objective evidence; and van Maanen had in effect superimposed them, measured the changes, and described the rotation that had given rise to these changes. By 1921 he had similar results for three more spirals. 'Congratulations on the nebulous results!', Shapley wrote. 'Between us we have put a crimp in the island universes, it seems, – you by bringing the spirals in and I by pushing the Galaxy out. We are indeed clever, we are.'

Another development concerned the radial velocities of spirals. As the number of known velocities grew, it became clear that Slipher had been premature in interpreting the evidence as revealing a drift of the Galaxy among the spirals. Spirals that were approaching the Galaxy proved to be exceptional; for the most part, the spirals were receding, in all directions. Shapley even suggested that they might be little more than wisps of matter driven off from the Galaxy by radiation pressure from the combined light of the galactic stars. And so the movement in favour of the 'island universe' theory of spirals lost a little of the momentum it had recently acquired.

The 'Great Debate'

In 1920 a meeting in Washington was arranged, at which Harlow Shapley and Heber D. Curtis were to argue for their contrasting positions on the structure of the universe. But their attentions were in fact focused in different directions. Shapley was committed to his Big Galaxy; the spirals were to him of minor importance, though they could scarcely be comparable galaxies with diameters of 300,000 light years. Curtis was studying spiral nebulae, which he believed to be island universes; the structure of the Galaxy was of secondary interest, though he required it to be of the manageable size accepted by nearly everyone except Shapley.

Though the meeting has since been built up into legend as 'The Great Debate', this is the result of confusion between what actually happened in Washington and what was published many months later. At the meeting itself, Shapley was uncomfortably aware of the presence of representatives from Harvard Observatory, whose vacant directorship he coveted; and so he made sure he did not suffer defeat, by reading a text so elementary as to verge on the trivial. Curtis, who spoke second, thought about abandoning his technical material, but decided it was then too late to change. In the published version of the 'debate', however, each man gave a powerful summary of the arguments in favour of his position. Astronomers found it hard to decide whose case was the stronger.

Since 1917 the 100-inch Hooker reflector had been in service at Mount Wilson. It had the largest mirror of any telescope in the world and this was of superb optical quality, and so the 100-inch was uniquely powerful for the photography of faint objects such as nebulae. In 1920 J. C. Duncan (1882–1967), director of the small observatory at Wellesley College in Massachusetts, had noticed three faint variable stars in the spiral nebula M 33, on photographs that included one taken with the 100-inch. They were encouraging indications of what might be achieved through a determined campaign with the 100-inch.

Astronomy in Southern California, 1920–60

In the remainder of this chapter we trace the widening interests of astronomers, from the study of the stars in the Galaxy (and that important study continued, and continues today), to embrace the study of galaxies in the larger universe. Astronomy has always been an international and an internationally-minded profession. Yet it may seem in what follows that one observatory and one small group of astronomers dominate the history for several decades, and a word of explanation is necessary.

It was part of the legacy of George Ellery Hale that the new and most powerful telescopes in the world were built on carefully selected sites in the coastal range mountains of southern California, with base offices and laboratories in the small town of Pasadena near Los Angeles. The Mount Wilson Observatory was operated by the Carnegie Institution of Washington, and the later Palomar Mountain Observatory by the California Institute of Technology. For some years from 1948 the two worked together in loose association, with shared access to facilities.

The 200-inch telescope was long in building. First planned as early as the 1920s when the 100-inch had proved successful, the pyrex glass mirror blank was not ready for grinding until 1936; the war years intervened, and after not unexpected technical problems in commissioning, the completed telescope came into regular use only in 1949. At about the same time, and also on Palomar Mountain, another less heralded but unique instrument became operational. This was a large camera, of nearly 50-inches aperture, of a type that had been invented by the Estonian-born optician Bernhard Schmidt (1879–1935) who was working in Hamburg in the 1930s. This was the most innovative telescope design for centuries; basically a reflecting telescope but with a weak single correcting lens, it combined large aperture with wide field. The Palomar 'Big Schmidt' embarked upon a photographic survey of the whole northern sky of previously undreamed-of quality; copies of the plates were made available to other observatories as the work slowly progressed.

Although visiting astronomers were always welcomed in Pasadena, they had only limited direct access to the telescopes that were then alone in the world in being powerful enough to tackle the more exacting observational problems of extragalactic research. For the most part the instruments were used by the relatively small established staff, who had opportunities for research denied to their colleagues elsewhere in the world. Several became leaders in the field.

Edwin Powell Hubble (1889–1953) was born in Missouri
and gained a Rhodes scholarship to Oxford University,
where he studied law, but he eventually turned to astronomy. He joined the Mount Wilson staff in 1919, and
although his egocentric and flamboyant personality did not
always endear him to colleagues, the discoveries he made
during his career there can be said to have changed our
picture of the universe. Sadly, his use of the 200-inch telescope to which he had so keenly looked forward was
restricted by his terminal illness.

Working with Hubble, and a near contemporary, was
Milton L. Humason (1891–1972), whose career was remarkable. With no university education but with an enthusiasm
for astronomy, he had worked for a time as a mule team
driver in the early days of Mount Wilson. He then joined
the assistant staff in a junior capacity, learned photography,
and eventually became a specialist in the long-exposure
spectroscopy of faint galaxies with the big telescopes.

Walter Baade (1893–1960) had a more conventional academic upbringing in Germany, gaining his doctorate in astrophysics at Göttingen in 1919 before joining the staff of the
Hamburg Observatory (where he came to know and encourage the optician Bernhard Schmidt). After a year as a visitor
at Mount Wilson in 1930 he joined the permanent staff in
1931, returning to Göttingen on his retirement in 1958.

Hubble's discovery of Cepheids in the Andromeda Nebula, 1923–24

Release from army service in 1919 enabled Hubble to take
up an earlier invitation to join the Mount Wilson staff. In
October 1923, Hubble embarked on a systematic search for
novae in the Andromeda Nebula, biggest of all the spirals
and presumably one of the most accessible. He found one
on the very first of his plates. This at least was his initial
verdict. But as he worked his way through the Mount
Wilson collection of photographs of the nebula, mostly
taken with the 60-inch and going back to 1909, he realised
that this star was no newcomer but a variable. When the
search was complete he had found over sixty plates on
which the star was present, sometimes fainter than nineteenth magnitude, at other times as bright as eighteenth.
The plates were numerous enough to allow him to plot the
pattern of the rise and fall of the star's light.

It began to look as if the variable was the greatest prize
possible: a Cepheid, one of the 'lighthouse' stars that
Shapley had used in his distance determinations of the
globular clusters. If so, its light should leap rapidly from

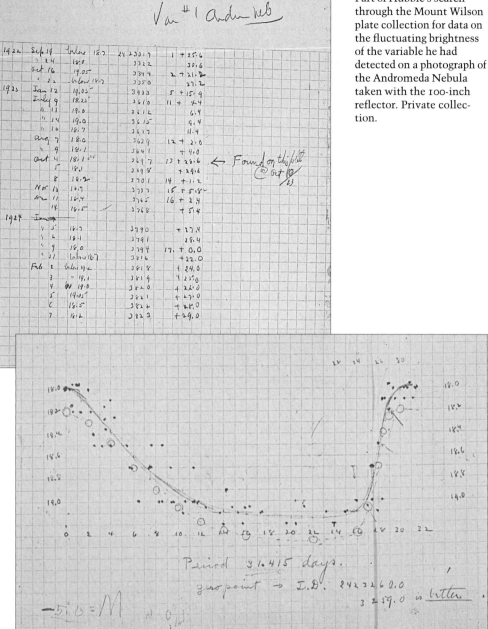

Part of Hubble's search through the Mount Wilson plate collection for data on the fluctuating brightness of the variable he had detected on a photograph of the Andromeda Nebula taken with the 100-inch reflector. Private collection.

The results of Hubble's search: the fluctuations were cyclic and the light curve was of characteristic Cepheid shape (see page 279), with period 31.415 days. Private collection.

minimum to maximum. To test the prediction, Hubble
took a series of plates during the first week of February
1924. The brightness of the star rose rapidly as expected
(see figure): it was indeed a Cepheid.

Hubble must have had difficulty in containing his excitement as he confirmed the characteristic shape of the light
curve and its lengthy period of over 31 days. The longer the
period of a Cepheid, the greater the luminosity of the star;
yet this star appeared as only eighteenth magnitude at best,
and was so faint that it had passed unnoticed by earlier
Mount Wilson investigators. For it to be so luminous and
yet appear so faint, its distance – and the distance of the
nebula of which it was part – must be enormous, indeed
approaching one million light years. Even on Shapley's estimates, the nebula lay far outside the Galaxy.

Furthermore, the Cepheid proved that the nebula contained not merely star-like objects of doubtful interpretation, but a true star that varied in familiar fashion. And not
just one such star: by the time Hubble was confident
enough to break silence on 19 February with a letter to
Shapley, he had found a second variable, as well as nine
novae.

It was a momentous step in the history of astronomy.
Hubble had used the same yardstick as Shapley, but in far
more convincing and direct fashion. Shapley, to whom the
Big Galaxy was paramount and the status of the spirals a
secondary consideration, immediately accepted the implications of Hubble's discovery.

Not so van Maanen, who was Hubble's colleague but no
friend – the pair of them communicated only via intermedi-
aries. Hubble hesitated to go into print, troubled by the
incompatibility of his results with van Maanen's alleged
rotations in spiral nebulae. Eventually, and only at the
insistence of Shapley and others, Hubble allowed a paper
summarising his discoveries to be read in his absence at a
meeting of the American Astronomical Society, on New
Year's Day, 1925. Those present knew the long debate had
ended: there are indeed many 'island universes'.

Hubble's demolition of van Maanen

Two serious anomalies remained: the van Maanen rotations, and Shapley's estimate for the diameter of our
Galaxy, which implied that if the Andromeda Nebula was
an island, the Galaxy was a continent by comparison. The
van Maanen problem became acute as the Dutchman point-
edly persevered with his comparisons of pairs of photographs of spirals, concluding in every case that the nebula

was rotating at a speed that made it physically impossible
for it to be an island universe.

Hubble, his irritation ripening into anger, finally decided
on an ingenious strategy. During 1932 and 1933 he retraced
van Maanen's steps, comparing the very same pairs of
photographic plates (though with significant differences in
technique). In addition, he had new plates specially taken
for the purpose. He also enlisted two experienced colleagues, who were to make their own independent measurements.

All three agreed that in the nebulae in question, 'internal
motions of the order predicted by van Maanen do not exist'.
It could hardly be supposed that Hubble and his colleagues
had all made errors in their measurements, errors that by
the most amazing coincidences invariably cancelled out the
'rotations' and left them with null results. Clearly, the rotations were illusory.

Hubble drafted papers for publication embodying his
demolition of van Maanen, but the director of Mount
Wilson would not permit a public display of the feuding
within his staff. A compromise was imposed, and in 1935
readers of the *Astrophysical Journal* were no doubt
intrigued to find there a two-page paper by Hubble deli-
cately outlining his conclusions, immediately followed by a
two-page paper by van Maanen, conceding that 'it is desirable to view the motions [the rotations in the spirals] with
reserve'.

The Galaxy and the Andromeda Nebula: comparability at last

The second anomaly created by Hubble's discovery
stemmed from Shapley's estimate of the size of the Galaxy.
According to Shapley, the Galaxy was some 300,000 light
years in diameter; Hubble's distance for the Andromeda
Nebula implied that in diameter it was only one-tenth as
big as the Galaxy, and so a mere one-thousandth in volume.
The discrepancy was diminished in 1930 by Trumpler's discovery of the obscuration that dims the light from distant
objects in the galactic plane, which had misled Shapley into
thinking galactic Cepheids more distant than they really
were. This reduced the Galaxy's diameter to 100,000 light
years. And meanwhile, increasingly sensitive photographic
plates showed that the diameter of the Andromeda Nebula
was more extensive than had been thought. Yet even so, the
Galaxy remained a super-system and the Andromeda
Nebula small by comparison.

The instincts of many astronomers made them

uncomfortable over any theory that privileged our location in the universe. The discrepancy was the more puzzling in that the Andromeda Nebula resembled the Galaxy in so many ways. Each contained numerous bright Cepheids, a system of globular clusters, and (in all probability) spiral arms outlined by layers of dust and highly luminous blue stars. Yet the novae in the Andromeda Nebula were supposedly fainter than their counterparts in the Galaxy, as were the most luminous stars, and the globular clusters.

The globular clusters were particularly puzzling, since here uncertainties as to the effects of loss of light through the presence of obscuring matter in interstellar space could play no part: obscuration would affect equally a globular cluster and the distance-measuring stars in it, and their relative brightnesses would be unaffected. Yet the globular clusters in the Andromeda Nebula were on current figures some four times fainter than the globulars in the Galaxy.

One of the few astronomers to point out repeatedly that these anomalies would disappear if the Andromeda Nebula were assigned a distance twice as large as that currently accepted was Knut Lundmark (1889–1958) of Lund in Sweden. Also, in 1922 the Estonian astronomer Ernst Öpik (1893–1985) used an ingenious new method to determine the distance of the Andromeda Nebula. A few years earlier F. G. Pease of Mount Wilson had used the Doppler Effect (see page 256) to measure the rotation velocities of different parts of the nebula. Öpik showed that if, as seemed reasonable, the mix of stars was the same as in our own Galaxy and their average ratio of mass to luminosity was the same, the Andromeda Nebula had to be about 1.5 million light years away if Pease's observations were to be explained. This was also much greater than Hubble's 1925 value of 900,000 light years, and would imply that the Andromeda Nebula was bigger than Hubble thought. But so much had been invested in the accuracy of the Cepheid method of establishing distances, that it had become almost an article of faith among astronomers, and Hubble's values prevailed.

One obstacle to progress was the lack of information about the bright central region of the Andromeda Nebula; its outer parts had been 'resolved' into stars on photographic plates, but not so the central condensation. With existing instrumentation the task seemed hopeless. The Mount Wilson 100-inch was the biggest reflector in the world, but even so, theoretical considerations put the resolution of the central bulge at the very limit of its powers. In any case, the city lights of nearby Los Angeles made the quest hopeless.

But as luck would have it, these lights were dimmed as a security measure when the United States entered the Second World War; and although most of the astronomers were away on war service, one who remained was the German Walter Baade (see page 292) who had a physical handicap (and who in any case had overlooked the formalities of taking out American citizenship, and so was exempt from military duty).

Baade was exceptionally skilled in long-exposure photography with the 100-inch, and his first efforts in the autumn of 1942 brought him to the verge of success. Unfortunately, although the city was relatively dark, his fast blue-sensitive plates registered the faint background glow of the night sky, and this limited the useful exposure time to about ninety minutes. Baade therefore turned to the newly-introduced red-sensitive plates. These would be less sensitive to the background glow, which he further reduced with coloured filters. They would also be more sensitive to red stars; and the fact that Hubble had resolved the spiral arms of the nebula, where there were very bright, blue stars, but had failed to resolve the central bulge suggested that red stars might predominate there.

In the autumn of 1943, Baade photographed the central region of the Andromeda Nebula with exposures of about four hours each. His control of the telescope had to be meticulous in the extreme, not least because changes in air temperature as the night progressed constantly threatened to throw the instrument out of focus. He was rewarded with photographs that showed stars in their 'thousands and tens of thousands', as was also the case with the two smaller elliptical companion nebulae.

Large numbers of the stars were indeed red, but they were much brighter than the red stars Trumpler had found in the open clusters within our Galaxy, and so they occupied different positions on the H–R diagram. It eventually occurred to Baade that the red stars in the Andromeda Nebula had, instead, positions on the diagram similar to the red stars in the globular clusters that surround our Galaxy.

Could the comparison be extended? Did the central bulge of the Andromeda Nebula also have short-period variables of the type known as RR Lyrae stars, similar to those so common in our globular clusters? Unfortunately, there was no way of answering the question, for such stars would be altogether too faint for detection with existing instrumentation.

But Hubble and Baade found an alternative way forward. Some years before, Shapley had announced the discovery of a new type of stellar system, examples of which were in the constellations Sculptor and Fornax. They were located outside our Galaxy, but were much smaller than the elliptical galaxies. Yet while they were much bigger than our globular clusters, their populations of stars resembled them: lots of red giants and RR Lyrae variables. Astronomers tended to think of them as overgrown globular clusters, but after Baade found numerous red giants in the centre of the Andromeda Nebula, the penny finally dropped: the Sculptor and Fornax systems were intermediate between the globular clusters and the elliptical nebulae, and similar populations of stars were to be found in all three.

In 1944, Baade announced his conclusion that stars belonged to one or other of two types. Type I stars populated the central plane of the Galaxy. They included stars like the Sun and most of its near neighbours, and the stars that are found in open clusters like the Hyades and the Pleiades. They had been formed from the interstellar material, gas and dust, that also occupied the central plane of the Galaxy, and their average chemical composition was similar to that of the Sun. The most luminous blue giants at the top end of the Main Sequence had formed only recently from the interstellar material, and all – stars, gas and dust – followed nearly circular orbits about the centre of the Galaxy.

Type II stars, on the other hand, were older stars, found in the gas- and dust-free elliptical galaxies, in the equally dust-free globular clusters associated with spiral galaxies, and in the central bulges of the spirals. In our Galaxy, the globular clusters and the isolated Type II stars move in elliptical orbits inclined at all angles to the galactic plane. The stars of Type II on average do not share the rapid circular motion of the Sun and the other Type I stars in the central plane of the Galaxy. When the orbit of a Type II star carries it into the neighbourhood of the Sun, its measured velocity consequently appears to be high.

The true complexity of nature had once again frustrated astronomers' attempts to impose a simple pattern. And in time, even Baade's two stellar populations would themselves prove to be an oversimplification.

Another anomaly that concerned Baade involved RR Lyrae variables. These variables were especially numerous in globular clusters, and for this reason were often spoken of as 'cluster variables'. They had much in common with the Cepheids of shorter periods, with which they had earlier been conflated. But whereas the luminosity of a Cepheid depended on the number of days in its period, the luminos-

ities of RR Lyrae stars in our Galaxy were all much the
same, irrespective of the length of their periods.

In his estimates of the distances of globular clusters (and
hence of the diameter of the Galaxy), Shapley had sometimes been forced by the lack of Cepheids into using RR
Lyrae stars instead, even though RR Lyrae variables with
periods of half a day were about two magnitudes fainter
than Cepheids with periods of 10 days. In photographs of
the Andromeda Nebula taken with the 100-inch, these
Cepheids appeared to have magnitude 20, in which case the
RR Lyrae variables would presumably have magnitude 22 –
beyond the reach of the 100-inch, but within reach of the
200-inch being built at Palomar Mountain, far enough
south of Los Angeles to avoid the worst of the glare of the
city lights.

Baade started to use the 200-inch as soon as it came into
service in 1949. He expected the telescope would reveal RR
Lyrae variables in the Andromeda Nebula, but, as he told
the 1952 meeting in Rome of the International
Astronomical Union, the first plates taken of the nebula
with the 200-inch 'showed at once that something was
wrong'. The reason was to be found in Baade's two populations of stars. The nearby Cepheids that Shapley had used
to calibrate the period–luminosity relationship were in the
spiral arms of the Galaxy and so of Type I, and these now
proved to be more luminous – and therefore more distant –
than had been thought. So too were the remote Cepheids
that Hubble had detected in the spiral arms of the
Andromeda Nebula, which likewise had to be removed to a
greater distance. But the Cepheids that Shapley had used to
determine the distances of the globular clusters, and hence
the diameter of the Galaxy, were of Type II, and their
brightnesses had been correctly assessed. It followed that
the diameter of the Galaxy remained as before, while the
distance, and therefore the diameter, of the Andromeda
Nebula was to be doubled. As Baade put it:

> Moreover, the error must be such that our previous esti-
> mates of extragalactic distances – not distances within our
> own Galaxy – were too small by as much as a factor 2. Many
> notable implications followed immediately from the cor-
> rected distances: the globular clusters in M 31 [the
> Andromeda Nebula] and in our own Galaxy now come out
> to have closely similar luminosities; and our own Galaxy
> may now come out to be somewhat smaller than M 31.

Our Galaxy thus lost the preeminence with which it had
been endowed by Shapley, and was given the status it has

today, that of a somewhat inferior sister to the Andromeda Nebula.

The expansion of the universe and the theories of relativity

In his address to the Rome meeting in 1952, Baade added a further comment: 'Above all, Hubble's characteristic time scale for the Universe must now be increased from about 1.8×10^9 years to about 3.6×10^9 years.' How had Hubble succeeded in assigning an age to the Universe?

To answer this question we must go back into the nineteenth century. In astronomy the application of Newton's inverse-square law of gravity, applied to increasingly difficult problems in celestial mechanics with ever increasing mathematical power, had been triumphantly successful. It worked in the context of a simple and familiar universe: a framework of three dimensions of space in which the geometry of Euclid applied, and in which a uniform flow of time – past, present and future – could be measured everywhere by clocks which, in theory if not in practice, were uniformly perfect. But one irritating observation remained: the anomalous advance of the perihelion of Mercury, discovered by Le Verrier in 1859 (see page 164), obstinately refused to yield to Newtonian theory.

In the physics laboratory, the middle decades of the nineteenth century had produced a vastly complex body of knowledge about the relations between electricity, magnetism and matter, which were largely explained by Clerk Maxwell in the 1860s, in a unifying theory that ranks in importance with Newton's theory of gravity. There were features of Maxwell's equations that made other theoretical physicists, like Ernst Mach and the Dutchman H. A. Lorentz (1853–1928), think deeply about the concepts of space and time and their relation; and mathematicians such as Gauss and his pupil G. F. B. Riemann (1826–66) had already explored geometries in curved spaces that were logically consistent even though they disregarded one of Euclid's postulates.

In 1880, the Polish-born American physicist Albert A. Michelson (1852–1931) was working at the Astrophysical Observatory in Potsdam, which had a vibration-free cellar. This fortunate circumstance allowed him to carry out a delicate optical measurement suggested by Maxwell's theory. The puzzling result was, in brief, that the electromagnetic theory of light required an aether, but the Earth could not be detected moving through it.

These and other investigations show that cracks were appearing in what had seemed a firmly-established structure of physics and mechanics, and that concepts that had once seemed beyond challenge were now being questioned. The greatest contributions to the creation of a new set of concepts of space and time were to be made by the German-born Albert Einstein (1879–1955) in the first two decades of the twentieth century. As Newton had said of himself, Einstein stood on the shoulders of giants, for Einstein developed his own ideas from the context of this new spirit of enquiry, by building on the work of Lorentz and numerous others.

Einstein's first important contribution came in 1905 with the recognition that the measurement of quantities like length and time were made, not in an absolute Newtonian framework, but relative to the observers. This Special Theory of Relativity explained, for example, the puzzling result of Michelson's experiment.

The General Theory, of 1915, went far beyond. It was essentially a theory of gravitation – not a modification of Newtonian theory, but a quite new way of thinking about the relation between matter, space, and time. The equations permitted the calculation of planetary motions (as Newton's had done) and almost immediately gave a new theoretical value for the motion of the perihelion of Mercury which now agreed with the observed one. They also predicted that a ray of light would be deflected when it passed near a massive body like the Sun by an amount that was significantly greater than the Newtonian value. The test was made by two expeditions from the Greenwich Observatory, under the direction of Eddington, at a particularly favourable eclipse of the Sun in 1919, when the new value was verified.

The redshift in the spectra of the nebulae

What concerns us here is that the equations also permitted various solutions (depending on the assumptions that were made, for example, about the amount of matter in the universe and the geometry of space-time) as to how the universe would behave 'in the large'. Some of these solutions predicted that the universe would not be static, but would expand.

This brings us back to the observations of the large radial velocities of the spiral nebulae by Slipher from 1912 onwards (see page 288). It is clear that at first Einstein was not aware of Slipher's work. Equally, the difficult concepts, and advanced mathematics, of Einstein's General Theory were beyond the reach of most practising astronomers. But from the middle 1920s, particularly following the theoretical work of Willem de Sitter (1872–1934) of Leiden University and A. A. Friedmann (1888–1925) of St Petersburg, it became apparent that cosmology was no longer a speculative subject but a new branch of science with both theoretical and observational aspects.

By 1925, a total of forty-five radial velocities of nebulae were available, most of them determined by Slipher. A few, however, had been checked at other observatories, and the reliability of Slipher's measurements was generally accepted. The largest were over 1,000 km/sec, suggesting that the nebulae were independent bodies outside the gravitational control of the Galaxy, consistent with Hubble's newly demonstrated theory of island universes.

The great range in the observed velocities made them difficult to interpret. In particular, it was difficult to disentangle the component of the observed motion that was due to the combined motions of the Sun and of our own Galaxy and not to that of the nebula under investigation. In 1918 Carl Wirtz (1876–1939) of Kiel, Germany, had attempted a method that had proved useful in the study of the motions of stars, and had introduced into the equations an additional term, known as the K term. It was in effect a quantity to be subtracted from all the measured velocities before one searched for the motion of the solar system. The K term for the nebulae also turned out to be enormously higher than that for stars, being measured in hundreds of kilometres per second rather than only a handful. It did not escape Wirtz's attention that the K term implied a remarkable expansion of the system of the nebulae, with nebulae rushing away in all directions.

In 1924, now using forty-two nebulae, Wirtz tried to see if the K term depended on the distance of the nebulae. These distances were of course unknown; but if the nebulae were supposed to belong to a single class of similar objects, then the smaller the apparent diameter, the greater the distance. A rough relation between distance and velocity away from the observer did emerge, in the sense that the more distant the nebula, the faster it was receding. The cosmological implications were now clearly understood: the title of Wirtz's paper (translated from the German of the *Astronomische Nachrichten* of 1924) is 'De Sitter's cosmology and the motion of the spiral nebulae'.

Hubble and the law of the redshifts
Hubble was well aware of Wirtz's work and acknowledges it freely in his writings. From 1925 he devoted a generous

ration of his observing time on the 100-inch telescope (and, in the very last years of his life, on the 200-inch) to this problem of observational cosmology – the relation between the velocity of recession and the distance of the spiral nebulae. Over the years the development of thinking about these objects is reflected in the term astronomers used for them: 'spiral nebulae' become 'extragalactic nebulae' – that is, nebulae outside our own Galaxy – and finally the modern 'galaxies', that is, objects akin to our own Galaxy.

Hubble shared the work with his colleague Milton Humason: Humason concentrated his observing skills on the determination of the radial velocities of ever more distant (and fainter) galaxies, and Hubble devised and put into practice methods of determining the actual distances from the brightness of the images on the photographic plates. But even with the 100-inch telescope, Cepheid variables were detectable only in the nearest galaxies. Hubble developed a 'stepping stone' method which later astronomers followed. In the galaxies where Cepheids were just visible, he identified the most luminous individual stars, which were some 50 or 100 times brighter than the Cepheids, and then used the same types of stars as standard objects in more distant galaxies. It was tediously slow work, in which small errors could easily multiply, as Hubble recognised.

By 1929 he had radial velocities and independent distance determinations for twenty-four galaxies, and he published a graph of the velocities (up to 1,100 km/sec) as a function of distance (up to 2×10^6 parsecs, or a little over 6×10^6 light years). Though the scatter was large, there was clearly a relation between velocity and distance; the best straight line showed that for every million parsecs (that is, 1 Mpc) the velocity of recession increased by 500 kilometres per second. The quantity for which Hubble gave this first estimate, 500 km/sec/Mpc, became known as Hubble's constant, and is one of the fundamental constants of cosmology. That the velocity of recession of galaxies is proportional to distance is known as Hubble's Law, or – since the velocities are determined from the Doppler shift to the red – the Law of the Redshifts.

How far out into the universe did the law hold? This had to be determined by a quite different method, for redshifts could be measured for galaxies in which even the most luminous stars could not be discerned. Hubble developed an ingenious method. Just as there are clusters of stars in the Galaxy, so also there are readily recognizable clusters of galaxies in the universe. Humason took examples of these clusters, ones whose size and apparent brightness showed they were at very different distances, and measured the redshift of the brightest galaxies in each of the clusters. Hubble recognized that a cluster of galaxies might contain one or two individual members that were freakishly bright, but he considered that the risk of such anomalies could be avoided if one measured in each cluster the apparent magnitude, not of the brightest galaxy, but of the fifth brightest. Working along these lines, Hubble and Humason showed that the law of the redshifts held to great distances. This implied that if Hubble's constant was known by other means, the redshift of a faint galaxy could be used to calculate its distance.

In 1935 Hubble was invited to give a course of lectures at Yale University, and he published them under the title of *The Realm of the Nebulae*. It clearly summarized the knowledge of observational cosmology at the time, and influenced a whole new generation of astronomers.

The interpretation of the redshift

It is interesting to note that the redshift was, from the first, interpreted as a Doppler shift – that is, due to the move-

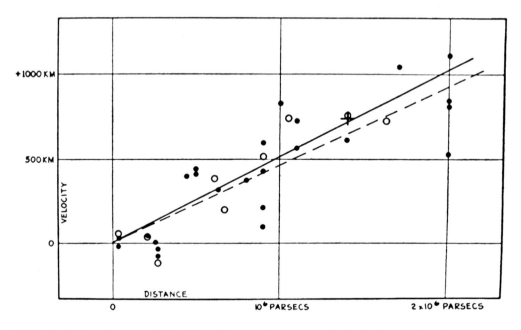

Hubble's velocity–distance relation, for galaxies at distances up to 2×10⁶ parsecs. The vertical axis represents radial velocities in kilometres per second (corrected for the motion of the solar system itself), the horizontal axis the distances in parsecs. The black discs and circles indicate the observational data, corrected by alternative methods of calculating the solar motion. The (alternative) lines indicate the simple relation inferred from the data. Private Collection.

ment of the galaxy away from the observer – and was expressed in kilometres per second. The implication was that the universe was expanding. There is no doubt that the nearly simultaneous detection of the redshifts and the derivation of solutions of Einstein's equations that suggested that the universe would be expected to expand greatly encouraged this interpretation.

It followed from the expansion interpretation that as one went back in time, the galaxies were closer together: at some early time (which we may simplistically term 'the beginning of the universe') the universe would have been extremely dense. The interval from then to the present could be termed 'the age of the universe', which in the simplest interpretation was inversely proportional to the value of Hubble's constant – the larger Hubble's constant, the younger the universe. This already presented something of a problem in the 1940s: the Hubble age of about 1.8×10^9 years was less than the accepted age of the Earth required by geologists.

We have already seen (on page 299) that Baade's version of the Cepheid calibration reduced the value of Hubble's constant by a factor of two, and a further downward revision to about 100 km/sec/Mpc followed in 1958 when Hubble's former pupil, Allan Sandage (1926–   ), showed that what Hubble had thought to be the individual brightest stars in galaxies were in fact aggregations of highly luminous stars embedded in gaseous nebulae. Both these reductions eased the conflict with the geological time scale.

It was already recognized, however, that this definition of the age of the universe was an oversimplification, for the value of Hubble's 'constant' would change with time if the expansion slowed down. In the 1950s and 1960s, Hubble and his successors expected these problems to yield to an early resolution. They did not, and uncertainties and disagreements remain to the present day.

9

Astronomy's widening horizons

Michael Hoskin and
Owen Gingerich

Where does history end and astronomy begin? Historians of every scientific discipline feel less and less secure as they approach the present day, and the historian of astronomy is no exception. The majority of astronomers who have ever lived are alive today, and the avalanche of publications is so great that to read even the titles of every article would be well-nigh impossible. To make matters worse for the historian, the written communications between astronomers in times past – documents that were often preserved for posterity – are being replaced by telephone calls, conversations in the conference coffee-queue, electronic mail messages, and so forth, few of which leave any trace. But in seeking to look at the present with an historian's eyes our task is not completely hopeless, for it is already clear that certain contemporary developments will be seen by future historians as of fundamental and lasting significance. Those that involve theory often require a good knowledge of physics for their understanding and so for the most part are beyond our present scope. But data must be obtained before they can be interpreted; and in the collection of data it may even be that readers – as taxpayers and voters – have themselves a role to play.

Extending the human senses: 'more light!'

Unlike many natural scientists, the astronomer does not have Nature on the rack – he cannot 'put Nature to the question', in the seventeenth-century phrase of Francis Bacon. The physicist or chemist can set up a controlled experiment, and like a torturer compel Nature to answer the questions put to her. Only occasionally is the astronomer able take the initiative. There have been visits to the Moon by so-called 'astronauts', and voyages of exploration by spacecraft to other planets in the solar system, while in 1986 no fewer than five spacecraft were sent to intercept Comet Halley – the European Space

Agency's *Giotto* came within 600 kilometres (375 miles) of
the nucleus and spent ten hours gathering data and taking
pictures. But for the most part the task of the astronomer is
as it always has been, to collect and make sense of whatever information Nature is prepared to supply to Earth-
based observers.

The invention of the telescope provided Galileo and his
contemporaries with an extension of the human eye, and by
the end of the eighteenth century, observers had become
acutely aware that the bigger the objective lens or primary
mirror of a telescope, the fainter the objects it could bring
into view. The clamour for ever-larger telescopes has grown
in volume ever since.

By the middle of the present century, American philan-
thropy had provided the more fortunate of American
astronomers with privileged access to the secrets of the universe. The astronomer with a 'state of the art' instrument
at his disposal has an advantage over his rivals, not only in
the disinterested pursuit of new knowledge, but in the
competition for status and salary within the astronomical
community; and in the last chapter we saw something of
the achievements of observers blessed with time on the
100-inch on Mount Wilson or, later, the 200-inch on
Palomar Mountain. Unsurprisingly, their colleagues in
other countries demanded comparable facilities, but the
cost of constructing and running a major telescope had
reached a level that often called for governmental, and
sometimes even international, funding.

There was a further problem, created by an unhelpful
Nature. Until the 1970s most of the major telescopes of the
world were located in the northern hemisphere, where
most of the manufacturers and most of the astronomers
were based. There are rich skies, however, south of the
celestial equator: they include the dense star clouds in the
direction of the centre of the Galaxy, and the two
Magellanic Clouds (which are irregular galaxies, and our
nearest intergalactic neighbours). It is as though the fates
had conspired to make astronomical research as difficult
and expensive as possible. However, as we shall see, in the
last quarter of a century the increasing speed and economy
of modern air travel and ease of communication have permitted the development of southern sites with facilities at
least equal to those in the north. But the costs have continued to escalate.

In the US one response was the foundation in 1957 of the
federally funded Association of Universities for Research in
Astronomy, to fund three major observatories: for the

northern skies, at Kitt Peak in Arizona, where a 3.8-metre
(150-inch) telescope began work in 1973; for the southern
skies, at Cerro Tololo in Chile, where a matching instrument came into service in 1976; and a solar observatory, at
Sacramento Peak in New Mexico.

Other countries in many cases relied on international
collaboration, often with time-sharing on the resulting
facilities. Thus a consortium of nations of continental
Europe sponsored the European Southern Observatory at La
Silla, north of Santiago in Chile. The 3.6-metre (147-inch)
telescope there came into service in 1971, and there are
now more than a dozen telescopes on the site. Such international collaboration involves bureaucracy, not to say
diplomacy; but in times of economic stringency, member
states find it harder to renege on their promises of funding
for international projects than they would with purely
domestic schemes.

The 1970s saw two other major southern telescopes
come into operation: the 3.9-metre (153-inch) Anglo-
Australian telescope at Siding Spring Mountain in New
South Wales, built by Australia and the United Kingdom,
and the 3.6-metre (141-inch) telescope on Mauna Kea in
Hawaii, created by a partnership between France, Canada
and Hawaii. Nor was the northern hemisphere neglected.
The United Kingdom, Spain, Denmark, and Sweden joined
forces to build an observatory on La Palma, in the Canary
Islands, where conditions were more favourable than anywhere on the mainland of Europe, and they were soon
joined by the Netherlands and Ireland.

All these instruments were located on sites chosen so as
to reduce as far as possible the limitations inflicted by the
presence of an atmosphere around the Earth: sites with less
frequent cloud cover, little air movement, a reduced thickness of the atmosphere, and also greater remoteness from
city lights – in brief, carefully selected mountain-tops. It is
more than a century since the establishment of the first
mountain-top observatory of note, Lick Observatory on
Mount Hamilton near Santa Cruz in California, where
observations began in 1881 at an altitude of over 4,000 feet.
Today the highest major observatory is on Mauna Kea, at an
altitude of nearly 14,000 feet, a height that poses a challenge to the physiology of the observers and staff. Needless
to say, the remoteness of such sites has added still further
to the costs.

Meanwhile, the collection and analysis of information
has been transformed for the better by the technological
revolution of recent decades. Photographic plates, which

near the end of the nineteenth century brought about their
own revolution in the way astronomers gathered data, are
in fact not very efficient as light collectors. The great
majority of photons striking a photographic plate are
reflected away, and at best only about 2 per cent serve to
impress an image on the sensitive surface. The rapid
development in the 1980s of photosensitive charge-coupled
devices, or 'CCDs' (which have wide-scale commercial
importance for video cameras), enabled astronomers to
assimilate photons with efficiencies exceeding 70 per cent.
This meant that a 30-inch telescope in 1990 could record
more photons than the 200-inch could in 1960.
Furthermore, by acquiring the images electronically,
observers could inspect progress as an observation was
being made, thereby eliminating the wasteful problems of
under- or over-exposure. And the CCDs provided yet
another advantage – they could be flown in planetary
probes or orbiting observatories such as the Hubble Space
Telescope, and the images could be transferred to Earth by
radio links.

The high speed of modern computers serves not only to
process electronic images, but also to drive telescopes and
to permit 'active optics'. One long-standing problem in the
construction of large reflectors has been the distortion of
the primary mirror as the instrument is tilted in different
directions and the mirror bends under its own weight; today
this distortion can be controlled mechanically, the pressure
at various positions on the back of the mirror being
continuously altered to compensate for changes in internal
stress. In a similar way the constant changes in the atmosphere above the telescope can be compensated for by introducing into the light path a thin, deformable mirror that
responds to these changes many times per second. Tracking
an object as the sky rotates is another job a computer can
undertake, and therefore equatorial mountings (see page
187) may now be dispensed with in favour of the simpler
and cheaper up/down, left/right of the altazimuth.

As a result, the days when amateur astronomers – men
like William Herschel or William Huggins – could be at the
forefront of the science are long gone; amateurs still have a
role to play, but it is in areas neglected by the professionals,
such as the search for arriving comets, the observation and
counts of meteors during meteor showers, or the monitoring of variable stars. Gone too are the days from the recent
past, when an astronomer on his observing nights could
have undisputed command of a great instrument. The
modern telescope depends for its continued operation on

teams of experts. Observing has been turned into a collaborative exercise, in which the astronomer finds himself in the hands of specialist support staff whose expertise is as likely to be in engineering or computing as in physics or astronomy. National observatories that once undertook their own research now find themselves occupied in maintaining facilities for a succession of visiting astronomers from academia.

The next generation telescopes

Most of the telescopes recently constructed have had mirrors with diameters a little less than the 200 inches of Palomar Mountain. In 1976 a 6-metre (236-inch) telescope was completed at Zenenchukskaya in what was then the USSR; but it proved a disappointment, and some thought the limit of useful mirror size (imposed by the movements in the atmosphere above the telescope) had already been reached. But experience of the new telescopes on high mountains showed that much of the disturbance came from air currents near and within the dome, and even within the instrument itself. Something could be done to alleviate these problems, and so the construction of telescopes with larger mirrors became attractive once more. Even so, for such a massive instrument to work effectively, it would need to be supported by the very latest technology.

The bulk of the daunting cost of one of these 'Next Generation Telescopes' (or NGTs) is incurred in the fabrication and shaping of the primary mirror, in the construction of machinery for controlling the telescope, and in the erection of the building with its rotating dome. The latter two are governed by the mirror's weight and by its focal length. Deeply hollowed mirrors of short focal length (and therefore requiring a relatively short tube and a dome of modest dimensions) are difficult to shape by conventional means; but a mirror can actually be cast with approximately the required shape, if the material of which it is made is rotated while still molten, and slowly allowed to cool. This is because a liquid under rotation in a circular container piles up towards the outside of the container, and its surface thereby acquires (approximately) the parabolic shape that is ideal for a telescopic mirror.

The largest NGT, the Keck I telescope on Mauna Kea, came into partial operation in 1990. Its mirror is 9.8 metres (387 inches) in diameter, and is formed of 36 hexagonal segments, nine of which were in place when the telescope first 'saw light'. A matching telescope, Keck II, is located nearby, and when used in conjunction the resolving power of the two instruments becomes equivalent to that of a telescope as large as the distance between the two domes.

The Hubble Space Telescope

Today's astronomer, however, is not wholly dependent on terrestrial observatories, for the use, not only of aircraft, but of rocketry, orbiting satellites and spacecraft – developments funded for military use or for reasons of national prestige – has opened the way to observation with instruments located above the atmosphere.

By far the most substantial of these efforts is the Hubble Space Telescope (or HST); launched in April 1990 from the Space Shuttle *Discovery*, it orbits 600 kilometres (370 miles) above the Earth. Its primary mirror is of diameter 2.4 metres (94 inches), medium-size by the standards of terrestrial instruments, but capable in outer space of a resolution far surpassing its Earth-based competitors. The smoothness of the HST's mirror far exceeded the standards of any previous large astronomical instrument, but unfortunately, the surface was made highly accurately but to an erroneous shape, and the telescope fell far short of meeting expectations. A servicing mission from a Space Shuttle in 1993 successfully placed correcting elements in the optical path. Subsequently, the exquisite smoothness of the mirror's surface has paid off in a spectacular series of high-resolution images, most notably allowing the discovery of hundreds of Cepheid variables in the distant Virgo cluster of galaxies.

The invisible universe of radio astronomy

Visible light, the electromagnetic radiation to which the human eye responds, is only one of the bands in which rays succeed in penetrating the atmosphere. Another, much larger band consists of radio waves with wavelengths between about 1 millimetre and 30 metres. The first clue that waves in this range were arriving from celestial bodies, as 'starry messengers' bearing new information, came in 1932. In that year, an engineer at the Bell Telephone Laboratory named Karl G. Janksy (1905–50) was using a steerable antenna to study interference on the recently inaugurated trans-Atlantic radio-telephone service. In addition to thunderstorms, Jansky met with a background hiss that varied in intensity with a period four minutes short of 24 hours – that is, the period of the rotation of the heavenly sphere. The signals turned out to come from the Milky

Way, and were strongest in the direction of the centre of the
Galaxy.

Jansky's results attracted little attention, but in 1937
another American radio engineer, Grote Reber (1911–   ),
built himself an antenna with a paraboloidal shape and
began to devote his spare time to following up Jansky's
work. For nearly a decade he produced contour maps of the
intensity of radio waves across the sky. This span of activity included the war years, and wartime radar operators on
the lookout for hostile activity also found themselves
encountering 'interference' coming from outer space. In
1942, James S. Hey (1909–   ) and his colleagues at the
British Army Operational Research Group were investigat-
ing what was thought to be German jamming of British
radar, when they found that the intense radio emission was
coming in fact from the Sun.

When the war ended, there were ex-service physicists
with the expertise to develop this new science of 'radio
astronomy', and discarded military equipment for them to
use. One notable early triumph of radio astronomy was the
demonstration that the Galaxy does indeed have a spiral
structure.

As we have seen (see page 274), speculations to this
effect had proved difficult to substantiate, not only because
the human inhabitants of the Galaxy would inevitably find
it hard to discern the shape of the star system to which
they themselves belonged, but also because dust clouds
concealed most of the galactic stars from sight. During the
Second World War, Dutch astronomers were deprived of
most of their facilities, and so found themselves forced to
concentrate on theoretical questions. Jan Oort of Leiden
Observatory realized that, because of their longer wavelengths, radio waves might be able to penetrate the dust
clouds, and he invited his student, Hendrik van de Hulst
(1918–   ), to investigate what radio waves might be
expected.

In a paper published in 1945, van de Hulst showed that
when the spin of a 'neutral' hydrogen atom reversed itself,
there would be an emission at a wavelength of about 21cm.
In any individual hydrogen atom, such a spontaneous reversal was to be expected only once in millions of years; but
hydrogen is the most abundant element in the universe,
and so there might be enough reversals occurring in the
Galaxy to be detectable on Earth.

It was 1951 before the prediction was confirmed, first at
Harvard (by use of a simple horn antenna built of plywood
and copper foil and mounted outside a laboratory window),
and later in the Netherlands and Australia. The latter two

groups then embarked on a collaborative program to map out the intensity and velocities of the 21-cm line in different directions in the Galaxy. It was laborious work – the Leiden group used an old German radar antenna which for nearly two years they had constantly to reposition by hand-cranking every 2¼ minutes – but in the resulting map the spiral arms of the Galaxy could be seen emerging at last.

Poignantly, William W. Morgan of Yerkes Observatory had recently completed an optical investigation of the structure of the Galaxy using the highly luminous stars of

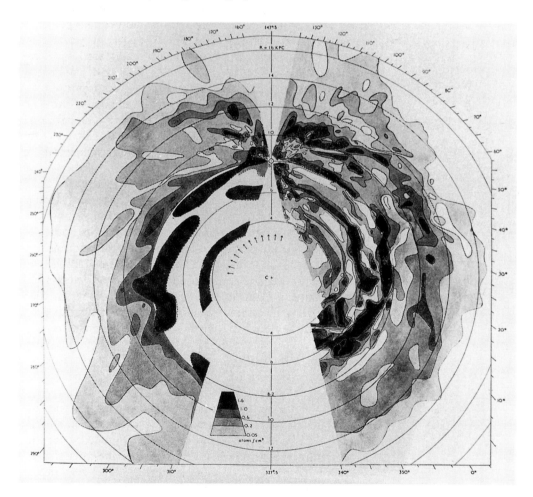

The composite diagram prepared in 1958 to show the spiral structure of the Galaxy, as determined by radio astronomers in the Netherlands and Australia from the distribution of neutral hydrogen. In the middle of the diagram is the galactic centre, and the position of the Sun is marked above it by S, at an assumed distance of 8,200 parsecs. The left of the diagram is mainly the work of the Sydney observers, while the more detailed picture to the right is based on the Dutch results. Where the two halves meet, there is reasonable agreement. Courtesy of Institute of Astronomy, Cambridge University.

types O and B to trace out the spiral arms, and his short presentation of his results to the Christmas 1951 meeting of the American Astronomical Society convinced the audience that the spiral structure of the Galaxy had at last been demonstrated. But Morgan fell ill the following spring, and when he was able to resume work his draft paper had been superseded by the achievements of the radio astronomers.

The first localized ('discrete') source of radio emission was found by Hey only a few months after the end of hostilities, and two years later Martin Ryle (1918–84) and F. Graham Smith (1923–   ) at Cambridge located the most powerful such source in the northern hemisphere. Could such 'radio stars' be identified with visual objects? Such identifications – or their lack – would clearly be fundamental to understanding the nature of radio stars. But there was a more specific reason: until identifications were made, radio astronomers would be ignorant of the third dimension, that of distance. The distance of a radio source would be unknown until it had been identified with a visual object, and the redshift (see pages 302–5) of the object's visual spectrum translated into distance, on the familiar theory that the greater the redshift, the greater the speed of recession, and the more distant the object.

Unfortunately the first radio telescopes were not sufficiently precise as to the position on the sky of a radio source to make such an identification possible. In 1949 came a significant advance, when three Australian radio astronomers, John G. Bolton, G. J. Stanley and O. B. Slee, succeeded in identifying three discrete radio sources with objects familiar to the optical astronomers; these included the Crab Nebula, which Edwin Hubble had earlier recognized as the remnant of the supernova explosion of 1054 (see page 59).

Two years later, Graham Smith was able to obtain an accurate position for the radio source known as Cygnus A; this allowed Walter Baade and Rudolph Minkowski at Palomar Mountain to identify it with a curious-looking object, which Baade interpreted as a distant pair of colliding galaxies. This interpretation had to be abandoned, but Cygnus A gave a foretaste of the cosmological potential of radio astronomy.

As the number of such identifications grew, it became clear that objects that lay at enormous distances, though very feeble emitters of light – so feeble that many may never be detected by optical telescopes – could nevertheless be strong sources of radio waves. Radio astronomers were 'seeing' these sources as they were a very long time ago,

when their light set out on its journey to us. This meant
that if the universe was indeed expanding from its origin in
a superdense beginning or 'Big Bang', as most cosmologists
believed, the radio astronomer could observe regions as
they were when significantly closer in time to the Big Bang
itself. This would permit evidence from radio astronomy to
be used to test theories of cosmology.

Radio astronomy and cosmological controversy

In 1948 three Cambridge astronomers, Fred Hoyle
(1915–), Hermann Bondi (1919–) and Thomas Gold
(1920–), none of them enamoured of the idea that the universe had an origin, proposed instead a theory whereby the
universe appeared broadly the same, at all times and from
all places. In this 'Steady State' cosmology there would be
no beginning, and no end. That the galaxies were receding
from each other was not the consequence of a Big Bang
(Hoyle's term); rather, the galaxies were simply moving
apart in all directions, much as the currants of a cake move
apart as the cake expands during cooking. To fill the spaces
left by the receding galaxies and so preserve the overall
appearance of the universe, the three friends postulated the
creation of matter in intergalactic space, matter that was
required to appear on a scale sufficient to form new galaxies.

If the Steady State cosmologists were right, then in the
remote past the distribution of the major components of the
universe was broadly the same as it is today; but if the Big
Bang theorists were right, then these components were
more concentrated when the universe was younger and
smaller. This was a question to which radio astronomy
might supply the answer.

A radio telescope that seemed suited to the task was also
in Cambridge, where Ryle had pioneered a technique
whereby several matching radio telescopes, one or more of
them mobile, could be teamed up to piece together a radio
map of the sky equivalent to that provided by a single 'dish'
of enormous, and quite impracticable, size (a method that
has culminated in the Very Large Array radio telescope near
Socorro, New Mexico, which consists of 27 dish antennas
shared between the three arms of a vast Y-shaped configuration). There was no love lost between Ryle and Hoyle; Ryle
more than once announced results of his surveys that
supposedly disproved Steady State, and Hoyle more than
once rejected them.

Eventually the question was decided – provisionally, at

least – by another route. In the late 1940s George Gamow
(1904–68) and colleagues from George Washington
University, Ralph A. Alpher (1921–   ) and Robert Herman
(1914–   ), predicted as one of the consequences of what
would later be termed the Big Bang, that throughout the
universe there would today be residual radiation; this
would be the same in all directions, and at a temperature of
some 5° above absolute zero (5K). Gamow pictured the
formation of the elements in the initial minutes of the
explosion as coming from a primordial mixture of radiation
and nuclear particles, which he called 'ylem', but this
scheme ran into difficulties because of the lack of a stable
element of mass 5. Although the study was well-publicized,
interest in these ideas declined because of the problems
with the proposed scenario for element formation.

Unaware of Alpher and Herman's forecast, Robert H.
Dicke at Princeton in 1964 proposed his own theory, which
also predicted a background radiation a few degrees above
absolute zero. Dicke's group set to work forthwith to build
an instrument to detect this radiation, and so test the theory.

Meanwhile, oblivious of these theoretical ideas remote
from their own concerns, two engineers at nearby Bell
Telephone Laboratories, Arno Penzias (1933–   ) and Robert
Wilson (1936–   ), were at work with a reflector originally
designed to test communication by satellite. They took
great pains to track down and eliminate all sources of interference, including two roosting pigeons, but they found that
from every direction in the sky there came an unaccounted-
for radiation at about 3K. They consulted Dicke as to the
possible cause, only to learn that he was actively searching
for what they had found.

In the eyes of most cosmologists, the discovery in
1964–65 of this 'microwave background radiation' tipped
the scales decisively against Steady State, which passed into
the limbo of discarded theories. Indeed, it had always been
motivated by views as to the form the universe ought to
have, rather than the form it does have. Its begetters had
pointed out that it was a theory that made clear predictions
and so had the scientific merit of being vulnerable to dis-
proof; and disproved it was. Yet historians remember earlier
arguments that polarized the astronomical community,
over whether the Earth or the Sun was at the centre of the
universe, or whether nebulae were star clusters or clouds of
luminosity – questions whose answers proved to be less
straightforward that was expected; and they feel uncomfortable when that community is polarized yet again between
opposing theories.

Both Steady State and Big Bang presupposed the recession of galaxies in all directions; yet even this interpretation of the evidence can awake unease. At root, the observational facts are that the light from small, faint galaxies, when turned into a spectrum, is found to be 'shifted' towards the red, and the more so, the fainter and smaller the galaxy. It is hard to doubt that 'small and faint' implies distant, but the redshift may be open to other interpretations. For example, it might be that the universe is static and that photons of light, in their long journeys across intergalactic space, somehow gradually lose energy and become of longer wavelength. However, all alternative theories are in some respect unsatisfactory, while the observed redshifts *behave* as if they represent an expansion, and for the present this continues to be the simplest and most satisfactory explanation of the observations.

There has indeed been something of a methodological dilemma: the theoretical cosmologist wanted the observing astronomer to produce 'facts' that would determine which was the correct model of the universe, but the very reduction of raw observations to produce such 'facts' required the assumption of a cosmological model which the observations were supposed to verify.

The universe hidden by the atmosphere

As we have seen, visible light and radio waves are the forms of electromagnetic radiation that can penetrate the atmosphere. Indeed, radio waves find bad weather no obstacle, and radio telescopes can be employed, day and night, at low level sites with poor climates. But other regions of the spectrum – those that lie between the near infra-red and radio waves, and the short-wavelength ultra-violet, X-rays and gamma rays – are effectively blanketed from view by the atmosphere. Today, as a result of developments in spacecraft, the whole of the spectrum is available for study; new areas of investigation have been spawned, and the integration of astronomy and physics has become ever stronger. The different wavebands provide, in part, pictures of the universe at different temperatures.

The first attempts at observation above the atmosphere took the form of instruments carried in rockets. As early as 1946, captured German V-2 rockets were used to obtain the first ultraviolet and X-ray observations of the Sun. V-2s were 14 metres (46 feet) tall and could climb as high as 150 kilometres (90 miles). Unfortunately the rocket would break up on impact, and so it was necessary either to

separate off the parts that contained the scientific instruments so that these could descend more slowly, or to trans-
mit the data by radio during flight. Though the V-2s were
superseded within a few years by rockets designed for space
research, a rocket is above the atmosphere for only a matter
of minutes, and so the data gathered by the early flights provided little more than crude snapshots that confirmed there
were windows on the universe as yet unutilized.

This naturally whetted the astronomical appetite.
Fortunately for would-be observers, the international
rivalry in space that followed the launching in 1957 of the
first Russian *sputnik* resulted in the provision of a whole
series of astronomical satellites, orbiting the Earth and
recording data in all the hitherto missing regions of the
spectrum. US 'orbiting astronomical observatories' were
launched in 1968 and 1972, and the International
Ultraviolet Explorer (a joint European–UK–US venture) in
1978. The American series of Small Astronomical Satellites
was inaugurated in 1970 with the launch from the coast of
Kenya of SAS-1, which generated a catalogue of 161 X-ray
sources; this was followed by SAS-2, launched in 1972,
which carried equipment for measurement of gamma rays,
and SAS-3, launched in May 1975 and carrying X-ray detec-
tors. The Netherlands–UK–US Infrared Astronomical
Satellite was launched in 1983, and in its eleven-month
active life assembled a sensitive and complete mapping of
the infrared sky. With the data from these and subsequent
satellites, astronomers have in a few short decades enormously enlarged the horizons of their science, and strength-
ened its links with mainstream physics.

Still more dramatic has been the use of space probes to
investigate from close range the other members of the solar
system. In 1959 the Soviet Luna 3 provided the first view of
the backside of the Moon; the most surprising discovery
was that the Moon's far side lacked the large dark basaltic
maria that characterize the near side facing the Earth. From
this beginning, direct exploration of the solar system has
blossomed, with spacecraft making close approaches to all
the planets except Pluto, as well as to satellites, comets,
and asteroids. These missions transmitted technical data of
gravitational and magnetic fields, and also the now familiar
'postcards from space' that have revealed in detail many
new worlds. Jupiter's satellite Io, with its active sulphurous
volcanoes, and Uranus's satellite Miranda, with its fascinating combination of old, heavily cracked terrain and a young
complex landscape with bright and dark scarps and ridges,
are among the strange small worlds disclosed by unmanned

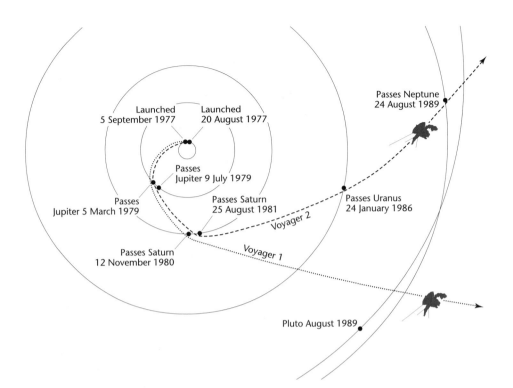

Launched
5 September 1977

Launched
20 August 1977

Passes
Jupiter 9 July 1979

Passes
Jupiter 5 March 1979

Passes Saturn
25 August 1981

Voyager 2

Passes Saturn
12 November 1980

Voyager 1

Passes Neptune
24 August 1989

Passes Uranus
24 January 1986

Pluto August 1989

instruments launched on lengthy journeys beyond the
Moon.

The first high definition close-ups of the Moon's surface
came in July 1964, with the US lunar probe Ranger 7, which
sent back thousands of televised images in preparation for
the manned Apollo landings that followed three years later.
American astronauts brought back hundreds of mineral
specimens, which helped to establish the age and complex
chemistry of the lunar origin.

In 1970 Venera 7 became the first spacecraft to send back
signals from the surface of another planet, and the first
actual pictures of Venus's landscape were produced by
Soviet probes in 1975. Meanwhile, the American Mariner
series documented the surfaces of Mercury and Mars. The
Viking probes, launched in 1975, gave magnificent views of
the immense volcanic calderas and the gigantic Valles
Marineris canyon system on Mars. The American Voyagers
transmitted remarkable details of the weather on Jupiter and
the ring and satellite systems of Saturn (in 1979), and then
of Uranus and Neptune (in 1986 and 1989 respectively).

These explorations revitalized planetary studies, which
had fallen into comparative disfavour earlier in the century
in competition with the more glamorous investigations
into cosmology and astrophysics.

The itineraries of Voyager
1 and Voyager 2 among the
outer planets.

The discovery of exotic objects

We have seen how already in the first half of the twentieth century, the scope of astronomy was enlarged until its range extended from complex subatomic processes to models of the universe involving General Relativity. At the same time it was increasingly realized that physical conditions in space were altogether more extreme than could be achieved in any laboratory, and that one could not rule out objects and processes that had seemed to belong rather to science fiction.

Yet the instincts of even the most forward-looking astronomers were conservative. Eighteenth-century specu-lators had discussed the characteristics of stars so dense that light would be prevented from leaving them by the strength of their gravitational attraction (see page 189); and according to Einstein's General Relativity, such bizarre objects (today's 'black holes') were theoretically possible as end-products of stellar evolution, provided the stars were massive enough for their inward gravitational attraction to overwhelm the repulsive forces at work. In 1935, Subrahmanyan Chandrasekhar correctly argued at a meeting of the Royal Astronomical Society that white dwarf stars more than half again as massive as the Sun could not withstand the intense tug of gravity and would collapse further; but Sir Arthur Eddington, even though he was himself a leading advocate of Relativity, poured scorn on the young man's proposals, finding it impossible to accept that Nature behaved in so extraordinary a fashion. Before the decade was out, J. Robert Oppenheimer (1904–67) and Hartland Snyder (1913–62) focused attention on the process of gravitational collapse itself, and showed that General Relativity was needed to describe the dynam-ics of the process, thus laying the groundwork for what is called 'black hole physics'.

It was in the 1960s that observers with radio telescopes recognized two examples of the near-incredible types of objects that exist in the astronomer's universe. On three occasions in 1962 the Moon passed in front of one of the radio sources found during Ryle's third Cambridge survey, and observations of these 'occultations' allowed observers at Parkes radio telescope observatory in New South Wales to show that the object was a double radio source, one com-ponent of which they identified with a blue, star-like object of the thirteenth magnitude. This prompted the Dutch-born astronomer Maarten Schmidt (1929–   ) to use the Palomar 200-inch reflector to study the spectrum of the star.

The spectrum was peculiar indeed: superimposed on a continuum were bright emission lines that seemed to be of an unknown element, but which Schmidt eventually recognized as the series of lines of hydrogen investigated as long ago as 1885 by the Swiss physicist Johann Balmer. But these lines had an extraordinarily large redshift, which implied that the star-like object lay far outside our Galaxy – provided, that is, that the redshift was interpreted in the usual way.

There was acrimonious debate over the nature of this 'quasi-stellar radio source' or 'quasar', for if it was at the vast cosmological distance implied by its large redshift, an enormous intrinsic luminosity was required. As time passed, quasars with greater and greater redshifts were discovered, including ones whose redshifts implied they were receding with over 90 per cent of the speed of light, and that they were situated thousands of millions of light years away. To be visible at such distances, quasars had to be perhaps a million times more luminous than a complete galaxy; yet they had also to be small, as their luminosity could sometimes vary by a factor of 2 in a matter of hours. Only a supermassive black hole offered any hope of a physical explanation of the processes involved. Small wonder that some astronomers refused to believe the redshifts had been correctly interpreted.

The second extraordinary type of object was discovered four years later. Antony Hewish (1924–   ), a Cambridge radio astronomer, and his student, Jocelyn Bell, had constructed a crude but effective radio telescope consisting of poles and wires strung out across a four-acre site. Their purpose involved the study of the impact of the interplanetary plasma (material formed of ions and electrons moving freely) on the waves from distant radio sources. In the course of their investigations, they chanced upon a weak source that eventually proved to be giving out regularly-spaced pulses with a period of just over a single second. It seemed that such a very rapid transmission must have been generated on Earth, perhaps at a military installation – or if not, by 'Little Green Men' elsewhere in the universe – but it eventually became clear that the source was celestial and natural.

The 'pulsating radio star' or 'pulsar' proved to be a rotating neutron star, a tiny but immensely dense star of a type whose existence had been predicted (see page 269) as long ago as 1934 by Fritz Zwicky and Walter Baade, shortly after the discovery of the neutron itself by James Chadwick. It was rotating once a second, and each time it did so its beam

of radiation swept the Earth, much as a lighthouse beam
sweeps the surrounding water. Hundreds of pulsars have
since been discovered, all rotating with astonishing speed:
that at the heart of the Crab Nebula has a period of only
$\frac{1}{30}$th of a second.

Pulsars have proved to be of immense significance.
These neutron stars can form only through the collapse of
less dense stars, proving that such collapses do take place.
In a neutron star, the stupendous pressure towards further
gravitational collapse is resisted by the forces between the
neutrons; but calculations show that in a star only a few
times more massive, these forces would be insufficient, and
the star would become a black hole.

Pulsars are the most stable clocks known in the universe, and this property is of great importance. A number of
pulsars are binary (two stars bound together by gravity, one
of which is a radio pulsar); in about half of these, the pulsar
rotates a hundred or more times every second. The first
binary pulsar was discovered in 1974 by Russell A. Hulse
and Joseph H. Taylor at the Arecibo Observatory in Puerto
Rico, where a radio telescope dish no less than 305 metres
(1,000 feet) in diameter is sited in a natural hollow. Its two
neutron stars move around each other in a little under 8
hours, in very eccentric orbits, which allows astronomers a
very detailed knowledge of the stars and their behaviour;
and the system is in effect a perfect clock in a rotating
frame of reference, just what is required for several tests of
General Relativity. In particular, there is a loss of kinetic
energy which Relativity ascribes to 'gravitational radiation';
this loss shrinks the orbit and speeds up the motion of the
stars, and as time goes on the loss accelerates. Over the
years, prediction and observation of this orbital decay in the
Hulse-Taylor pulsar have matched each other to better than
1%, in what has been described as 'the most far-reaching
and thorough test of General Relativity ever performed'.

The violent universe

The discoveries of the 1960s and 1970s gave astronomers an
entirely different view of the pace of astronomical events.
The notion that celestial evolution moved at an ever-stately
march gave way to a cosmos filled with sudden happenings.
From the split-second oscillations of the pulsars to the
comparatively swift light variations of the quasars, cosmic
events took place far more rapidly than the billion-years-
evolutionary cycle of stars like the Sun. New stars were
observed in formation, some of which would last only a few

hundred thousand years, and by the 1980s astronomers realized that the cataclysmic final collapse of a supernova core could be reckoned in milliseconds. Even the staggering explosion of the Big Bang, with the formation of hydrogen and helium, was clocked at about three minutes.

Astrophysicists began to appreciate that much of the unexpected radiation outside the visible portion of the spectrum arose from non-thermal processes. That is, the radiation could not be associated with the random motion of hot atoms, but instead with highly ordered motions such as electrons moving in magnetic fields. For two or three decades astronomers were fond of calling their new vision 'the violent universe', but gradually the term fell out of use, while the expression 'high energy astronomy' gained increasing currency.

At the same time that astronomers were becoming familiar with such violent and rapid events as 'the first three minutes' of creation or the implosion-explosion of supernovae, they discovered that these diverse phenomena were linked in the explanation of how the elements have come to be. In the intense non-equilibrium interactions in the initial minutes of the Big Bang, photons of pure energy were transformed into protons and electrons, which soon combined into hydrogen and helium, but the absence of a stable mass 5 isotope prevented heavier elements from forming. In contrast, in the slow equilibrium processes in the cores of evolving stars, it was possible to build up carbon, oxygen, and some higher elements, while in the catastrophic collapse of supernovae many heavier elements could be formed and blown out into space; this cosmic debris then became the material for a subsequent generation of stars. This theoretical picture of element formation was conceived largely by Geoffrey and Margaret Burbidge (1925– , 1919–), Fred Hoyle, and Willy Fowler (1911–95) and independently by A. G. W. Cameron (1925–). One of the most profound insights of astronomy in the second half of the twentieth century, it has provided a solid basis for observational investigations of the evolving chemical composition of the heavens, and dramatic confirmation that our universe has a long history of slow evolutionary change.

In conclusion

This short final chapter has done no more than mention some of the instruments, observational techniques, and theories that have been developed in the second half of the twentieth century, and offer samples of the extraordinary objects that populate the universe of the contemporary astronomer.

The gulf that separates the modern astronomer from the general public is nothing new. Those at the frontiers of astronomy, with the learning required to understand the heavens, have always been an elite, set apart from the rest of society to whom this esoteric knowledge has filtered down in much-diluted forms. This was no doubt true in pre-historic times, and it was certainly true in China and in Mesoamerican cultures such as that of the Maya. In the Middle Ages, Ptolemy's *Almagest*, with its epicycles, defer-ents and equants, was a closed book to all but a very few. Copernicus's *De revolutionibus* proved equally impenetra-ble, and Kepler's achievements went unappreciated by intel-lects of the calibre of Galileo and Descartes. The implications of Newton's *Principia* were left to a tiny band of brilliant mathematicians. With Einstein, the gulf became still wider: in 1920, the organiser of the 'Great Debate' (see page 290), C. G. Abbot, rejected relativity as a possible topic, saying 'I pray to God that the progress of science will send relativity to some region of space beyond the fourth dimen-sion, from whence it may never return to plague us'. And since then, the technical and conceptual barriers facing the student of astronomy have grown steadily more daunting.

Perhaps only once in history has there been substantial harmony between the scholar's universe and that of the ordinary people. In the Latin Middle Ages, God was in his heaven with the angels and the blessed, and mankind was on *terra firma* at the centre of the cosmos, no place of honour but a position that fitted man's significance in the moral order. Yet even then, the epicycles of the astronomers caused dissention in the Faculties of Arts; and it would not be long before Copernicus's reform would lead the English poet, John Donne (1573–1631), to write:

> The New Philosophy calls all in doubt,
> The Element of fire is quite put out;
> The Sun is lost, and th'Earth, and no man's wit
> Can well direct him where to look for it . . .
> 'Tis all in pieces, all coherence gone;
> All just supply, and all Relation.

Earlier in this book we discussed Plato's challenge to his contemporaries, in effect inviting them to answer the ques-tion of whether the universe is an intelligible cosmos. Today, as down the centuries, the answer might be: 'Yes, but only just.'

Chronology

3500BC
Construction of Newgrange.

3000 Construction of Stonehenge I. Egyptian religious calendar.
Egyptian administrative calendar.

2500

2000

1500 Old Babylon: Venus records (place-value numerical system).

1000 Babylonian *Enuma* takes definitive form.

Babylonians identify Metonic cycle.

Eratosthenes, size of the Earth.

Hipparchus, models for the Sun and the Moon, star catalogue, precession.

Systematic Babylonian records   700

600

Greek cosmology   600

500

Greeks recognize sphericity of Earth.   400
Plato poses problem of planets.
Eudoxus, models of spheres.
Aristotle's world picture.   300
Aristarchus, distances of the Sun and the Moon.

Apollonius, eccentres, epicycles, deferents.   200

Greek planetary models based on Babylonian records   100BC

AD100

Ptolemy, *Almagest, Planetary Hypotheses, Tetrabiblos.*   200

Medieval Christendom   300

Martianus Capella.   Macrobius.   Calcidius.   400

Boethius.   500

600

Bede.   700

800

Medieval Islam

632 Death of Muhammad.

House of Wisdom in Baghdad.   *Translations from Greek*   900
al-Battani.

al-Sufi.
ibn Yunus.   ibn al-Haytham.   Astrolabe enters West.   1000

al-Zarqali.   1100
1120–25 Cairo Observatory.
Averroës.   *Translations from Greek and Arabic*
al-Bitruji.   1200

1259 Maragha Observatory founded for al-Tusi.   Sacrobosco: *Sphere.*
Anonymous: *Theory of the Planets.*   1300

al-Shatir.   1400

1420 Samarkand Observatory built by Ulugh Beg.
Invention of printing.   1500
1474 Peurbach, *New Theories of the Planets.*   1496 *Epitome* of the *Almagest.*
1543 Copernicus: *De revolutionibus.*

1575–80 Istanbul Observatory.

325

1570
1572 'Tycho's Nova'.
1577 Tycho finds comet to be celestial.   *1576–97 Tycho Brahe's*
1580   *observatory on Hven*

1590

1596 Kepler, *Cosmographic Mystery*.
1600
1602 Tycho's star catalogue published.   1604 'Kepler's nova'.
1610 1609 Kepler, *New Astronomy*; first two laws. Galileo's first telescopic observations.   1610 Galileo, *Starry Messenger*.
1611 Kepler designs 'astronomical' telescope.   1613 Galileo, *Letters on Sunspots*.
1620 1619 Kepler's third law.

1627 Kepler, *Prutenic Tables*.
1630
1632 Galileo, *Two Great World Systems*.

1640 1640 Gascoigne's micrometer.
1644 Descartes, *Principles of Philosophy*.

1650

1660 1659 Huygens elucidates Saturn's rings.
1663 Gregory designs reflector.   1665 *Philosophical Transactions* begun.
1667 Paris Observatory founded. Mira's period
1670   identified.
1672 Newton shows white light composite.
1675 Greenwich Observatory founded.   1677–78 Halley at St Helena
1680 1679 *Connoissance des Temps* founded.

1687 Newton, *Principia*.
1690   1690 Hevelius's star catalogue.
1695 Halley's Comet recognized as periodic.
1700

1710

1720 1721 Halley's papers on symmetric star system.
1725 Flamsteed's *British Catalogue*.
1730 1728 Newton, *System of the World*.   1729 Bradley's paper on aberration of light.

1740

1750 1750 Wright, *An Original Theory*.   1751–53 Lacaille at Cape of Good Hope .
1757 Michell argues most double stars binary.   1758 Achromatic lens described by Dollond.   1759 Return of Halley's Comet.
1760 1761 Trial begins of Harrison's H4 chronometer. Transit of Venus widely observed.
1767 *Nautical Almanac* founded. Transit of Venus widely observed.
1770

1780 1781 Messier's final catalogue of nebulae. W. Herschel discovers Uranus.
1783 Goodricke (and Pigott) suggest Algol eclipsing binary.   1785 W. Herschel's cross-section of Galaxy.   *1783–1802*
1790 1789 W. Herschel's 40-foot reflector completed.   *W. Herschel's*
1796 Laplace, *Exposition*.   *sweeps for nebulae*
1800   1799 Laplace, *Traité*, vol. 1.   1800 *Monatliche Correspondenz* founded.   1801 Piazzi discovers Ceres.
1803 W. Herschel confirms binary stars.

1810
1814–15 Fraunhofer maps lines in solar spectrum.
1818 Bessel, *Fundamenta astronomiae*.

1820 Foundation of Royal Astronomical Society.   1820

1823 *Astronomische Nachrichten* begun.
                                        1824 9½-inch Dorpat refractor mounted.

                                                                                                     1830

                                                                              1834–38 J. Herschel
1837 Struve announces parallax of Vega.   *at Cape of Good Hope*
                                          1838 Bessel announces parallax of 61 Cygni.
1839 Pulkovo Observatory founded; 15-inch refractor.   1840

1842 Corona and prominences observed during solar eclipse.
1844 Bessel argues Sirius and Procyon have unseen companions.   1843 Schwabe announces sunspot cycle.
1845 'Leviathan of Parsonstown' reflector completed. Spiral structure of nebula recognized. First daguerreotype of Sun.
1846 Supposed resolution of Orion Nebula. Neptune (and satellite) discovered.
1847 Harvard 15-inch refractor completed. W. Struve's layered model of the Galaxy.
                                                           1849 Gould's *Astronomical Journal* founded.
                                                                                                     1850
1852 Sabine announces sunspots linked with magnetic storms.

1856 Pogson's proposed magnitude scale published. First silver-on-glass astronomical mirrors.
1857 Clerk Maxwell shows Saturn's rings made of particles.   1858 De la Rue photographs sunspots by collodion process.
1859 Tempel finds nebulosity around Merope in Pleiades. Bunsen and Kirchhoff associate elements with
      spectral lines in laboratory. Publication of *Bonner Durchmusterung* begins.   1861–2 Kirchhoff's map of solar spectrum with
                                                                                                     1860
1862 Ångstrom shows hydrogen present in solar atmosphere.   elements identified.
      Clark observes companion of Sirius.   1863 Astronomische Gesellschaft founded.
1864 Donati examines light of comet with spectroscope. Huggins shows a nebula to be formed of gas.
1866 Schiaparelli links August meteors with comet.
1868 Secchi describes four types of stellar spectra.
1870 Young observes reversing layer during solar eclipse.   1870
1872 Draper photographs spectrum of Vega.

1876 Dry gelatine plates used in photography.
                                          1877 Hall discovers two moons of Mars.

                                                                                                     1880
1882 Gill's photograph of comet contains numerous stars.

1885 Nova S Andromedae encourages opposition to 'island universes'.

1887 Pickering initiates study of spectroscopic binaries. *Carte du Ciel* inaugurated.

1890 Lockyer, *Meteoritic Hypothesis*.   1890

1897 Yerkes 40-inch refractor completed.

1900 *Astronomischer Jahrsbericht* first published, with literature for 1899.   1900

1904 International Union for Co-operation in Solar Research founded.
                                                   1905 Hertzsprung suspects existence of a class
1906 Kapteyn's *Plan of Selected Areas*.   of 'giant' stars.
1908 Mt Wilson 60-inch reflector completed.

1910

1912 Leavitt's period–luminosity relation for Cepheids in Small Magellanic Cloud.

1913 Russell presents first 'H–R diagram' for field stars (stars not in clusters).

1914 Slipher announces large radial velocities of spiral nebulae. Shapley's theory of pulsating stars. Adams and Kohlschütter establish method of spectroscopic parallaxes.

1916 Van Maanen claims evidence of rotation in M 101.

1917 Mt Wilson 100-inch reflector completed.

1918 Shapley proposes 'Big Galaxy'. Publication begins of Henry Draper catalogue of stellar spectra.

1919 International Astronomical Union founded.

1920

1920 'Great Debate' between Shapley and Curtis. Saha publishes theory of ionization in stellar atmospheres.

1923 Hubble finds Cepheid variable in Andromeda Nebula.

1925 Hubble demonstrates Andromeda Nebula is independent galaxy. Payne, *Stellar Atmospheres*.

1926 Eddington, *The Internal Constitution of the Stars*.

1927 Oort analyses stellar motions to study structure of Galaxy.

1929 Hubble shows galaxies receding in conformity with 'law of redshift'.

1930

1930 Trumpler demonstrates existence of interstellar dust in plane of Galaxy. Discovery of Pluto.

1931 Chandrasekhar investigates structure of white dwarf stars.

1932 Jansky constructs antenna, detects radio waves from Milky Way.

Radio astronomy's early years

1934 Baade and Zwicky publish theory of neutron stars; show novae fall into two classes.

1937 Reber constructs 9-metre dish steerable in elevation.

1939 Outbreak of World War II leads to intensive development of radar.

1939 Bethe's detailed theory of nuclear source of stellar energy.

1940

1942 Detection of radio waves from the Sun.

1944 Baade announces discovery of stellar populations.

1945 End of World War II releases radar equipment and personnel for scientific work.

1946 V2 rockets make ultraviolet observations of Sun.

Early space research

1948 '48-inch' (1.25-metre) Schmidt telescope completed, Palomar Mountain (USA). 200-inch telescope completed, Palomar Mountain (USA).

1948 V2 rockets detect solar X-rays; Aerobee rockets introduced.

1949 First identification of radio sources with optical objects.

1950

1952 Baade announces revision of distance scale.

Ground-based optical telescopes since 1950

Radio astronomy since 1950

1951 Detection of 21-centimetre line.

1952 Optical identification of Cygnus A.

1955

1957 Jodrell Bank (UK) 250-foot (fully) steerable dish.

1959 120-inch telescope, Mt Hamilton (USA).

1960

1961 Parkes (Australia) 64-metre steerable dish.

1963 Arecibo (Puerto Rico) 305-metre dish in natural hollow. First quasars identified.

1964 Cambridge (UK) one-mile telescope. Detection of microwave background radiation.

1965 Jodrell Bank/Malvern (UK) VLBI, 127-kilometre baseline. 1965

1967 Canadian VLBI, 3,074-kilometre baseline.

1968 Announcement of discovery of first pulsar.

1970 Westerbork (Netherlands) 3-kilometre telescope. 1970

1972 Cambridge (UK) 5-kilometre telescope. Effelsberg (Federal Republic of Germany) 100-metre steerable dish.

1973 3.8-metre telescope, Kitt Peak (USA). 1.24-metre UK Schmidt telescope, Siding Spring (Australia).

1975 3.9-metre Anglo-Australian Telescope, Siding Spring (Australia).

1975 First observations with Very Large Array, Socorro (USA). 1975

1976 4-metre US telescope, Cerro Tololo (Chile). 6-metre telescope, Mt Pastukhov (USSR).

1977 3.6-metre European Southern Observatory Telescope, La Silla (Chile).

1978 3.8-metre UK Infrared Telescope, Mauna Kea, Hawaii.

1979 3.6-metre Canada–France–Hawaii Telescope, Mauna Kea, Hawaii.
3-metre US NASA Infrared Telescope Facility, Mauna Kea, Hawaii.
Multiple Mirror Telescope, Mt Hopkins (USA).

1980 Very Large Array fully operational. 1980

1984 3.5-metre telescope, Calar Alto (Spain).

1985

1986 James Clerk Maxwell Telescope, 15-metre millimetre and submillimetre dish, Mauna Kea, Hawaii.

1987 4.2-metre UK telescope, Canary Islands.

1989 3.5-metre European Southern Observatory's New Technology Telescope, La Silla (Chile).

1989 Sweden–European Southern Observatory 15-metre submillimetre dish, La Silla, Chile.

1990 US Caltech Submillimeter Observatory, 10.4-metre submillimetre dish, Mauna Kea, Hawaii. 1990

1991 '10-metre Keck I' (9.8-metre) US Next Generation Telescope, Mauna Kea, Hawaii.

1993 US Very Long Baseline Array VLBI, baseline up to 8,000 kilometres.

VLBI: Very Long Baseline Interferometer. ESA: European Space Agency.

Space research since 1950

1955

1957 First USSR sputnik launched.

1959 USSR Luna 3 gives first images of back of Moon.
1960

1962 US Aerobee rocket detects X-ray star.

1964 US Ranger 7 gives images of Moon's surface.
1965

1968 US Orbiting Astronomical Observatory OAO-2 launched.
1969 First human on Moon.
1970 1970 USSR Venera 7 sends signals from Venus. US Small Astronomical Satellite SAS-1 launched.

1972 European Space Research Organization ultraviolet satellite TD-1 launched. US Orbiting Astronomical Observatory OAO-3 ('Copernicus') launched. US Small Astronomical Satellite SAS-2 launched.

1974 Astronomical Netherlands Satellite (Netherlands and US) launched. US Mariner 10 photographs Mercury, Venus.
1975 1975 USSR Venera 9 photographs surface of Venus. US Small Astronomical Satellite SAS-3 launched. ESA gamma ray satellite
1976 US Viking probes photograph surface of Mars.   Cos-B launched.
1977 US Voyagers 1 and 2 launched.
1978 International Ultraviolet Explorer (ESA, UK, USA) launched. US High Energy Astrophysical Observatory HEAO-2 ('Einstein
1979 Voyagers 1 and 2 visit Jupiter.   Observatory') launched.
1980 1980 Voyager 1 visits Saturn.
1981 Voyager 2 visits Saturn.

1983 Infrared Astronomical Satellite (Netherlands, UK, USA) launched. ESA X-ray satellite EXOSAT launched.

1985
1986 Voyager 2 visits Uranus. Giotto (ESA) and other spacecraft intercept Halley's Comet.

1989 Voyager 2 visits Neptune. US Galileo probe to Jupiter launched. ESA Hipparchos astrometric satellite launched.
1990 1990 US Hubble Space Telescope launched. Röntgenstrahlen Satellit (X-ray satellite) ROSAT (Germany,
1991 US Compton Gamma Ray Observatory launched.   also UK and USA) launched.
1992 US Extreme Ultraviolet Explorer EUVE launched.
1993 Hubble Space Telescope optics repaired. Japan Asca X-ray satellite launched.

1995 Galileo visits Jupiter. ESA Solar and Heliospheric Observatory launched. ESA Infrared Space Observatory launched.

Glossary

An *asterisk before a word indicates another entry.

A

aberration of light the small, constantly varying displacement in the observed position of a star, caused by the velocity of the Earth-based observer's orbit around the Sun.

absolute brightness see *luminosity.

absolute magnitude see *magnitude.

absorption band (line) a dark band (line) superimposed on a continuous spectrum, caused by the absorption of light from an incandescent source in passing through a gas of lower temperature.

achromatic lens a lens of two or more components, designed to eliminate as far as possible the effects of *chromatic aberration.

active optics techniques for making rapid corrections in the shape of a telescopic mirror or radio dish, in response to temporary changes from its designed shape.

alidade a sighting bar in an astrolabe or other instrument.

altazimuth the mounting of a telescope or other instrument so that it may be moved independently in *altitude or *azimuth about a horizontal and a vertical axis.

altitude the angle of a celestial body above the horizontal.

angular velocity the rate of change of an angle.

annual parallax the small, constantly varying displacement in the observed position of a star, caused by the displacement of the Earth-based observer from the centre of the solar system (see also *parallax).

aphelion the position in the orbit of a planet or comet where it is furthest from the Sun.

apogee the position in the orbit of the Moon where it is furthest from Earth.

apparent brightness see *brightness.

apparent magnitude see *magnitude.

arc minute one-sixtieth of a degree (1′).

arc second one-sixtieth of an arc minute; $\frac{1}{3600}$th of a degree (1″).

armillary sphere a skeletal sphere with graduated rings representing circles on the celestial sphere, used for instruction or observation.

asteroid a minor planet; a small body in orbit about the Sun, mostly in near-circular orbits in the gap between Mars and Jupiter.

astrolabe a medieval instrument for measuring *altitudes and for computations involving movements of celestial bodies (see pages 63–7).

astronomical unit the distance of the Earth from the Sun, usually taken as half the major axis of its elliptical orbit.

astrophysics the study of celestial bodies by analysis of their light.

atmospheric refraction the bending of the path of light from a celestial body by the Earth's atmosphere.

azimuth angle in the horizontal plane, usually measured eastwards from the north point.

B

backstaff a navigator's instrument for measuring the *altitude of the Sun.

Big Bang the moment in the past when, according to certain models, the universe began to expand from an initial condensed state.

binary star two stars held together by their mutual attraction, each in orbit about their common centre of gravity.

black hole a celestial body so massive that its gravitational pull prevents light from escaping from it.

brightness the *apparent* brightness of celestial object as seen by an observer from the Earth; for *absolute* brightness, see *luminosity.

C

CCD see *charge-coupled device.

Cepheid a pulsating *variable star whose prototype is Delta Cephei.

charge-coupled device an electronic detector sensitive over a wide range of wavelengths, in recent years replacing the photographic plate in many uses.

chromatic aberration a defect in a refracting telescope in which not all wavelengths of light are brought to the same focus, causing a coloured fringe around the image of a star.

chromosphere part of the outer gaseous layers of the Sun, briefly visible during a total eclipse.

climate a climatic zone, a belt of the Earth's surface between two parallels of *latitude.

collapse, gravitational the process whereby a star, star system or *nebula collapses into itself as a result of the mutual gravitational pull of its constituents.

collodion process a photographic process involving a solution of guncotton etc. in a mixture of alcohol and ether.

colour excess a change in the *colour index of a star; the amount of reddening suffered by the starlight in passing through interstellar dust.

colour index of a star, the difference between its *photographic magnitude and *visual magnitude.

complement of an angle, the angle which, when added to the given angle, makes one right angle of 90°.

concave of a lens, one that diverges a beam of light, being thinner at the centre than the edge, in contrast to *convex.

concentric spheres in Greek astronomy, a geometrical model formed of a nest of spheres with a common centre, used to reproduce a planet's motion.

conjunction the time when two bodies of the solar system have the same celestial *longitude, especially the apparent near approach of two planets as one overtakes the other.

convex of a lens, one that converges a beam of light, being curved in the manner of a magnifying glass, in contrast to *concave.

corona the outermost parts of the Sun's atmosphere, visible in a total eclipse.

cosmos the universe, considered as organized and lawlike rather than at the mercy of chance.

cotangent the inverse of *tangent.

crepe ring one of the inner (and relatively faint) rings of Saturn.

cross-staff an instrument used in the Middle Ages for measuring the angle between two objects.

culmination the passage of a celestial body across the observer's *meridian.

D

decan in Egyptian astronomy, one of 36 stars or groups of stars used for telling the time at night (and possibly participating as divinities in rituals).

declination angle north or south of the celestial *equator, the counterpart of *latitude on Earth.

deferent in Greek astronomy, a circle that 'carries' on its circumference either the Sun or the centre of another circle (*epicycle).

dip of a magnetic needle swinging in a vertical plane, the angle by which it departs from the horizontal.

discrete radio source a radio source located in a precisely-defined direction.

divided-lens micrometer an accessory in a *reflector, for measuring small angles; the *objective of a *refractor when divided along a diameter for the same purpose.

Doppler Effect a change in the observed frequency of light (or other radiation) caused by the motion of the body towards or away from the observer.

double star two stars in the sky so close that at first they appear to be one star.

doublet lens a lens consisting of two elements, e.g. an *achromatic lens.

dwarf star one of the commonest stars, of low mass (like the Sun), as distinct from the rarer giant stars.

E

eccentric circle in Greek astronomy, a circle that is not centred on the Earth.

eccentricity of an ellipse, a measure of how much the foci depart from the centre.

eclipsing binary a close *binary star in which at least one star periodically eclipses its companion, usually also a *spectroscopic binary.

ecliptic the observed annual path of the Sun in a *great circle around the sky.

electromagnetic radiation a flow of energy (for example, light and radio waves) produced when electrically charged particles are accelerated.

electron a stable elementary particle that is a constituent of all atoms.

elementary in Aristotelian cosmology, the region below the Moon, occupied by the four elements (earth, water, air and fire).

elongation the angle separating a planet from the Sun; the angle planet–Earth–Sun

emission band (line) a band (line) in the *spectrum of a glowing gas under low pressure.

ephemeris a table of (daily) positions of a heavenly body.

epicycle in Greek astronomy, a small circle whose centre moves on a *deferent and which carries a planet or another small circle.

equant point in Greek astronomy, the seat of uniform angular motion.

equator, celestial the circle in which the heavenly sphere is cut by the plane of the Earth's equator.

equatorial a form of mounting for a telescope or other instrument in which one axis is directed to the celestial North or South Pole.

equinox one of the two occasions in the year (in March and September) when the Sun crosses the celestial *equator.

extragalactic nebula a term used earlier this century for *galaxy.

F

faculae active regions in the upper part of the *photosphere of the Sun that appear unusually bright.

fixed star in Greek astronomy, a star that always maintains its position relative to the other stars, in contrast to a planet; a star, in the modern sense of the term.

focal length in a telescope, the distance between the *objective lens or mirror and the point where the light comes to a focus.

focus of an ellipse, two points A and B on the major axis such that $AP + BP$ is a constant for all points P on the ellipse.

focus, empty in an (elliptical) planetary orbit, the focus not occupied by the Sun.

Fraunhofer lines narrow, dark *absorption lines cutting across the continuous *spectrum of the Sun or a star.

Fraunhofer spectrum the spectrum of the Sun when seen in detail.

fundamental particles the simplest forms of matter.

G

galaxy a large aggregation of stars, star clusters, and other objects, analogous to our Galaxy or Milky Way system.

gamma rays electromagnetic radiation with the shortest wavelengths (less than about 10^{-12} metres).

Giant Branch a region of an *H-R diagram populated by large and highly luminous stars.

globular cluster a spherical system of hundreds of thousands of stars.

graduated marked with divisions (of an arc or circle in an instrument for measuring angles).

great circle in the sky, a circle centred on the observer, dividing the sky into two equal halves.

H

H-R diagram a graph correlating the *luminosities and *spectral types of stars, developed independently by Hertzsprung and Russell.

heliacal rising the first reappearance of a star, star cluster or planet in the dawn sky, after some weeks lost in the glare of the Sun.

heliacal setting the last appearance of a star, star cluster or planet in the evening sky, before some weeks lost in the glare of the Sun.

heliometer a telescope with a divided object-glass micrometer for the accurate measurement of angles.

Hertzsprung-Russell diagram see *H-R diagram.

hippopede the planetary path in the form of a figure-of-eight, generated by two of a nest of *concentric spheres.

hour circle on an astrolabe, a circle to indicate time by the division of day and night each into twelve equal parts.

HST Hubble Space Telescope.

I

impetus the theory developed in the fourteenth century to provide a cause for the motion of projectiles.

inclination the angle between the *ecliptic and the plane of a planet's orbit.

inequality in the study of lunar and planetary orbits, a departure from a simple pattern.

inertia in Newtonian mechanics, the tendency of a body at rest to stay at rest, or of a body in movement to continue to move in the same direction and with the same speed.

infrared radiation radiation at wavelengths between those of visible light and radio waves.

intercalary in a given calendar, the days, weeks or months added when necessary to keep in step with the natural cycle.

intrinsic brightness see *luminosity.

inverse-square law Newton's law of gravitational attraction, where the force diminishes with the square of the distance.

ionization a process whereby an atom or molecule loses one or more electrons.

K

kinetic energy the energy of a body by virtue of its motion, equal to the work it could do in coming to rest.

L

latitude degrees north (or south) of the Earth's equator; on the sky, the angular distance of an object north or south of the *ecliptic.

light year the distance travelled by light in one year (nearly 10^{13} km).

light-gathering power W. Herschel's term for the size of a telescopic mirror or *objective lens.

longitude on Earth, degrees east or west of a standard location (today, Greenwich); on the sky, the angle of an object measured eastwards along the *ecliptic from the spring *equinox point.

luminosity the total energy emitted by a star summed over all the wavelengths; used loosely to mean absolute *magnitude (absolute brightness, intrinsic brightness).

M

macrocosm the cosmos (the universe as a whole), in contrast to a microcosm or individual living body.

Magellanic Clouds two irregular *galaxies, visible to the naked eye in the southern sky, and our nearest extragalactic neighbours.

magnification the amount by which a telescope magnifies.

magnitude a scale for measuring the brightness of celestial objects in which a star of magnitude 1.0 is 100 times brighter than a star of magnitude 6.0; *apparent* magnitude, magnitude as observed from Earth; *absolute* magnitude, magnitude as the same star would appear if at a standard distance of 10 *parsecs.

Main Sequence the region in an *H-R diagram occupied by the majority of the stars.

mean a precise measure of 'average'.

mechanical philosophy the doctrine that explanations of natural phenomena must be in terms of impacts between bodies.

meridian the circle in the sky passing through the celestial North and South Poles and the observer's zenith.

meteor a fragment of fragile inter-planetary material, observed as a 'shooting star'.

meteorite a fragment of solid inter-planetary material large enough to penetrate our atmosphere and reach the surface of the Earth.

Meteoritic Hypothesis Lockyer's hypoth-esis that stars originated out of solid particles rather than gas.

Metonic cycle a calendric cycle based on the near equality between 19 years and 235 lunar (*synodic) months.

microcosm a living body, seen as orga-nized analogously to the cosmos or macrocosm.

micrometer a telescope accessory for measuring small angles.

microwave background radiation a weak radio signal, the same from all direc-tions, interpreted as a relic of the *Big Bang.

mural quadrant a *quadrant mounted on a north–south wall and used to measure the altitudes of celestial bodies at their *culmination.

muwaqqit a timekeeper in a mosque.

N

nebula a celestial body observed as an ill-defined patch of light.

neutron a particle present in the nuclei of all atoms except hydrogen.

neutron star a highly dense star that has undergone gravitational collapse, as a result of which its core is composed primarily of neutrons.

Next Generation Telescope a modern telescope whose design features make use of the most recent technology.

NGT a *Next Generation Telescope.

nocturnal a sixteenth-century (and later) instrument for telling the time at night, especially at sea.

nova a newly-appeared star (*nova stella*).

nutation the 'nodding' of the Earth's axis (period 18.6 years) caused mainly by the pull of the Moon on the non-spher-ical Earth.

O

objective (lens or 'object glass') the prin-cipal lens of a refracting telescope.

obliquity of the ecliptic the angle between the ecliptic and the celestial *equator.

Olbers's Paradox the problem of explain-ing the observed darkness of the night sky, which contrasts with what would be expected in an infinite universe with stars (or galaxies) regularly distributed.

open cluster a loosely-defined cluster of stars (such as the Pleiades), found in the disc of the Galaxy.

opposition the location of the Moon or

one of the outer planets, when opposite to the Sun in the sky.

P

parabola the curve formed by the intersection of a circular cone and a plane parallel to its axis.

parallax the change in the observed position of an astronomical object caused by a change in the observer's position (see also *annual parallax).

parameter a measured quantity used in calculations.

parsec a unit of distance used in stellar astronomy, the distance of a star with an *annual parallax of 1 *arc second; 1 parsec=3.26 *light years.

perigee the position of the Moon in its orbit where it is nearest the Earth.

perihelion the position in the orbit of a planet or comet where it is nearest the Sun.

period–luminosity relationship a correlation between the periods and mean *luminosities of *Cepheids, useful as a distance indicator.

periodic variation a regular change, especially in the orbit of the Moon or a planet or in the *luminosity of a variable star.

perturbations an irregularity in a planetary orbit caused by the attractive pull of another planet.

photographic magnitude a measure of *magnitude as recorded on a photograph.

photosphere the visible surface layer of the Sun, from which its white light is emitted.

place-value the writing of numbers in which significance is given to the positions of the component symbols (as in the arabic numerals we use today).

planet in Greek astronomy, a 'wandering' star (Sun, Moon, Mercury, etc.); since Copernicus, a major satellite of the Sun.

planetary nebula a celestial body with the light of a nebula and the disc of a planet (W. Herschel).

polar circles the Arctic and Antarctic Circles.

precession the motion of the Earth's axis about the pole of the *ecliptic in a period of 25,800 years, caused by the pull of Sun and Moon on the non-spherical Earth.

projection in the *astrolabe, the association of a point in the sky with a point on the astrolabe plate by means of lines drawn from the celestial South Pole.

prominences streamers of glowing gas visible in the outer layers of the Sun.

Q

qibla in Islamic practice, the direction towards Mecca.

quadrant an instrument with a graduated arc of 90°, usually of brass, used for measuring angles

quadrature the Moon near its quarters when the angle Sun–Earth–Moon is a right angle, and similarly for exterior planets when they are at a *longitude 90° east or west of the Sun.

quadrivium in medieval studies, the four mathematical subjects (arithmetic, music, geometry, astronomy).

quantum mechanics, quantum theory
modern atomic theory developed from
the hypothesis that radiation from
electrons is emitted in discrete
packets or *quanta*.

quasar a 'quasi-stellar object', the optical
counterpart of a strong, though
extremely distant, radio source.

quintessence in Aristotelian cosmology,
the (fifth) element of which the
heavens are composed.

R

radial velocity velocity towards or away
from the observer ('line of sight velocity').

radio source a source of radio waves from
outside the solar system.

radio star a star with an unusually strong
emission at radio wavelengths (originally: any *discrete radio source).

radius vector in Kepler's second law, the
line from the Sun to a planet.

redshift the displacement to longer wavelengths of features in a spectrum, particularly of distant *galaxies, thought
to result from the *Doppler Effect.

reduction the correction in the observed
position of a heavenly body to eliminate effects due to instrumental
errors, *aberration, *nutation, etc..

reflector a telescope in which a mirror is
used to collect the rays of light.

refractor a telescope in which a lens
(*objective) is used to collect the rays
of light.

retrogression the interval during which a
planet reverses its normal west-to-east
motion across the sky.

right ascension the counterpart on the

sky of terrestrial longitude, measured
eastwards from the intersection of the
*ecliptic and the celestial *equator in
Aries.

RR Lyrae star a type of pulsating star
similar to the regular *variable star so
named.

S

secular variation a change in the orbit of
the Moon or planet that accumulates
indefinitely.

selective absorption the scattering of
starlight by interstellar dust, with con-
sequent changes in the colour of the
star.

sextant an instrument with a graduated
circle of 60°, used for measuring
angles; a portable navigational instrument in which mirrors effectively
double the 30° arc.

sidereal time time measured by the stars
rather than by the Sun (the sidereal
day is about 4 minutes shorter than
the twenty-four hours of the solar
day).

sine of an angle in a right-angled triangle,
the length of the opposite side divided
by the hypotenuse.

solar apex the position in the sky
towards which the solar system is
travelling (in the constellation
Hercules).

solstices the dates in June and December
when the Sun reaches its maximum
distance from the celestial *equator
and so passes overhead inhabitants in
one of the *tropics.

spectral type the classification of a star
by the features in its *spectrum.

spectrograph, spectroscope an instrument for studying the *spectrum of light.

spectroheliogram a photograph of the Sun in the light of one line in the spectrum (for example, the red hydrogen line), revealing *faculae and *prominences.

spectroscopic binary a *binary star whose nature is revealed by the study of its *spectrum

spectroscopic parallax the distance of a star inferred from its *spectral type and apparent *magnitude.

spectrum the spreading of a beam of light or other radiation according to wavelength, as in the colours of the rainbow.

spherical aberration the blurring in a telescopic image formed by a spherical mirror or lens, caused by the failure of the light rays to converge to a single point.

spiral nebula a nebula of spiral form, but used sometimes as a general name for *galaxies.

Steady State a cosmological model in which the universe appears broadly the same, everywhere and at all times.

sublunary in Aristotelian cosmology, in the terrestrial region ('below the Moon').

supernova the cataclysmic explosion of a star at the end of its evolution.

synodic month the interval of time between successive *conjunctions of the Sun and Moon.

T

tangent of an angle in a right-angled triangle, the side opposite the angle divided by the side adjacent.

transit (i) of a celestial body, when it crosses the *meridian; (ii) the passage of one of the inner planets across the face of the Sun.

trigon three signs of the zodiac forming an equilateral triangle.

trivium in medieval studies, the three literary subjects (grammar, rhetoric, logic).

tropics (i) the circles in the sky marking the extreme positions north and south of the celestial *equator reached by the Sun at the *solstices; (ii) the corresponding circles on Earth.

U

ultraviolet radiation radiation at wavelengths 'beyond the violet' end of the visible *spectrum, in the range from about 10^{-8} to 10^{-7} metres.

V

variable star a star whose apparent *brightness varies significantly, either periodically or irregularly.

visual magnitude a measure of *magnitude as seen with the human eye, as distinguished from *photographic magnitude.

volvelle in Ptolemaic astronomy, a component of a cardboard replica of a planetary model, used to calculate positions or to demonstrate the Ptolemaic system.

vortex in the cosmology of Descartes, a whirlpool of matter surrounding the Earth, the Sun, etc.

W

white dwarf a small, dim star of high density, at the end of its evolution.

X

X-rays radiation at shorter wavelenths
than the ultraviolet, in the range from
about 10^{-11} to 10^{-8} metres.

Z

zenith the point in the sky directly overhead the observer.

zij in Islamic astronomy, a table for calculating planetary positions.

zodiac the belt around the sky in which
the Sun, Moon, and principal planets
are always found, conventionally
divided since Antiquity into twelve
equal regions or 'signs', each 30° in
extent and named after a contiguous
constellation.

Further reading

General

The present text derives from that of *The Cambridge Illustrated History of Astronomy* (Cambridge University Press, Cambridge, 1997); the *Illustrated History* makes extensive use of colour, and so can offer more on subjects, such as astronomical instruments, where much can be learned from illustration.

A survey of the history of astronomy that discusses some topics in greater detail than in the present work is *The Fontana/Elsevier History of Astronomy and Cosmology* by John North (Fontana, London, and Elsevier, New York, 1995). It includes an extended bibliographical essay covering both popular and specialist writings.

One other survey of the history of astronomy, by a leading astronomer, is still of value, though somewhat dated: *A History of Astronomy*, by A. Pannekoek (publ. in Dutch in 1951; English transl., George Allen & Unwin, London, 1961; reprinted in paperback, Dover Publications, New York, 1990).

For accounts of the lives and work of all the major figures in history of astronomy (except those still living at the time of publication), see the *Dictionary of Scientific Biography*, edited by C. C. Gillespie (16 vols, Charles Scribner's Sons, New York, 1970–80, plus later additions). In many cases the entry in the *DSB* is quite simply the best discussion available,

and the work is available in most major libraries.

An in-depth study of the history of astronomy down to 1950 is contained in *The General History of Astronomy*, under the general editorship of Michael Hoskin (4 vols in 7 parts, Cambridge University Press, Cambridge, 1984– , in progress); the relevant volumes and parts are listed below in the reading lists for individual chapters.

Many history of science journals contain occasional articles in history of astronomy, but a journal dedicated to the subject is *Journal for the History of Astronomy*, ed. by Michael Hoskin (quarterly plus annual *Archaeoastronomy* supplement, Science History Publications, 16 Rutherford Road, Cambridge CB2 2HH, UK).

Entries on most important topics in the history of astronomy are to be found in *History of Astronomy: An Encyclopedia*, ed. by John Lankford (Garland Publishing Inc., New York, 1997).

An anthology of interesting essays on topics ranging over the whole period of this book is *The Great Copernicus Chase, and Other Adventures in Astronomical History*, by Owen Gingerich (Sky Publishing Corporation, Cambridge, Mass., and Cambridge University Press, Cambridge, 1992). Another anthology, this time mainly on topics in astronomy since the invention of the telescope, is *The Astronomical Scrapbook: Skywatchers, Pioneers, and Seekers in Astronomy*, by Joseph Ashbrook (Sky Publishing Corporation, Cambridge, Mass., and Cambridge University Press, Cambridge, 1984).

Chapter 1: Astronomy before history

Aveni, Anthony F., *Ancient Astronomies* (St Remy Press, Montreal and Smithsonian Books, Washington DC, 1993)

Aveni, Anthony F., *Empires of Time: Calendars, Clocks and Cultures* (Basic Books, New York, 1989)

Aveni, Anthony F., *Skywatchers of Ancient Mexico* (University of Texas Press, Austin and London, 1980)

Aveni, Anthony F. (ed.), *The Sky in Mayan Literature* (Oxford University Press, Oxford and New York, 1992)

Aveni, Anthony F., and Gary Urton (eds), *Ethnoastronomy and Archaeoastronomy in the American Tropics* (New York Academy of Sciences, New York, 1982)

Burl, Aubrey, *Prehistoric Astronomy and Ritual* (Shire Publications, Princes Risborough, Aylesbury, HP17 9AJ, U.K., 1983)

Hodson, F. R. (ed.), *The Place of Astronomy in the Ancient World* (Oxford University Press, for the British Academy, London, 1974)

Ruggles, C. L. N., *Astronomy in Prehistoric Britain and Ireland* (Yale University Press, New Haven, 1999)

Ruggles, C. L. N. (ed.), *Archaeoastronomy in the 1990s* (Group D Publications, 81 Park Road, Loughborough, LE11 2HD, U.K., 1993)

The *Archaeoastronomy* supplement to *Journal for the History of Astronomy* is devoted to the subject matter of this chapter.

Chapter 2: Astronomy in Antiquity

Clagett, Marshall, *Ancient Egyptian Science*, vol. 2: *Calendars, Clocks, and Astronomy* (American Philosophical Society, Philadelphia, 1995)

Crowe, Michael J., *Theories of the World from Antiquity to the Copernican Revolution* (Dover Publications, New York, 1990)

Heath, Thomas L., *Greek Astronomy* (Dent, London, 1932; reprinted by Dover Publications, New York, 1991)

Hodson, F. R. (ed.), *The Place of Astronomy in the Ancient World* (Oxford University Press, for the British Academy, London, 1974)

Neugebauer, Otto, *The Exact Sciences in Antiquity* (2nd edn, Yale University Press, Providence, R.I., 1957; reissued by Dover Publications, New York, 1969)

Pedersen, Olaf, [and Mogens Pihl], *Early Physics and Astronomy* (Macdonald, London, and American Elsevier, New York, 1974; 2nd edn, Cambridge University Press, Cambridge, 1993)

Thurston, Hugh, *Early Astronomy* (Springer-Verlag, New York, 1993)

Van der Waerden, B. L., *Science Awakening II: The Birth of Astronomy* (Noordhoff International, Leiden)

Van Helden, Albert, *Measuring the Universe: Cosmic Dimensions from Aristarchus to Halley* (University of Chicago Press, Chicago and London, 1985)

Walker, Christopher (ed.), *Astronomy before the Telescope* (British Museum Press, for the Trustees of the British Museum, London, 1996)

See also the five extended essays dealing with astronomy in Egypt, Mesopotamia and India, in vol. xv of *Dictionary of Scientific Biography*, ed. by C. C. Gillispie (Charles Scriber's Sons, New York, 1978)

Chapter 3: Islamic astronomy

King, David A., *Astronomy in the Service of Islam* (Variorum, Aldershot, 1993)

King, David A., and George Saliba, *From Deferent to Equant* (New York Academy of Sciences, New York, 1987)

Saliba, George, *A History of Arabic Astronomy: Planetary Theory during the Golden Age of Islam* (New York University Press, New York and London, 1994)

Chapter 4: Medieval Latin astronomy

Crowe, Michael J., *Theories of the World from Antiquity to the Copernican Revolution* (Dover Publications, New York, 1990)

Gingerich, Owen, *The Eye of Heaven: Ptolemy, Copernicus, Kepler* (American Institute of Physics, New York, 1993)

Grant, Edward, *Planets, Stars and Orbs: Medieval Cosmology, 1200–1687* (Cambridge University Press, Cambridge, 1993)

Lindberg, David C. (ed.), *Science in the Middle Ages* (The University of Chicago Press, Chicago and London, 1978)

Pedersen, Olaf, [and Mogens Pihl], *Early Physics and Astronomy* (Macdonald, London, and American Elsevier, New York, 1974; 2nd edn, Cambridge University Press, Cambridge, 1993)

Rosen, Edward, *Three Copernican Treatises* (Columbia University Press, New York, and Oxford University Press, London, 1939; reprinted Dover Publications, New York, 1959, Octagon Books, New York, 1971)

Van Helden, Albert, *Measuring the Universe: Cosmic Dimensions from Aristarchus to Halley* (University of Chicago Press, Chicago and London, 1985)

Walker, Christopher (ed.), *Astronomy before the Telescope* (British Museum Press, for the Trustees of the British Museum, London, 1996)

Westman, Robert S. (ed.), *The Copernican Achievement* (University of California Press, Berkeley and Los Angeles, 1975)

Chapter 5: From geometry to physics: astronomy transformed

Aiton, Eric J., *The Vortex Theory of Planetary Motions* (Macdonald, London, and American Elsevier, New York, 1972)

Caspar, Max, *Kepler* (2nd English edn, Dover Publications, New York, 1993)

Drake, Stillman, *Discoveries and Opinions of Galileo* (Doubleday, Garden City, N.Y., 1957)

Drake, Stillman, *Galileo at Work: His Scientific Biography* (University of Chicago Press, Chicago and London, 1978)

Galileo, *Dialogue Concerning the Two Chief World Systems*, transl. by Stillman Drake (University of California Press, Berkeley and Los Angeles, 1953; rev. edn, 1967)

Galileo, *Sidereus Nuncius, or The Sidereal Messenger*, transl. by Albert Van Helden (University of Chicago Press, Chicago, 1989)

Jardine, Nicholas, *The Birth of History and Philosophy of Science* (Cambridge University Press, Cambridge, 1984)

Sharratt, Michael, *Galileo, Decisive Innovator* (Blackwell, Oxford, 1994; now published by Cambridge University Press)

Stephenson, Bruce, *Kepler's Physical Astronomy* (Spring Verlag, New York, 1987; paperback, Princeton University Press, Princeton, NJ, 1994)

Stephenson, Bruce, *The Music of the Heavens: Kepler's Harmonic Astronomy* (Princeton University Press, Princeton, NJ, 1994)

Thoren, Victor E., *The Lord of Uraniborg: A Biography of Tycho Brahe* (Cambridge University Press, Cambridge, 1990)

Van Helden, Albert, *Measuring the Universe: Cosmic Dimensions from Aristarchus to Halley* (University of Chicago Press, Chicago and London, 1985)

Westfall, Richard S., *Essays on the Trial of Galileo* (Vatican Observatory, 1989; distributed by The University of Notre Dame Press, Notre Dame, Indiana)

Wilson, Curtis, *Astronomy from Kepler to Newton: Historical Studies* (Variorum, Aldershot, 1989)

See also *The General History of Astronomy*, vol. 2: *Planetary Astronomy from the Renaissance to the Rise of Astrophysics*, Part A: *Tycho Brahe to Newton*, ed. by R. Taton and C. Wilson (Cambridge University Press, Cambridge, 1989)

Chapter 6: Newton and Newtonianism

Armitage, Angus, *Edmond Halley* (Nelson, London, 1966)

Bennett, J. A., *The Divided Circle: A History of Instruments for Astronomy, Navigation and Surveying* (Phaidon Christie's, Oxford, 1987)

Christianson, Gale E., *In the Presence of the Creator: Isaac Newton and his Times* (Free Press, New York, 1984)

Cook, Alan, *Edmond Halley: Charting the Heavens and the Seas* (Clarendon Press, Oxford, 1998)

Coyne, G. V., M. Heller and J. Zycinski (eds), *Newton and the New Direction in Science* (Vatican Observatory, 1988; distributed by Libreria Editrice Vaticana, Vatican City State)

Grant, Robert, *History of Physical Astronomy* (Henry G. Bohn, London, 1852)

Grosser, Morton, *The Discovery of Neptune* (Harvard University Press, Cambridge, Mass., 1962)

King, Henry C., *The History of the Telescope* (Charles Griffin, London, 1955)

Koyré, Alexandre, *Newtonian Studies* (Harvard University Press, Cambridge, Mass., 1965)

Westfall, Richard S., *Never at Rest: A Biography of Isaac Newton* (Cambridge University Press, Cambridge, 1980)

See also *The General History of Astronomy*, vol. 2: *Planetary Astronomy from the Renaissance to the Rise of Astrophysics*, Part A: *Tycho Brahe to Newton*, and Part B: *The Eighteenth and Nineteenth Centuries*, ed. by R. Taton and C. Wilson (Cambridge University Press, Cambridge, 1989, 1995)

Chapter 7: The astronomy of the universe of stars

Armitage, Angus, *William Herschel* (Thomas Nelson, London, 1962)

Buttmann, Günther, *The Shadow of the Telescope: A Biography of John Herschel* (Lutterworth Press, Guildford and London, 1974)

Crowe, Michael J., *Modern Theories of the Universe from Herschel to Hubble* (Dover Publications, New York, 1994)

Ferris, Timothy, *Coming of Age in the Milky Way* (The Bodley Head, London, 1988)

Hoskin, Michael, *Stellar Astronomy: Historical Studies* (Science History Publications, Cambridge, 1982)

Hoskin, Michael, *William Herschel and the Construction of the Heavens* (Oldbourne, London, and American Elsevier, New York, 1963)

King, Henry C., *The History of the Telescope* (Charles Griffin, London, 1955)

Lubbock, Constance A. (ed.), *The Herschel Chronicle: The Life-story of William Herschel and his Sister Caroline Herschel* (Cambridge University Press, Cambridge, 1933)

Whitney, Charles A., *The Discovery of Our Galaxy* (Angus & Robertson, London, 1971)

See also *The General History of Astronomy*, vol. 3: *Stellar Astronomy, Instrumentation and Institutions from the Renaissance to the Mid-nineteenth Century* (Cambridge University Press, Cambridge, in preparation)

Chapter 8: The message of starlight: the rise of astrophysics

Berendzen, Richard, Richard Hart and Daniel Seeley, *Man Discovers the Galaxies* (Neale Watson Academic Publications, New York, 1976)

Brush, Stephen G., *A History of Modern Planetary Physics* (3 vols, Cambridge University Press, Cambridge, 1996)

Christianson, Gale E., *Edwin Hubble: Mariner of the Nebulae* (Farrar, Straus & Giroux, New York, 1995)

Clerke, Agnes C., *A Popular History of Astronomy During the Nineteenth Century* (Adam & Charles Black, Edinburgh and London, 1st edn 1885, 2nd edn 1887, 3rd edn 1893, 4th edn 1902)

Clerke, Agnes C., *Modern Cosmogonies* (Adam & Charles Black, London, 1905)

Clerke, Agnes C., *The System of the Stars* (Longmans, Green, London, 1890; 2nd edn, Adam & Charles Black, London, 1905)

Crowe, Michael J., *Modern Theories of the Universe from Herschel to Hubble* (Dover Publications, New York, 1994)

Ferris, Timothy, *Coming of Age in the Milky Way* (The Bodley Head, London, 1988)

Hearnshaw, J. B., *The Analysis of Starlight: One Hundred and Fifty Years of Astronomical Spectroscopy* (Cambridge University Press, Cambridge, 1986)

Hearnshaw, J. B., *The Measurement of Starlight: Two Centuries of Astronomical Photometry* (Cambridge University Press, Cambridge, 1996)

Hermann, Dieter B., *The History of Astronomy from Herschel to Hertzsprung* (Cambridge University Press, Cambridge, 1984)

Hubble, Edwin, *The Realm of the Nebulae* (Yale University Press, New Haven, Conn., 1936; reprinted by Dover Publications, New York, 1958)

Kragh, Helge, *Cosmology and Controversy: The Historical Development of Two Theories of the Universe* (Princeton University Press, Princeton, N.J., 1997; distributed in the U.K. by Wiley, London)

Krisciunas, Kevin, *Astronomical Centers of the World* (Cambridge University Press, Cambridge, 1988)

Lang, Kenneth R., and Owen Gingerich (eds), *A Source Book in Astronomy and Astrophysics, 1900–1975* (Harvard University Press, Cambridge, Mass., and London, 1979)

Leverington, David, *A History of Astronomy from 1890 to the Present* (Springer, London, 1995)

North, J. D., *The Measure of the Universe: A History of Modern Cosmology* (Oxford University Press, Oxford, 1965; Dover Publications, New York, 1990)

Paul, Erich Robert, *The Milky Way Galaxy and Statistical Cosmology, 1890–1924* (Cambridge University Press, Cambridge, 1993)

Osterbrock, Donald E., *James E. Keeler: Pioneer American Astrophysicist* (Cambridge University Press, Cambridge, 1984)

Osterbrock, Donald E., John R. Gustafson and W. J. Shiloh Unruh, *Eye on the*

Sky: Lick Observatory's First Century
(University of California Press,
Berkeley and Los Angeles, 1988)

Smith, Robert, *The Expanding Universe:
Astronomy's 'Great Debate'
1900–1931* (Cambridge University
Press, Cambridge, 1982)

Struve, Otto, and Velta Zebergs,
Astronomy of the 20th Century
(Macmillan, New York and London,
1962)

See also *The General History of
Astronomy*, vol. 4: *Astrophysics and
Twentieth-century Astronomy to
1950*, ed. by Owen Gingerich
(Cambridge University Press,
Cambridge;, Part A, 1984, Part B, in
preparation)

*Chapter 9: Astronomy's widening
horizons*

Edge, David O., and Michael J. Mulkay,
*Astronomy Transformed: The
Emergence of Radio Astronomy in
Britain* (Wiley-Interscience, New
York, 1976)

Hey, J. S., *The Evolution of Radio
Astronomy* (Elek Science, London,
1973)

Hirsch, Richard F., *Glimpsing an
Invisible Universe: The Emergence of
X-ray Astronomy* (Cambridge
University Press, Cambridge, 1983)

Lang, Kenneth R., and Owen Gingerich
(eds), *A Source Book in Astronomy
and Astrophysics, 1900–1975* (Harvard
University Press, Cambridge, Mass.,
and London, 1979)

Lightman, Alan, and Roberta Brawer,
*Origins: The Lives and Worlds of
Modern Cosmologists* (Harvard
University Press, Cambridge, Mass.,
1990)

Needel, Allan (ed.), *The First 25 Years in
Space* (Smithsonian Institution Press,
Washington, 1989)

Smith, R. W., *The Space Telescope: A
Study of NASA, Science, Technology
and Politics* (Cambridge University
Press, Cambridge, 1989)

Sullivan, W. T. III (ed.), *The Early Years of
Radio Astronomy* (Cambridge
University Press, Cambridge, 1984)

Verschuur, Gerrit L., *The Invisible
Universe: The Story of Radio
Astronomy* (The English Universities
Press, London, and Springer-Verlag,
New York, 1974)

Index